Recent Developments in Geopolymers and Alkali-Activated Materials

Recent Developments in Geopolymers and Alkali-Activated Materials

Editor

Sujeong Lee

Basel • Beijing • Wuhan • Barcelona • Belgrade • Novi Sad • Cluj • Manchester

Editor
Sujeong Lee
Resources Utilization
Devision
Korea Institute of Geoscience
and Mineral Resources
Daejeon
Korea, South

Editorial Office
MDPI
St. Alban-Anlage 66
4052 Basel, Switzerland

This is a reprint of articles from the Special Issue published online in the open access journal *Materials* (ISSN 1996-1944) (available at: www.mdpi.com/journal/materials/special_issues/geopolymers_alkali-activated_materials).

For citation purposes, cite each article independently as indicated on the article page online and as indicated below:

Lastname, A.A.; Lastname, B.B. Article Title. *Journal Name* **Year**, *Volume Number*, Page Range.

ISBN 978-3-7258-0014-8 (Hbk)
ISBN 978-3-7258-0013-1 (PDF)
doi.org/10.3390/books978-3-7258-0013-1

© 2024 by the authors. Articles in this book are Open Access and distributed under the Creative Commons Attribution (CC BY) license. The book as a whole is distributed by MDPI under the terms and conditions of the Creative Commons Attribution-NonCommercial-NoDerivs (CC BY-NC-ND) license.

Contents

About the Editor vii

Preface ix

Sujeong Lee
Special Issue: "Recent Developments in Geopolymers and Alkali-Activated Materials"
Reprinted from: Materials 2024, 17, 245, doi:10.3390/ma17010245 1

Muhammad Ahmed, Piero Colajanni and Salvatore Pagnotta
A Review of Current Research on the Use of Geopolymer Recycled Aggregate Concrete for Structural Members
Reprinted from: Materials 2022, 15, 8911, doi:10.3390/ma15248911 3

Laura Ricciotti, Alessio Occhicone, Claudio Ferone, Raffaele Cioffi, Oreste Tarallo and Giuseppina Roviello
Development of Geopolymer-Based Materials with Ceramic Waste for Artistic and Restoration Applications
Reprinted from: Materials 2022, 15, 8600, doi:10.3390/ma15238600 20

Chengjie Zhu, Ina Pundienė, Jolanta Pranckevičienė and Modestas Kligys
Effects of Na_2CO_3/Na_2SiO_3 Ratio and Curing Temperature on the Structure Formation of Alkali-Activated High-Carbon Biomass Fly Ash Pastes
Reprinted from: Materials 2022, 15, 8354, doi:10.3390/ma15238354 37

Sujeong Lee and Arie van Riessen
A Review on Geopolymer Technology for Lunar Base Construction
Reprinted from: Materials 2022, 15, 4516, doi:10.3390/ma15134516 60

Muhammad Nasir Amin, Kaffayatullah Khan, Muhammad Faisal Javed, Fahid Aslam, Muhammad Ghulam Qadir and Muhammad Iftikhar Fara
Prediction of Mechanical Properties of Fly-Ash/Slag-Based Geopolymer Concrete Using Ensemble and Non-Ensemble Machine-Learning Techniques
Reprinted from: Materials 2022, 15, 3478, doi:10.3390/ma15103478 71

Byoungkwan Kim, Sujeong Lee, Chul-Min Chon and Shinhu Cho
Setting Behavior and Phase Evolution on Heat Treatment of Metakaolin-Based Geopolymers Containing Calcium Hydroxide
Reprinted from: Materials 2022, 15, 194, doi:10.3390/ma15010194 91

Muhammad Ibraheem, Faheem Butt, Rana Muhammad Waqas, Khadim Hussain, Rana Faisal Tufail, Naveed Ahmad, et al.
Mechanical and Microstructural Characterization of Quarry Rock Dust Incorporated Steel Fiber Reinforced Geopolymer Concrete and Residual Properties after Exposure to Elevated Temperatures
Reprinted from: Materials 2021, 14, 6890, doi:10.3390/ma14226890 103

Taewan Kim, Choonghyun Kang and Kiyoung Seo
Development and Characteristics of Aerated Alkali-Activated Slag Cement Mixed with Zinc Powder
Reprinted from: Materials 2021, 14, 6293, doi:10.3390/ma14216293 127

Juan María Terrones-Saeta, Jorge Suárez-Macías, Francisco Javier Iglesias-Godino and Francisco Antonio Corpas-Iglesias
Development of Geopolymers as Substitutes for Traditional Ceramics for Bricks with Chamotte and Biomass Bottom Ash
Reprinted from: *Materials* **2021**, *14*, 199, doi:10.3390/ma14010199 . . . . **144**

Piotr Rożek, Paulina Florek, Magdalena Król and Włodzimierz Mozgawa
Immobilization of Heavy Metals in Boroaluminosilicate Geopolymers
Reprinted from: *Materials* **2021**, *14*, 214, doi:10.3390/ma14010214 . . . . **164**

About the Editor

Sujeong Lee

Sujeong Lee is a Principal Research at KIGAM and has diverse experience as a researcher in Geopolymer and Applied Mineralogy, particularly in various electron microscopy analyses such as SEM (Scanning Electron Microscopy), TEM (Transmission Electron Microscopy), MLA (Mineral Liberation Analysis), and electron tomography. Sujeong Lee has collaborated with geologists, metallurgists, ceramic experts, and materials engineers at prominent universities and governemment-backed research organizations. Sujeong Lee has developed experimental skills, as well as an understanding various minerals, alloys, industrial byproducts, geopolymers, nanomaterials, and ceramic materials through cooperation with expers in various fields. She has taken part in various research projects as a PI or a member on coal cleaning, mineral sands, Pb ores, Zn ores, Cu-Co ores, and Ti-V ores projects at KIGAM.

Preface

This reprint compiles eight recent research papers and two review papers to provide insights into the latest trends in geopolymer and alkali-activated materials (AAM). It introduces various raw materials and manufacturing methods, covering diverse applications, such as lunar exploration bases. Geopolymer cement, as opposed to conventional cement, is an alternative material with superior properties, including heat resistance and acid resistance. Through this reprint, the understanding and utilization of geopolymer and AAM as low-carbon emission alternatives to traditional cement are expected to expand.

Sujeong Lee
Editor

Editorial

Special Issue: "Recent Developments in Geopolymers and Alkali-Activated Materials"

Sujeong Lee [1,2]

[1] Resources Utilization Division, Korea Institute of Geoscience and Mineral Resources, 124 Gwahang-no, Yuseong-gu, Daejeon 34132, Republic of Korea; crystal2@kigam.re.kr

[2] Resources Engineering Department, Korea National University of Science and Technology, 217 Gajeong-ro, Yuseong-gu, Daejeon 34113, Republic of Korea

Citation: Lee, S. Special Issue: "Recent Developments in Geopolymers and Alkali-Activated Materials". *Materials* **2024**, *17*, 245. https://doi.org/10.3390/ma17010245

Received: 17 December 2023
Revised: 20 December 2023
Accepted: 28 December 2023
Published: 2 January 2024

Copyright: © 2024 by the author. Licensee MDPI, Basel, Switzerland. This article is an open access article distributed under the terms and conditions of the Creative Commons Attribution (CC BY) license (https://creativecommons.org/licenses/by/4.0/).

As efforts toward global sustainability converge with the imperative to reduce the environmental impact of construction materials, extensive research and development is underway in the field of geopolymers and alkali-activated materials (AAMs). This Special Issue aims to comprehensively present the latest research findings, methodologies, and crucial insights from leading researchers and practitioners in this field.

Our contributors, comprising outstanding researchers, scholars, and industrial experts, have each brought forth new perspectives to enhance our collective understanding of geopolymers and AAMs. Geopolymers and AAMs are gaining significant attention not only for their potential to replace traditional materials but also for surpassing performance expectations and serving as low-carbon, environmentally friendly technologies to address climate change. The unique properties of geopolymers, such as their high strength, durability, and fire resistance, reinforce their position as eco-friendly alternatives. Moreover, AAMs broaden the horizon of sustainable construction materials by incorporating a wide range of raw materials, such as industrial by-products, waste, and biomass ash. This not only makes a significant contribution to environmental sustainability but also enhances the durability and overall performance of AAMs.

While past research primarily focused on understanding and improving the chemical and physical properties of geopolymers and AAMs, recent studies actively explore the utilization of various industrial by-products, waste materials, and biological sources as raw materials. Research efforts are also directed toward predicting material properties using machine learning, optimizing mix designs, and forecasting the characteristics of geopolymers and AAMs. These materials find applications not only in construction but also in diverse fields such as nanomaterials, aerospace, ceramics, and space exploration. Despite these advancements, there is an ongoing need for fundamental research, standardization, and specification improvements for geopolymers and AAMs.

This Special Issue features eight research papers and two review papers covering a spectrum of topics. Ricciotti et al.'s [1] study demonstrates the production of metakaolin geopolymer-based plaster by recycling waste generated in the production process of porcelain stoneware products, with excellent adhesion suitable for the restoration and preservation of artworks. Zue et al. [2] showcase the synthesis of low-temperature C-S-H using the alkali activation of high-carbon biomass fly ash, a challenging material to recycle. Amin et al. [3] employ machine learning to predict the mechanical properties of 156 geopolymer concrete samples, exploring the efficiency of machine learning in enhancing the production process of geopolymers. Kim et al. [4] investigate the positive influence of Ca additives on the early strength of geopolymers, providing advanced insights into phase evolution through a Rietveld refinement analysis. Ibraheem et al. [5] explore the diversification of raw materials in AAMs by adding quarry rock dust and steel fiber, presenting effective ways to recycle waste resources and enhance strength. Kim et al. [6] manufacture a porous alkali-activated material using zinc powder as a foaming agent, introducing a novel reaction pathway compared to commonly used aluminum powder. Rozek et al. [7] report on the performance of boroaluminaosilicate geopolymers in immobilizing heavy metals, surpassing traditional cement concrete in metal fixation capabilities. Terrones-Saeta et al. [8] demonstrate the production of ceramics with sufficient strength using chamotte as an alumino-silicate raw material and potassium-rich biomass bottom ash as an alkali activator. Ahmed et al. [9] provide a specialized analysis of geopolymer recycled aggregate concrete made from recycled aggregates and geopolymers in terms of structural applications. Lee and Riessen [10] analyze the feasibility of manufacturing geopolymers from lunar-simulant-based materials originating from lunar regolith, highlighting the growing interest in lunar exploration.

This Special Issue highlights the synergy between academia, industry, and research institutions, confirming their pivotal role in the continuous advancement of this field. The collaborative efforts showcased on these pages signify the scientific community's dedication to adapting to ongoing global changes.

We extend our sincere gratitude to all the authors who contributed to this Special Issue. Their invaluable research contributions made this Special Issue possible, and we genuinely appreciate their efforts in advancing the sustainable development of construction materials.

In conclusion, we hope that this Special Issue serves as a comprehensive resource, fostering innovation and dialogue among researchers, engineers, policymakers, and industry experts in this dynamic field.

Conflicts of Interest: The author declares no conflict of interest.

References

1. Ricciotti, L.; Occhicone, A.; Ferone, C.; Cioffi, R.; Tarallo, O.; Roviello, G. Development of geopolymer-based materials with ceramic waste for artistic and restoration applications. *Materials* **2022**, *15*, 8600. [CrossRef] [PubMed]
2. Zhu, C.; Pundienė, I.; Pranckevičienė, J.; Kligys, M. Effects of Na_2CO_3/Na_2SiO_3 ratio and curing temperature on the structure formation of alkali-activated high-carbon biomass fly ash pastes. *Materials* **2022**, *15*, 8354. [CrossRef] [PubMed]
3. Amin, M.N.; Khan, K.; Javed, M.F.; Aslam, F.; Qadir, M.G.; Faraz, M.I. Prediction of mechanical properties of fly-ash/slag-based geopolymer concrete using ensemble and non-ensemble machine- learning techniques. *Materials* **2022**, *15*, 3478. [CrossRef] [PubMed]
4. Kim, B.; Lee, S.; Chon, C.-M.; Cho, S. Setting behavior and phase evolution on heat treatment of metakaolin-based geopolymers containing calcium hydroxide. *Materials* **2022**, *15*, 194. [CrossRef]
5. Ibraheem, M.; Butt, F.; Waqas, R.M.; Hussain, K.; Tufail, R.F.; Ahmad, N.; Usanova, K.; Musarat, M.A. Mechanical and microstructural characterization of quarry rock dust incorporated steel fiber reinforced geopolymer concrete and residual properties after exposure to elevated temperatures. *Materials* **2021**, *14*, 6890. [CrossRef] [PubMed]
6. Kim, T.; Kang, C.; Seo, K. Development and characteristics of aerated alkali-activated slag cement mixed with zinc powder. *Materials* **2021**, *14*, 6293. [CrossRef]
7. Rożek, P.; Florek, P.; Król, M.; Mozgawa, W. Immobilization of heavy metals in boroaluminaosilicate. *Materials* **2021**, *14*, 214. [CrossRef]
8. Terrones-Saeta, J.M.; Suárez-Macías, J.; Iglesias-Godino, F.J.; Corpas-Iglesias, F.A. Development of geopolymers as substitutes for traditional ceramics for bricks with chamotte and biomass bottom ash. *Materials* **2021**, *14*, 199. [CrossRef]
9. Ahmed, M.; Colajanni, P.; Pagnotta, S. A review of current research on the use of geopolymer recycled aggregate concrete for structural members. *Materials* **2022**, *15*, 8911. [CrossRef] [PubMed]
10. Lee, S.; van Riessen, A. A review on geopolymer technology for lunar base construction. *Materials* **2022**, *15*, 4516. [CrossRef] [PubMed]

Disclaimer/Publisher's Note: The statements, opinions and data contained in all publications are solely those of the individual author(s) and contributor(s) and not of MDPI and/or the editor(s). MDPI and/or the editor(s) disclaim responsibility for any injury to people or property resulting from any ideas, methods, instructions or products referred to in the content.

Review

A Review of Current Research on the Use of Geopolymer Recycled Aggregate Concrete for Structural Members

Muhammad Ahmed, Piero Colajanni * and Salvatore Pagnotta

Department of Engineering, University of Palermo, 90128 Palermo, Italy
* Correspondence: piero.colajanni@unipa.it

Abstract: Geopolymer cement (GPC) is a sustainable alternative to ordinary Portland cement (OPC) that considerably cuts the emission of carbon dioxide linked to the building of concrete structures. Over the last few decades, while a large number of papers have been written concerning the use of GPC with natural aggregates and OPC with recycled aggregates, few papers have been devoted to investigating the use of Geopolymer Recycled Aggregate Concrete (GRAC) in structural members. Most of them show more interest in the mechanical strength of the material, rather than the structural behavior of RC members. This review critically compiles the present and past research on the behavior of structural members cast with different types and compositions of GRAC. The focus is on the few research studies investigating the structural behavior of GRAC elements, with an analysis of the load-bearing capacity, the load-deflection mechanism, shear behavior, tensile and flexural strength, and ductility of GRAC structural members. This review aims to indicate the research and experimental tests needed in the future for characterizing the behavior of structural members made up of GRAC.

Keywords: geopolymer cement; recycled aggregate; structural behavior; compressive strength; flexural strength

1. Introduction

Ordinary Portland Cement (OPC) concrete is the world's most common and most widely used binding constructional material. It has a number of advantages as it is easily available, has a low cost, and is durable. However, this cement is criticized due to the emission of carbon dioxide during its manufacturing process. The production of OPC accounts for 5% of the world's CO_2 emissions; one ton of ordinary Portland cement releases approximately 0.9 tons of CO_2 [1,2].

Moreover, the production of normal concrete requires a lot of natural aggregates, and there are environmental impacts as a result of aggregate extraction, including conversion of land use, erosion, loss of habitat of different species, etc. [3]. At the same time, a massive amount of construction waste is produced every year due to the demolition of buildings and other concrete structures [4].

To mitigate the environmental pollution due to the construction industry and the excessive use of natural resources to produce OPC, the need arises for an alternative to OPC made using natural waste materials, especially industrial by-products (i.e., slag). In addition, it must also be efficient in terms of cost and characteristics [5].

The scientific community and the development sector are credited with developing Geopolymer Cement (GPC) and the latest research shows that it is now one of the most reliable alternatives to conventional constructional binding materials, i.e., OPC [6,7].

The material used for manufacturing GPC is mainly of geological origin and has semicrystalline or amorphous aluminosilicate polymeric network structures. However, different compositions and types of materials can be used for GPC production [8]. GPC production involves mixing an optimum quantity of source materials and alkali activators, then curing the prepared mixture at low or high temperatures. The source materials can be metakaolin or coal-fired fly ash [9].

GPC can also be produced when waste material such as fly ash, Ground Granulated Blast Furnace Slag (GGBS), and clay containing aluminosilicate minerals are treated with an alkali solution such as sodium hydroxide. The alkali solution helps reduce the setting time of geopolymer cement [10].

GPC is promising as an effective alternative to OPC as it could limit the emission of CO_2 gases and help reduce construction demolition waste [11]. Apart from being eco-friendly, the use of GPC instead of OPC results in lower production costs with comparable mechanical properties [12].

Geopolymer cement with aggregates produces geopolymer concrete. Due to the compelling properties of geopolymer concrete, it is now used for various applications in the construction sector, i.e., multifunctional plastering, and for thermal insulation [13]. It can also be used for soil stabilization or coastal infrastructures [14].

The concept of a circular economy, which is playing an increasingly prominent role in the definition of sustainable construction techniques and materials, promotes the use of Recycled Aggregates (RA) for concrete production. The use of RA obtained from Construction and Demolition Wastes (CDW) in the concrete industry can help preserve natural aggregate resources and will reduce the need for landfill space, making the construction industry more environmentally friendly and sustainable. The use of recycled aggregates taken from CDW has more than two decades of tradition in the field of construction with OPC concrete [15], as has been proven by various researchers [16].

However, the properties of recycled aggregate are not as effective as those of natural aggregate. Indeed, the use of recycled aggregates for concrete structural members is hindered by the attached mortar since it has some negative effects on the strength of the mixture. It increases both the porosity of the recycled aggregate and the development of two different Interfacial Transition Zones (ITZ)s, between the recycled aggregate and new mortar and between the new mortar and attached mortar (Figure 1). The ITZ between new and old mortar is the weak zone that causes the reduction in strength [17].

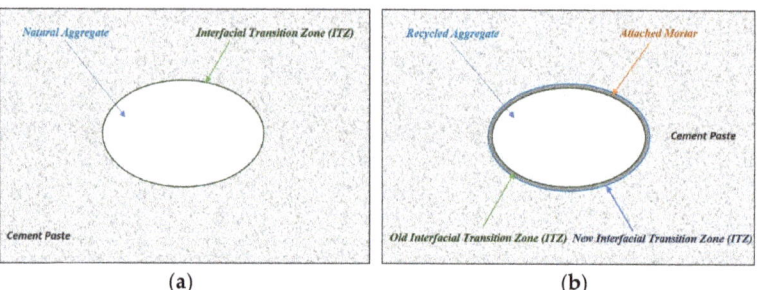

Figure 1. (a) Interfacial Transition Zone between NA and cement paste; and (b) Interfacial Transition Zone in the case of RA.

That is the reason behind the greater uncertainty about the mechanical properties of concrete with RA, compared to that with NA, which limits the use of RA in OPC for non-structural applications or structural use with a low replacement ratio. However, RA efficiency can be significantly improved by treating the recycled aggregate with chemicals, heat, and abrasion [18].

Geopolymer concrete makes it possible to use a larger amount of RA, as demonstrated by research conducted in [19], where, with 100% use of recycled aggregates, concrete compressive strength values in excess of 45 MPa were achieved, and thus it is certainly compatible with the structural use of Geopolymer Recycled Aggregate Concrete (GRAC).

Le and Bui (2020) determined that the old concrete obtained from demolished construction waste can be mechanically crushed, sieved, cleaned, and sometimes also chemically treated in order to obtain RA for structural GRAC [17]. These recycled aggregates can be used in concrete as a partial substitute or full substitute depending upon the requirement [20].

In this regard, it has to be pointed out that there is no unanimous consensus on the influence of the compressive strength of the concrete used for the extraction of recycled aggregates on the strength of the GRAC to be obtained, as some authors state that the latter depends on the strength of the original concrete [21], and others only on the quality of the aggregates [22,23].

To study the environmental benefits of GPC, Almutairi et al. (2021) conducted comprehensive research and found that the use of geopolymer cement in the construction industry will reduce 80% of carbon dioxide emissions associated with the production of concrete. It will also be helpful in reducing the cost of raw materials required to produce concrete [24].

There is already significant ongoing research on the use of OPC with RA in structural members, and some on GPC with NA, but there is very little research on the behavior of structural members made of GRAC. The available data also provide conflicting results and assessments on the efficiency of GRAC. To begin, this review paper briefly describes the past and current research developments on the characterization of the mechanical properties of GRAC and then focuses on the few experimental research papers investigating the behavior of GRAC structural members, rather than the material itself.

2. Research Development on the Mechanical Properties of GRAC

Although research on making concrete from GPC has been going on for almost a decade, no guidelines are yet available for making geopolymer concrete, and currently, no empirical model is available for the reliable prediction of the compressive and tensile strength of GRAC, since they depend on several factors such as binder types, aggregate types, the molarity of the alkaline solutions, mixing procedure, casting temperature, and environmental conditions [25].

This section analyzes some of the research conducted by various authors on the mechanical strength of GRAC, focusing more on the use of different materials and the RA replacement ratio for improving the main mechanical characteristics of GRAC that influence the behavior of structural elements, such as the compressive strength, workability, and tensile and flexural strength. The studies are presented according to the type of material used for the preparation of GPC.

2.1. Fly Ash (FA)

Early GPC production techniques were based upon the use of FA, which is an efficient binder, due to its richness in silica and alumina, and provides cement with high compressive strength, especially when it is cured at a high temperature.

Uddin Ahmed Shaikh (2016) did an experimental study to discover the durability and mechanical properties of GRAC made with FA (17% By Weight (BW)), sodium silicate (5% BW), and sodium hydroxide 8 M (2% BW). RA was used as a partial replacement (15%, 30%, and 50%) of NA. GPC with 100% NA was used as a reference for comparison. The test results showed that with an increase in RA content, the compressive strength, indirect tensile strength, and elastic modulus of geopolymer concrete decreased whether the test was performed after 7 or 28 days (Figure 2a) [25].

Moreover, it was found that the existing empirical models for OPC ((AS3600) [26]) and for GPC (Ryu et al. [27], Diaz Loya et al. [28]) containing natural aggregates underestimate the indirect tensile strength and overestimate the elastic modulus of GRAC (Figure 2b).

Nuaklong et al. (2016) performed a similar experimental study to find the effect of NaOH concentration (8 M, 12 M, and 16 M) on GPC prepared with FA and sodium silicate. Six different cylindrical samples with a 100 mm diameter and 200 mm height were prepared. Three of the samples had 100% limestone (as the natural aggregate) and three samples had 100% RA. The mechanical properties of GRAC prepared with FA (19% By Weight (BW)), sodium silicate (6.9% BW), and sodium hydroxide 8 M, 12 M, and 16 M (4.5% BW) were investigated.

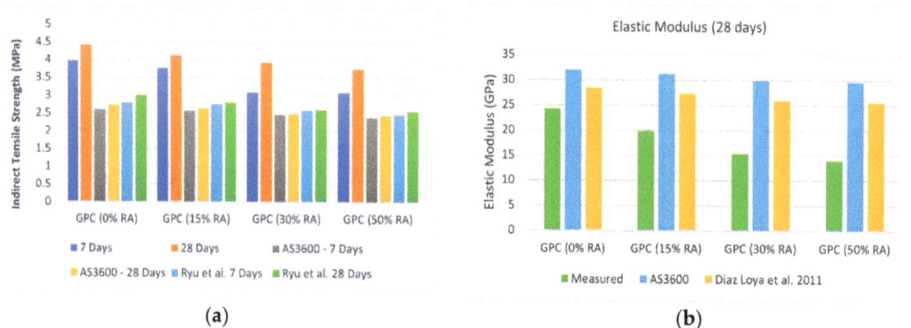

Figure 2. (a) Indirect tensile strength of GPC containing different percentages of NA and RA [27]; and (b) elastic modulus of GPC containing NA and RA [25,28].

The results showed that RA can be used in GPC with high calcium fly ash content. In the case of 8 M NaOH, the compressive strength of GRAC was found to be approximately 76.93% of GPC with NA, and 93% and 91% for 12 M NaOH and 16 M NaOH, respectively. In the case of 8 M NaOH, the flexural strength of GRAC was found to be approximately 95% of GPC with NA, while an increment of 4% and 7% was observed for 12 M NaOH and 16 M NaOH, respectively. Finally, 12 M NaOH was found to be most appropriate for GPC with high calcium fly ash content [29].

Wongsa et al. (2020) conducted comparative research to discover the physical properties of Pressed Geopolymer Concrete (PGC) made with GPC. The GPC was prepared with lignite coal fly ash, a sodium hydroxide solution (NaOH), and a sodium silicate solution. The GPC was mixed with RA obtained from demolition waste, Recycled Concrete Block Aggregate (RB), and limestone dust to obtain concrete to be used for concrete blocks. The RA and RB meshed into fine aggregates having a 4.75 mm diameter.

The results showed that: (a) pressed geopolymer concrete made up of limestone dust exhibited more compressive strength, less porosity, and water absorption than concrete made up of RB and RA; (b) the compressive strength of pressed GPC made up of RA and RB was nearly equal to the strength of moderate strength lightweight concrete prepared according to ACI 213 [30]; and (c) it was recommended that the pressed GPC with RA can be used not only for structural applications but also to make hollow geopolymer-based concrete blocks with better thermal insulation than cement-based concrete blocks [31].

Le and Bui (2021) studied the use of GRAC and the effect of the ratio between alkaline activated solution (AAS) and FA and the use of the lignosulfonate superplasticizer. AAS was taken as a combination of NaOH and a sodium silicate solution. Three different AAS/FA values (0.4, 0.45, and 0.5) were used for testing. An alkali-activated binder (geopolymer binder) was made using low calcium FA, a sodium silicate solution, a sodium hydroxide solution, and a lignosulfonate superplasticizer. Specimens were cured both at 60 °C and environmental temperature.

The results showed encouraging behavior of the GRAC specimen when the use of low calcium fly ash was joined with curing at 60 °C, even if a 100% replacement NA with RA was made, as a much lower decrease in strength was observed [32].

2.2. FA+GGBS (Flyash Combined with GGBS)

The use of GGBS, in addition to FA, increases the amount of aluminosilicate content, increasing the compressive strength and modulus of elasticity, and making them less dependent on the Water to Binder ratio (W/B). Moreover, it decreases the negative effect of weak ITZ that affects RA strength, resulting in a higher compressive strength.

Hu et al. (2019) performed an experimental study to find the flexural strength of GRAC members using different percentages of Ground Granulated Blast Furnace Slag (GGBS) and recycled aggregates. Twelve different mixtures were prepared. NA replacement was prepared at ratios of 0%, 50%, and 100% by weight. Specimens of size $100 \times 100 \times 400$ mm^3 were prepared and tested under a three-point loading test.

It was found that the use of GGBS provides an increment of compressive strength irrespective of the replacement ratio of RA, up to 100% when 30% of GGBS was used. This was due to the increase in high calcium content after GGBFS addition, which ultimately resulted in the formation of a gel phase that reduced the porosity.

The flexural strengths of the mixtures with recycled aggregate were found to be lower than that of the corresponding mixtures with natural aggregate. A further decrease in strength was observed with an increase in the RA percentage. The reason for the lower flexural strength was the poor quality of the recycled aggregate and the low bonding strength between the RA and the GPC. However, the flexural strength tended to increase with the addition of GGBS. After the inclusion of 10%, 20%, and 30% GGBS, the flexural strengths increased by 54%, 78%, and 92%, respectively, for the mixtures with 100% RA, and 51%, 60%, and 64%, respectively, for the mixtures with 50% RA [33], stressing that the addition of GGBS is more efficient when concrete with a 100% RA replacement ratio is considered.

Xie et al. (2019) conducted an experimental investigation to find the combined effects of FA and GGBS and considered the effect of W/B on the fresh and hardened properties of GRAC. A total of 100% RA was used for the preparation of GRAC, with different percentages of GGBS, FA, and the W/B ratio. The results revealed that the combination of GGBS and FA provides encouraging results regarding workability and the mechanical performance of GRAC.

Replacing the OPC matrix with FA/GGBS-based geopolymer improved the strength of recycled aggregate concretes. The compressive strength of GRAC was found to increase with a decrease in the W/B ratio, and with the inclusion of more GGBS content, the compressive strength increased. The compressive strength of GRAC with 50% and 75% GGBS content was found to be 50% and 180% higher than that of normal concrete [34].

Srinivas and Abhignya (2020) observed that by using FA and GGBS as a replacement for cement and RA as a replacement for NA, GRAC beams, and columns, performed much better than conventional reinforced concrete beams and columns referring to compressive and flexural strength. The optimum replacement percentage of RA was found to be 30% because with this replacement the ductile nature of both geopolymer and conventional concrete beams was almost the same [35].

Moulya and Chandrashekhar (2022) performed an experimental study to find the effect of the recycled aggregate replacement ratio on the strength of GRAC. The geopolymer concrete was made with a fixed ratio of GGBS to FA (50:50), and a sodium hydroxide concentration of 8 M was used. NA was replaced with RA at percentages of 0%, 10%, 20%, 30%, 40%, 50%, 60%, 70%, 80%, 90%, and 100%, and tests were performed after 3, 7, 14, and 28 days.

The results indicated that as the percentage substitution of RA increased, the compressive strength of the GRAC was reduced, but it was also found that the strength of GRAC increased with age/casting days. The maximum compressive strength of GRAC was found to be 60.02 MPa after 28 days with 70% RA replacement, which was very close to 100% natural aggregate geopolymer concrete. Hence, in this experimental analysis, it was recommended that a combination of FA and GGBS be used, with 8 M NaOH and 70% RA replacement for precast construction with environmental curing [36].

2.3. Metakaolin (MK)

Recently, the use of MK as a binder in GPC has been growing, since it is characterized by a high alumina and silica content and high reactivity due to its pozzolanic nature. Its high reaction speed with calcium hydroxide produces calcium aluminates hydrates and silicon aluminates that reduce the percentage of voids in concrete, improving its mechanical behavior.

In (2018) Nuaklong et al. (2018) modified the former mixture for GPC using MK with fly ash-based geopolymer concrete and performed a comparative assessment of the use of NA and 100% RA concrete. Two different schemes were adopted using: (1) Limestone as the natural aggregate in geopolymer concrete; (2) 100% recycled aggregate in geopolymer concrete. It was found that when the metakaolin amount was increased, the compressive strengths of GRAC with metakaolin (0, 10, 20, and 30%) were 32.9, 40.4, 45.0, and 47.2 MPa, respectively. GRAC mixtures with metakaolin achieved approximately 15–34% higher compressive strength than concrete without metakaolin. This enhancement of compressive strength was due to increased geo-polymerization and denseness of the microstructure. Increasing metakaolin from 10% to 30% also led to an increase in the splitting tensile strength from 2.9 to 5.4 MPa for GPC with NA and from 2.7 to 3.5 MPa for GRAC. The strength of geopolymer concrete also increased in both types (1 and 2) since the compressive strength of (1) was almost 7–19% higher than (2). Moreover, the researchers stressed that usually, the formation of geopolymer occurs by casting geopolymer slurry in the mold with a significant amount of alkali solution, increasing the chances of high porosity. These pores can act as a point of stress concentration and mechanical failure. In this context, the application of pressure reduces porosity [37].

Muduli and Mukharjee (2019) conducted an experimental study on the flexural strength of members made of GRAC. A total of 15 samples of size $100 \times 100 \times 500$ mm^3 were prepared with different percentages of MK and RA. Flexural strength tests were performed 28 days after casting. It was found that the sample without RA and MK had a flexural strength of 4.59 MPa, which decreased to 4.22 MPa and 3.9 MPa with the addition of 50% and 100% RA, respectively, in the concrete mix. The reason for the reduction was found to be poor bonding between RA and GPC and the presence of loose residual mortar attached to RA. MK proved to be efficient for the increase in flexural strength. For members having 50% RA, flexural strength gains of 2.8%, 8.1%, 9%, and 4.7% were observed with the incorporation of 5%, 10%, 15%, and 20% metakaolin, while 4.6%, 11.3%, 13.8% and 10% flexural strength gains were detected for concrete with 100% RA [38]. These results prove that MK is more efficient in increasing tensile strength, even with the higher replacement ratio of recycled aggregates.

Lee et al. (2020) stressed that superplasticizers can be used to improve the workability of concrete, but when they are applied to calcium-rich, alkali-activated materials, they give inconsistent results; by contrast, the use of Methyl Isobutyl Carbinol (MIBC) and polycarboxylate superplasticizers is found to be effective in improving the workability and strength of MK-based geopolymers, especially after 7 days [39].

Berhanul et al. (2021) examined the effect of metakaolin as a cement replacement on the properties of fresh and hardened recycled aggregate concrete and natural aggregate concrete. The recycled aggregates were obtained from first-hand cast laboratory cubes whose compressive strength was already known.

Different concrete mixtures were prepared and tested with different percentages of recycled aggregates and MK. Namely, 0%, 6%, 12%, 18%, and 24% replacement of OPC with MK, and 0%, 25%, 50%, 75%, and 100% replacement of NA with RA was made. The results showed that the use of metakaolin as a cement replacement improved the strength of GRAC. In the case of GRAC with 100% replacement with RA, the optimum content for OPC replacement with MK was found to be 6% [40].

Xu et al. (2021) reviewed the current research on the mechanical properties of GRAC, confirming that GPC in GRAC is an ideal substitute for cement and, similarly, RA is also an ideal substitute for NA because of its environment-friendly effects. The strength of GRAC depends on many factors such as the type of geopolymer, the casting temperature, the type of aggregates, etc. But these strength-influencing factors are similar to those for GPC, and there is a lack of research on the use of GRAC so the focus must be on the practical use of GRAC [41].

2.4. Mixing Procedure

Treatments of RA to reduce the unfavorable influence on GRAC mechanical properties lie beyond the scope of this review. Here, we will just mention that to mitigate the risk of failure due to the ITZ of recycled aggregates, Liang et al. (2013) previously proposed two different mixing procedures, named the Mortar Mixing Approach (MMA) and the Sand-Enveloped Mixing Approach (SEMA).

The schematic diagrams of both methods are given in (Figure 3). In the SEMA method, RA underwent pre-surface treatment 7 days before mixing, obtaining a higher 28-day compressive strength as compared to that of recycled aggregate concrete made with the MMA method. Due to these mixing processes, an additional layer of cement formed on the aggregate surface, which decreased its porosity as well as reduced its high-water absorption; this ultimately resulted in an improvement in strength. The results obtained showed that using MMA improved the compressive strength of concrete made with 100% coarse RA. However, these methods generate an increase in cost and casting time [42].

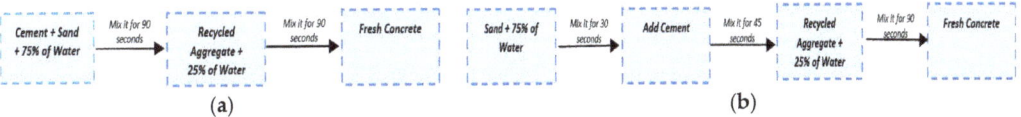

Figure 3. Schematic diagram of (**a**) the Mortar Mixing Approach (MMA) method; and (**b**) the Sand-Enveloped Mixing Approach (SEMA) method [42].

Recently, Alqarni et al. (2021) proposed a new two-stage mixing approach with silica fume and cement. In this treatment process, a cement-silica fume slurry solution was prepared by mixing the cement and silica fume in different percentages with water by weight. RA was dried in an oven for a day and then cooled. After that, the RA was mixed with a cement-silica fume slurry solution for about 30 min; this treatment was found to increase the compressive strength of concrete [43].

2.5. Comparison of Results

Figure 4 shows a comparison of the results reported by three different researchers, namely, Nuaklong et al., 2018 [37], Muduli and Mukharjee (2019) [38], and Berhanul et al., 2021 [40], showing the effect of metakaolin increases and/or the percentage of recycled aggregate replacement on the compressive strength of GRAC.

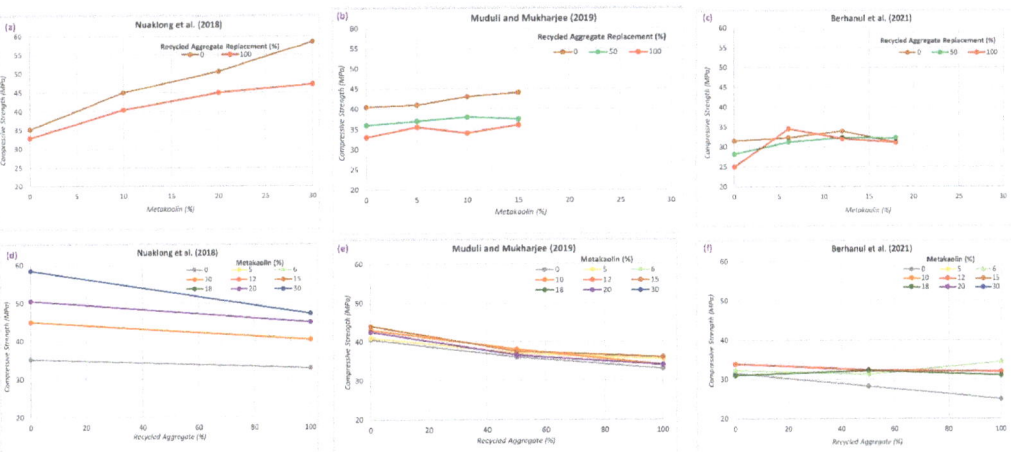

Figure 4. Relationship between the compressive strength of GRAC with different percentage contents of RA and MK. (**a**) compressive strength variation with different percentage of MK for 0% and 100% RA replacement [37]; (**b**) compressive strength variation with different percentage of MK for 0%, 50% and 100% RA replacement [38]; (**c**) compressive strength variation with different percentage of MK for 0%, 50% and 100% RA replacement [40]; (**d**) compressive strength relation with different percentage of RA for different percentage of MK [37]; (**e**) compressive strength relation with different percentage of RA for different percentage of MK [38]; (**f**) compressive strength relation with different percentage of RA for different percentage of MK [40].

Figure 4a,b shows that the larger the replacement with MK, the larger the increase in the compressive strength; surprisingly, the increase is more evident for NA than for RA. By contrast, the results of Berhanul et al. (2021) [40], shown in Figure 4c, show an optimum MK replacement ratio, and the larger the recycled aggregate replacement ratio, the smaller the value of the optimum MK replacement ratio.

This trend is not reflected in most of the other research in the literature. Figure 4d,e shows the generalized reduction in the compressive strength with an increase in the RA replacement ratio; interestingly, the larger the MK content, the smaller the compressive strength reduction due to the use of RA.

A more comprehensive comparison of the different approaches to compensating the strength reduction due to the incorporation of significant amounts of recycled aggregates is reported in Figure 5, where the results of tests on both compressive and flexural strength (green line) are reported for different binders and recycling aggregate replacement ratios. From the discussion above and by critically analyzing the data represented in Figure 5, it can be seen that in the past, the strength and properties of GPC were not enough to incorporate a large percentage of recycled aggregates. Hence, low-quality GRAC was produced with less strength. However, with the modification in the chemical composition of GPC, i.e., by the addition of MK mixed with a NaOH solution with different concentrations and GGBS, a significant improvement was observed in the strength of GRAC, and encouraging results were obtained. In most studies, it was found that the strength was increased with the addition of more MK and GGBS, while it was reduced with the addition of more RA. However, in Nuakalong et al. (2016) [29], an optimal value of NaOH was found to be 12% while in Berhanul et al. (2021) [40], with MK, it was found to be 6%. The above results are qualitatively represented in Figure 6.

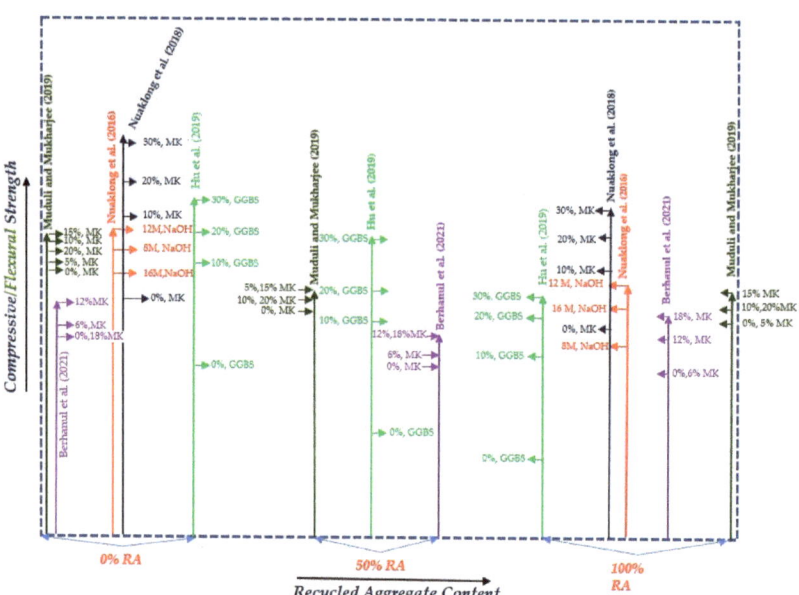

Figure 5. Comparative overview of previous research on GRAC strength with the inclusion of different percentages of RA and other materials (MK and GGBS) [29,33,37,38,40].

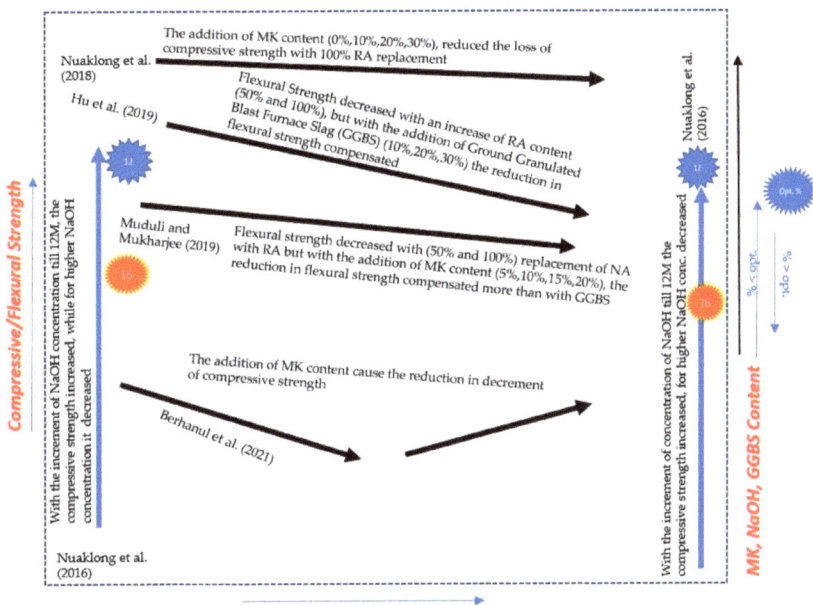

Figure 6. General overview of previous research on GRAC [29,33,37,38,40].

3. Research Progress on the Structural Members Made up of GRAC

This section focuses on the structural behavior of GRAC members, reviewing recent research developments on flexural strength, the load-bearing capacity, the load-displacement relationship, and the shear behavior of GRAC structural members. The results are discussed on the basis of the results reviewed in the previous section. Most of the research is devoted to investigating the flexural behavior of GRAC-reinforced beams with fixed concrete properties and the RA replacement ratio being varied, while only one paper investigated the behavior of a shear critical beam [44], and another investigated the behavior of a reinforced column under an axial load. Lastly, one paper (Romanazzi et al. (2022) [45]) is devoted to investigating the bond-slip relationship of reinforced GRAC elements.

Thangamanibindhu and Murthy (2015) carried out research on the behavior of environment-cured GRAC beams. The GPC was prepared using GGBS, FA, and sodium silicate solution. Sodium hydroxide was used as an alkali activator, and a superplasticizer was used to improve the mechanical characteristics. A total of nine beams were cast having a dimension of $100 \times 200 \times 1200$ mm^3, longitudinal and hanger reinforcement #2 with $\varnothing = 10$ mm, stirrups of $\varnothing = 6$ mm with a pitch of 100 mm, and tested in flexure. Three beams were prepared with conventional concrete mixes having a 0%, 10%, and 30% replacement of RA (1). Six beams had varying proportions of FA (12.6–8.33% BW), GGBS (4.2–8.5% BW), and recycled coarse aggregates (0%, 10%, and 30%) (2). A four-point loading scheme was used in the test. It was found that the average ultimate loads for GRAC beams ranged from 65 kN to 103.55 kN, while for (1) it ranged from 38.1 kN to 55.6 kN. Moreover, the cracking load of GRAC beams was found on average to be 30% more than that of conventional reinforced concrete beams. The load-carrying capacity of all the beams decreased when a larger quantity of recycled aggregates was incorporated. The same load-deflection characteristics were obtained for ordinary reinforced cement concrete beams and geopolymer concrete with 10% replacement of RA. The deflection for (2) ranged between 5.28 mm and 7.04 mm and for (1) it ranged between 3.35 mm and 4.54 mm. The failure behavior of geopolymer concrete beams was found to be similar to that of cement concrete beams, as both types of beams failed initially due to the yielding of the tensile steel; then concrete crushing occurred [46].

Kathirvel and Kaliyaperumal (2016) conducted an experimental study to investigate the influence of RA obtained from demolished construction waste on the flexural behavior of GRAC beams. The casting of GRAC occurred at room temperature. GGBS (19.71% BW), NaOH (3.28% BW), sodium silicate (6.5% BW), and superplasticizers were used to achieve high strength. A total of six beams with dimensions of 1.5 m × 0.1 m × 0.15 m, having #2 @ 12 mm ⌀ longitudinal bars, #2 @ 8 mm ⌀ hanger bars, and stirrups of 6 mm @100 mm c/c were cast. Five beams were of geopolymer concrete having 0%, 25%, 50%, 75%, and 100% replacement of RA, and one beam was normal concrete. All beams were tested under a four-point flexure load scheme similar to that indicated by ASTM C1161 [47]. RA was pre-wetted to mitigate the consequence of the rapid reduction in concrete workability.

The results revealed that with an increase in RA content, there was a slight decrease in initial stiffness. Due to pre-wetting and the inclusion of plasticizers up to 50% RA replacement, the compressive strength and water absorption characteristics improved (Figure 7a), while with the replacement of NA with RA up to 75% the load-bearing capacity of beams increased (Figure 7b). By contrast, it started to decrease after the replacement exceeded 75% [19].

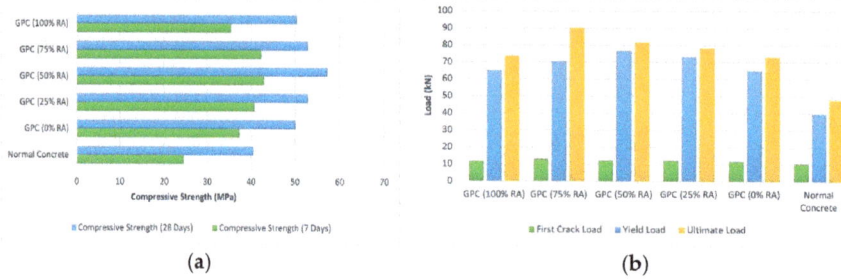

(a)   (b)

Figure 7. (a) The compressive strength of cylindrical specimens with different percentages of RA after 7 and 28 days; (b) the load-carrying capacity of beams at various RA replacement ratios [19].

Deepa and Jithin (2017) performed an experimental study to find the strength and behavior of GRAC beams. RA taken from demolition waste were used as coarse aggregate.

The ingredients of GPC were low-calcium FA (Class F), sodium silicate alkaline solutions, and a sodium hydroxide solution. Coarse aggregate, fine aggregate, and superplasticizer were used with GPC to prepare GRAC. NA were replaced with RA with the following percentages: 20%, 30%, 40%, 50%, and 60%. The optimum replacement ratio was found to be 40% on the basis of workability. Beams were cast of dimensions 175 mm × 150 mm × 1200 mm, having #2 @ 10 mm ⌀ and #1 @ 6 mm ⌀ longitudinal bars, #2 @ 6 mm ⌀ hanger bars, and stirrups of 6 mm @100 mm c/c. The beams were then subjected to a bending test. From the experimental study, it was concluded that there was a slight reduction in strength and deformability with the addition of RA (Figure 8b). The flexural strength of a GRAC beam with 20% RA replacement was 4% lower than that of a geopolymer concrete beam with 100% NA, while the reduction was 31% for the GRAC beam with 60% RA replacement. It was also observed that the GRAC beams showed a larger size and increased number of cracks as compared to normal geopolymer concrete beams (Figure 8a). This is due to the porous structure of the recycled aggregates which produces a reduction in the tensile strength of GRAC [48].

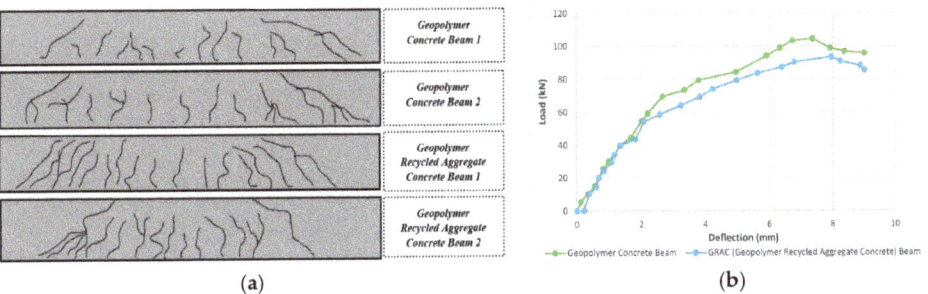

Figure 8. (a) The crack pattern of geopolymer concrete beams and GRAC beams; and (b) the load-deflection curves of geopolymer concrete beams and GRAC beams [48].

The main aim of the experimental study performed by Srinivas and Anhignya (2021) was to determine the optimum percentage of RA for GRAC and to study the behavior of structural members (beams and columns) when subjected to axial compression or bending. Recycled aggregates were obtained from demolition waste and mixed with geopolymer cement to produce GRAC. In the production of GRAC, GPC was prepared using fly ash (13.68% BW), GGBS (2.41% BW), and an alkaline solution (sodium silicate and sodium hydroxide). Naphthalene sulphonate formaldehyde and superplasticizers were used for better strength of GPC. In this experimental study, 20%, 30%, 40%, 50%, and 60% replacement of recycled aggregates was made. Three beams and three columns of dimensions 150 mm × 150 mm × 1200 mm were prepared. The beams were tested under a four-point loading test and the columns were tested under axial loading.

The results revealed that based on mechanical properties and workability, the optimum replacement percentage of RA was 40%. The crack pattern and failure mode of GRAC beams and geopolymer concrete beams were the same. GRAC columns with 40% of RA in axial compression behaved in the same way as geopolymer concrete beams. Due to the inclusion of naphthalene sulphonate formaldehyde, it was also found that almost 8% more ultimate load strength was obtained for GRAC as compared to geopolymer concrete beams; it was thus suggested that GRAC is a practical and eco-friendly solution [49].

Zhang et al. (2021) conducted a comparative study on GRAC and ordinary recycled aggregate concrete beams (OPC with RA). Static loading tests were conducted on three ordinary recycled aggregate concrete beams and seven GRAC beams. Metakaolin-based fly ash geopolymer and alkaline solution were used in the preparation of GPC. MK was used at 5.65% BW, while fly ash and potassium silicate were used at 5.65% BW. The test variables included the RA replacement ratio, the replacement pattern, and the reinforcement ratio. Three replacement ratios (30%, 70%, and 100%) of RA were taken. The conventional aggregate replacement pattern was to replace the same percentage of all particle sizes but in a new, larger replacement pattern; (up to 19 mm) NA particles were replaced with RA, and a 70% replacement ratio was used/set in both replacement patterns. Ten reinforced concrete beams with dimensions (1800 mm (L) × 100 mm (W) × 250 mm (H)) having the same geometry but different concrete types and replacement ratios were made. The bottom longitudinal reinforcement of eight beams was #2 @ 14 mm ⌀, one beam was #2 @ 10 mm ⌀, and one was #2 @ 18 mm ⌀. For all beams, the hanging bars were #2 @ 10 mm ⌀ and stirrups were 6 mm ⌀ @ 100 mm spacing.

The tests revealed that the geopolymer concrete has the same compressive strength as ordinary concrete but with a smaller elastic modulus (e.g., 28.9 GPa for ordinary and 10.2 GPa for geopolymer) because Young's modulus of geopolymer concrete is affected by a microstructure based on speciation of the alkali silicate activating solutions as well as the properties of the aggregates. Because of this, GRAC beams have a lower height of the neutral axis and more deflection than ordinary recycled aggregate concrete beams at the same loading (Figure 9a), depending on replacement patterns. The ultimate deflection was found to be 17.9 mm for CC14-100 (ordinary concrete beam with longitudinal bars of 14 mm ø with 100% RA), 19 mm for GC14-100 (geopolymer concrete beam with longitudinal bars of 14 mm ø with 100% RA), 12.7 mm for CC14-70-L (ordinary concrete beam with longitudinal bars of 14 mm ø with a 70% replacement of large natural aggregates), and 21.9 mm for GC14-70-L (geopolymer concrete beam with longitudinal bars of 14 mm ø with a 70% replacement of large natural aggregates) (Figure 9b). GRAC beams also had a slightly lower cracking load, ductility, and bending capacity. It was also found that the cracking load and cracking moment of the GRAC with 100% RA were found to be approximately 23% lower compared to the ordinary concrete beam with 100% NA. In this study, a high alkali solution was used which reduced the elastic modulus. When the alkali concentration was reduced, the geopolymer concrete showed better results [50].

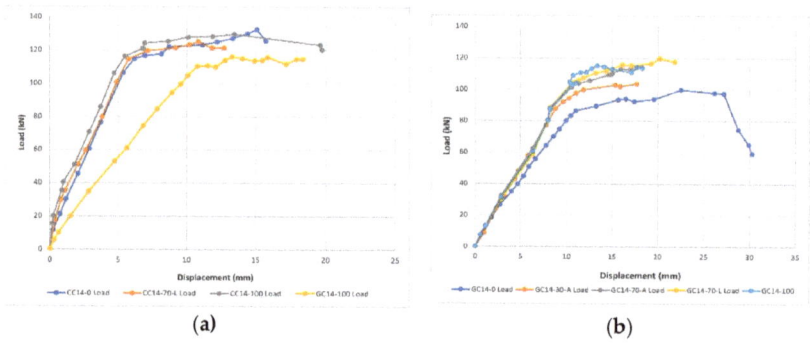

Figure 9. (a) The load-displacement curves of geopolymer and ordinary Portland cement beams; and (b) the load-displacement curves of GRAC with different ratios and replacement patterns [50].

Aldemir et al. (2022) were the only researchers who investigated the shear behavior and structural performance of GRAC beams in detail. A new type of geopolymer concrete was prepared from demolition wastes. Roof tiles, red clay, hollow bricks, and concrete rubble were used as wastes along with slag, fly ash, sodium hydroxide, and sodium silicate. In this study, four different mixes were prepared: (1) GRAC, (2) geopolymer natural aggregate concrete, (3) ordinary recycled aggregate concrete, and (4) normal concrete. In previous studies, the authors found that with the addition of RA both in ordinary concrete and in GPC, workability and compressive strength decreased but, in this study, it was assessed that, when the same Water Cement Ratio (W/C) ratio is used, in GRAC the porosity of the concrete decreased because part of the water is absorbed by the RA, thus increasing the strength of the concrete. Three shear span-to-effective depth ratios (a/d = 0.5, 1, 1.65) were used to examine the different failure modes. Four beams of dimension 150 × 250 × 1100 mm^3 were cast for each shear span-to-effective depth ratio and each concrete mix type. Then, 4-point bending tests were performed to determine the shear behavior of the beams. Parameters including load-deflection curves, moment curves, and crack propagation were used to assess the mechanical performance of the beams. It was also found that the compressive strength of the members made of this GRAC was 3% higher than that of conventional concrete members. The results indicated that the beams made up of geopolymer concrete exhibited a similar performance to normal concrete beams of the same grade. However, when recycled aggregates were used, then the failure mechanism shifted from flexure-dominated to shear-dominated. This shift was more common in the beams with a larger span to an effective depth ratio [44].

Raza et al. (2021) performed an experimental study on the structural performance of GRAC columns with glass fiber-reinforced composite bars and hooks subjected to a compressive axial load. Nine mid-scale circular columns of dimensions 250 mm × 1150 mm, having six, eight, or ten longitudinal reinforcing bars (reinforcement ratio of 1.57%, 2.11%, and 2.6%), and pitch hooks at 75 mm, 150 mm, and 250 mm (corresponding to a transversal reinforcement volumetric ratio of 1.42%, 0.71%, and 0.50%, respectively) were tested. The mix design of GPC was chosen with the following proportion by weight: RA 50.13%, sand 21.18%, water 5.20%, sodium hydroxide solution (14 M) 1.65%, FA 10.236%, GGBS 6.89%, superplasticizer Sika ViscoCrete-3425 0.16%, and sodium silicate 4.44%.

GRAC was prepared with the 100% replacement of NA. The axial force-displacement curves in Figure 10 stress the influence of the GFRP longitudinal reinforcement ratio and circular hoop spacing on the strength and deformation capacity of the GRAC. The authors noted that an increase in the number of longitudinal GFRP bars up to eight improved the axial load capacity of GRAC members, while a further increase to ten bars reduced the axial load capacity of the specimens. Reducing the hoop pitch from 250 mm to 150 mm produced an average increase in the axial load capacity by 6.3%, while a further reduction in the spacing to 50 mm produced a total gain of 1.13%. A noticeable increase in the ultimate deflection, i.e., the deflection of the post-peak softening branch at which the specimen attains 85% of its load capacity, was only found when ten vertical reinforcing bars were put in place (21% and 25% for pitch reduction from 250 mm up to 150 mm and 50 mm, respectively), while a clear trend was not revealed for six and eight longitudinal rebars [51]. The results prove that the effect of confinement provided by transversal reinforcement can be fully exploited to prevent the buckling of a longitudinal bar, i.e., the longitudinal bar diameter is large enough to avoid buckling phenomena with the chosen hook pitch.

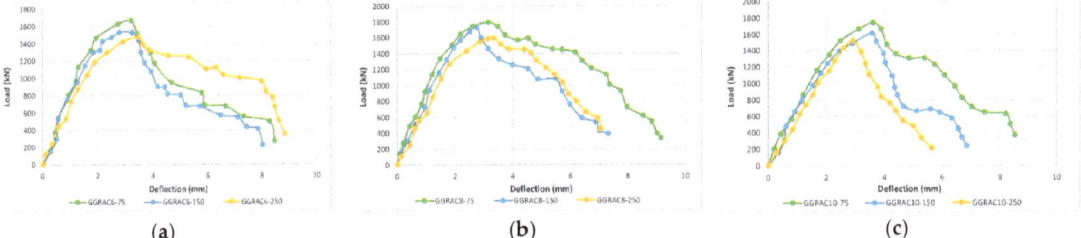

Figure 10. The load-displacement curves of GRAC columns with different reinforcement from glass fiber reinforced composite bars. (**a**) With 6 mm diameter bars and a hook pitch of 75 mm, 150 mm, and 250 mm, (**b**) with 8 mm diameter bars and a hook pitch of 75 mm, 150 mm, and 250 mm, and (**c**) with 10 mm diameter bars and a hook pitch of 75 mm, 150 mm, and 250 mm [51].

The bond strength between the concrete and steel reinforcement is essential for the ultimate strength of the structural members. Romanazzi et al. (2022) performed an experimental investigation to examine the bond behavior of GPC with steel bars and sand-coated Glass Fiber Reinforced Polymer (GFRP) bars using a pull-out test. The test results revealed that the adhesion bond characterizing the behavior of GPC is stronger than that shown with OPC, both for sand-coated GFRP bars and deformed steel. However, the ultimate bond strength of GPC with steel bars was two to three times higher than that of sand-coated GFRP bars. This is due to the fact that there is adequate mechanical interlocking and a good bond strength between the GPC and steel bars, irrespective of the bar diameter. Thus, in the case of GFRP bars, the predominant mechanisms are those of adhesion and friction, while between concrete and steel bars, the predominant contribution is that of mechanical interlocking [45].

4. Discussion

- The use of geopolymer concrete with recycled aggregate is a complex topic. The efficiency of GRAC depends on several factors: the type and composition of GPC, the molarity of the alkaline solutions, the mixing procedure, the curing temperature and environmental conditions, and the mechanical and chemical characteristics of the RA and ITZ; how these factors relate to the amount and characteristics of the attached

- mortar; the extraction process from the demolition construction waste; and, lastly, the replacement percentage of RA with NA, etc.
- The type of geopolymer cement created is one of the most important aspects, as different amounts and types of chemicals are used in the production of GPC. In older studies, GPC was prepared with fly ash and the results were not very encouraging (the recycled aggregate replacement ratio was limited to 30–40%).
- In recent research, it was found that when metakaolin and GGBS-based GPC were used, larger values of replacement ratios (up to 100%) can occur without a significant reduction in the flexural strength of the structural member. The tensile strength of GPC concrete can increase the cracking strength of the beam, which decreases with an increase in the RA replacement ratio; however, OPC with NA-reinforced beams and GRAC-reinforced beams usually exhibit a similar failure mode and cracking pattern. Only a shear-critical beam can exhibit premature failure when GRAC with a large RA replacement ratio is used.
- A large amount of RA can cause workability issues, but from the latest research, it was also found that the use of polycarboxylate superplasticizers and methyl isobutyl carbinol (MIBC) can improve workability, allowing for a reduction in the W/B ratio and increasing the strength of metakaolin-based geopolymers.
- In this regard, it must be pointed out that most of the research was performed using RA in a saturated surface condition which reduces the compressive strength and elastic modulus of the concrete.
- The excessive use of alkali activators reduces the elastic modulus of concrete, causing an increase in beam deflection. Hence, a precise quantity of alkali activators in relation to other materials should be used in GPC production. There is not yet a unanimous consensus on the exact quantity and type of material; therefore, the quantity and type of GPC should be chosen depending on the characteristics of available RA.
- An Interfacial Transition Zone (ITZ) develops between the attached mortar and the new cement paste. This is one of the weakest zones, so proper chemical and mechanical treatment is advised before using RA. It was found in the latest research that the addition of fillers and fly ash is helpful to fill the pores of RA, thus reducing the vulnerability of failure along the ITZ; to this aim, the Two Stage Mixing Technique (TSMA) can also be adopted, in which a cement coating forms on the surface of the recycled aggregate, thus filling up the cracks before actual mixing of concrete.
- From the literature review, it is seen that there are no general limits to the use of coarse RA in a concrete mixture. Some of the former researchers recommended a maximum 30% replacement of NA with RA, while recent researchers suggest that the RA replacement can be up to 50% or 100% if the mix design, batching methodology, and moisture condition of the RA are properly handled. In most of the research, the use of a 70% RA replacement ratio does not significantly affect the load capacity and only slightly affects the deformability of the GRAC beam loaded in flexure.
- In general, much research is still needed to identify the optimal mix design and to optimize the production methods and rules for setting the geopolymer concrete and its mechanical properties, particle size distributions, and aggregate processing for the production of GRAC. However, current knowledge already makes it possible to produce GRAC with a predetermined class of compressive strength, while its tensile strength and related characteristics, such as the bond between GRAC and reinforcement, related cracking phenomena, and ductility that can be conferred through confinement, are more uncertain.
- Particular attention must be paid when GRAC is to be used in conjunction with GFRP bars, since the confinement of the transverse reinforcement, which is able to provide an increase in compressive strength, is not always effective in increasing the deformation at the decay of resistance to 85% of the maximum value, and, more generally, ductility and toughness, a condition which can only be achieved in the presence of an adequate

number of longitudinal bars capable of transferring the confinement action exerted by the transverse reinforcement more uniformly.
- Regarding the bond between GRAC and reinforcement, GRAC has exhibited promising behavior when used in conjunction with steel cold-form reinforcement, while bond strength is reduced up to 33% when used with sand-coated GFRP bars. This is because the predominant influence ensuring the bond between concrete and steel bars is mechanical interlocking, while in the case of GFRP bars, the predominant mechanisms are those of adhesion and friction.
- It is also found that no research has been conducted on the beam-to-column joint made up of GRAC. The beam-column joint is one of the most vulnerable structural members belonging to moment-resisting frames made of cast-in-situ concrete. Thus, in order to prove the effectiveness of GRAC, this aspect should also be analyzed.
- Similarly, a research gap has been found regarding the seismic behavior of GRAC structural members. The seismic assessment of GRAC structural members should be carried out in order to evaluate the ductility and energy dissipation capacity of reinforced GRAC for construction in seismic areas.
- Therefore, from the above literature review, it can be concluded that there is still a need for experimental tests that study the behavior of structural members made up of GRAC, characterizing the phenomenon of bond, the strength and ductility of members subjected to bending with or without axial load, and the shear strength of members with and without transverse reinforcement. Moreover, there is a need for studies focusing on the influence of the casting and curing conditions on the mechanical strengths of the structural member.

Funding: This research received no external funding.

Institutional Review Board Statement: Not applicable.

Informed Consent Statement: Not applicable.

Data Availability Statement: The data presented in this study are available on request from the corresponding author.

Conflicts of Interest: The authors declare no conflict of interest.

Abbreviations

GRAC	Geopolymer Recycled Aggregate Concrete
OPC	Ordinary Portland Cement
GPC	Geo Polymer Cement
GGBS	Ground Granulated Blast Furnace Slag
RA	Recycled Aggregates
ITZs	Interfacial Transition Zones
NA	Natural Aggregate
MK	Metakaolin
HCFA	High Calcium Fly Ash
PGC	Pressed Geopolymer Concrete
RB	Recycled Concrete Block Aggregate
FA	Fly ash
MMA	Mortar Mixing Approach
SEMA	Sand-Enveloped Mixing Approach
GFRP	Glass Fiber Reinforced Polymer
TSMA	Two-Stage Mixing Technique
BW	By Weight
W/B	Water to Binder ratio
MIBC	Methyl Isobutyl Carbinol
W/C	Water Cement Ratio

References

1. Imbabi, M.S.; Carrigan, C.; McKenna, S. Trends and developments in green cement and concrete technology. *Int. J. Sustain. Built Environ.* **2012**, *1*, 194–216. [CrossRef]
2. Malhotra, V.M.; Mehta, P.K. *High-Performance, High-Volume Fly Ash Concrete: Materials, Mixture Proportioning, Properties, Construction Practice, and Case Histories*; Supplementary Cementing Materials for Sustainable Development: Ottawa, ON, Canada, 2002.
3. Langer, W.H. *Managing and Protecting Aggregate Resources*; Open-File Report 02-415; U.S. Department of the Interior, U.S. Geological Survey: Reston, VA, USA, 2002.
4. Akhtar, A.; Sarmah, A.K. Construction and demolition waste generation and properties of recycled aggregate concrete: A global perspective. *J. Clean. Prod.* **2018**, *186*, 262–281. [CrossRef]
5. Verian, K.P.; Ashraf, W.; Cao, Y. Properties of recycled concrete aggregate and their influence in new concrete production. *Resour. Conserv. Recycl.* **2018**, *133*, 30–49. [CrossRef]
6. La Scalia, G.; Saeli, M.; Adelfio, L.; Micale, R. From lab to industry: Scaling up green geopolymeric mortars manufacturing towards circular economy. *J. Clean. Prod.* **2021**, *316*, 128164. [CrossRef]
7. Thomas, B.S.; Yang, J.; Bahurudeen, A.; Chinnu, S.; Abdalla, J.A.; Hawileh, R.A.; Hamada, H.M. Geopolymer concrete incorporating recycled aggregates: A comprehensive review. *Clean. Mater.* **2022**, *3*, 100056. [CrossRef]
8. Srividya, T.; Kannan Rajkumar, P.R.; Sivasakthi, M.; Sujitha, A.; Jeyalakshmi, R. A state-of-the-art on development of geopolymer concrete and its field applications. *Case Stud. Constr. Mater.* **2022**, *16*, e00812. [CrossRef]
9. Lee, S.; Van Riessen, A. A Review on Geopolymer Technology for Lunar Base Construction. *Materials* **2022**, *15*, 4516. [CrossRef]
10. Kim, B.; Lee, S.; Chon, C.-M.; Cho, S. Setting Behavior and Phase Evolution on Heat Treatment of Metakaolin-Based Geopolymers Containing Calcium Hydroxide. *Materials* **2022**, *15*, 194. [CrossRef]
11. Pelisser, F.; Silva, B.V.; Menger, M.H.; Frasson, B.J.; Keller, T.A.; Torii, A.J.; Lopez, R.H. Structural analysis of composite metakaolinbased geopolymer concrete. *Rev. IBRACON Estrut. Mater.* **2018**, *11*, 535–543. [CrossRef]
12. Matsimbe, J.; Dinka, M.; Olukanni, D.; Musonda, I. Geopolymer: A Systematic Review of Methodologies. *Materials* **2022**, *15*, 6852. [CrossRef]
13. Saeli, M.; Mikael, R.; Seabra, M.P.; Labrincha, J.A.; La Scalia, G. Selection of Novel Geopolymeric Mortars for Sustainable Construction Applications Using Fuzzy Topsis Approach. *Sustainability* **2020**, *12*, 5987. [CrossRef]
14. Nawaz, M.; Heitor, A.; Sivakumar, M. Geopolymers in construction—Recent developments. *Constr. Build. Mater.* **2020**, *260*, 120472. [CrossRef]
15. Wang, B.; Yan, L.; Fu, Q.; Kasal, B. A Comprehensive Review on Recycled Aggregate and Recycled Aggregate Concrete. *Resour. Conserv. Recycl.* **2021**, *171*, 105565. [CrossRef]
16. Senaratne, S.; Lambrousis, G.; Mirza, O.; Tam, V.W.Y.; Kang, W.-H. Recycled Concrete in Structural Applications for Sustainable Construction Practices in Australia. *Procedia Eng.* **2017**, *180*, 751–758. [CrossRef]
17. Le, H.B.; Bui, Q.B. Recycled aggregate concretes—A state-of-the-art from the microstructure to the structural performance. *Constr. Build. Mater.* **2020**, *257*, 119522. [CrossRef]
18. Kapoor, K.; Bohroo, A.U.R. Study on the Influence of Attached Mortar Content on the Properties of Recycled Concrete Aggregate. In *Sustainable Engineering*; Lecture Notes in Civil Engineering; Agnihotri, A., Reddy, K., Bansal, A., Eds.; Springer: Singapore, 2019; Volume 30. [CrossRef]
19. Kathirvel, P.; Kaliyaperumal, S.M. Influence of recycled concrete aggregates on the flexural properties of reinforced alkali activated slag concrete. *Constr. Build. Mater.* **2016**, *102*, 51–58. [CrossRef]
20. Nikmehr, B.; Al-Ameri, R. A State-of-the-Art Review on the Incorporation of Recycled Concrete Aggregates in Geopolymer Concrete. *Recycling* **2022**, *7*, 51. [CrossRef]
21. Tabsh, S.W.; Abdelfatah, A.S. Influence of recycled concrete aggregates on strength properties of concrete. *Constr. Build. Mater.* **2009**, *23*, 1163–1167. [CrossRef]
22. Pani, L.; Francesconi, L.; Rombi, J.; Mistretta, F.; Sassu, M.; Stochino, F. Effect of parent concrete on the performance of recycled aggregate concrete. *Sustainability* **2020**, *12*, 9399. [CrossRef]
23. McNeil, K.; Kang, T.H.K. Recycled Concrete Aggregates: A Review. *Int. J. Concr. Struct. Mater.* **2013**, *7*, 61–69. [CrossRef]
24. Almutairi, A.L.; Tayeh, B.A.; Adesina, A.; Isleem, H.F.; Zeyad, A.M. Potential applications of geopolymer concrete in construction: A review. *Case Stud. Constr. Mater.* **2021**, *15*, e00733. [CrossRef]
25. Uddin Ahmed Shaikh, F. Mechanical and durability properties of fly ash geopolymer concrete containing recycled coarse aggregates. *Int. J. Sustain. Built Environ.* **2016**, *5*, 277–287. [CrossRef]
26. *AS 3600*; Design of Concrete Structures. Standards Australia, SAI Global: Sydney, Australia, 2009.
27. Ryu, G.S.; Lee, Y.K.; Koh, K.T.; Chung, Y.S. The mechanical properties of fly ash based geopolymer concrete with alkaline activators. *Constr. Build. Mater.* **2013**, *47*, 409–418. [CrossRef]
28. Diaz-Ioya, E.I.; Allouche, E.N.; Vaidya, S. Mechanical properties of fly ash based geopolymer concrete. *ACI Mater. J.* **2011**, *108*, 300–306.
29. Nuaklong, P.; Sata, V.; Chindaprasirt, P. Influence of recycled aggregate on fly ash geopolymer concrete properties. *J. Clean. Prod.* **2016**, *112*, 2300–2307. [CrossRef]
30. ACI Committee 213. *Guide for Structural Lightweight Aggregate Concrete*; American Concrete Institute: Farmington Hills, MI, USA, 2003.
31. Wongsa, A.; Siriwattanakarn, A.; Nuaklong, P.; Sata, V.; Sukontasukkul, P.; Chindaprasirt, P. Use of recycled aggregates in pressed fly ash geopolymer concrete. *Environ. Prog. Sustain.* **2020**, *39*, e13627. [CrossRef]

32. Le, H.-B.; Bui, Q.-B.; Tang, L. Geopolymer Recycled Aggregate Concrete: From Experiments to Empirical Models. *Materials* **2021**, *14*, 1180. [CrossRef]
33. Hu, Y.; Tang, Z.; Li, W.; Li, Y.; Tam, V.W.Y. Physical-Mechanical Properties of Fly ash/GGBFS Geopolymer Composites with Recycled Aggregates. *Constr. Build. Mater.* **2019**, *226*, 139–151. [CrossRef]
34. Xie, J.; Wang, J.; Rao, R.; Wang, C.; Fang, C. Effects of combined usage of GGBS and fly ash on workability and mechanical properties of alkali activated geopolymer concrete with recycled aggregate. *Compos. B Eng.* **2019**, *164*, 179–190. [CrossRef]
35. Srinivas, T.; Abhignya, G. A Review on Geopolymer RCC Beams made with Recycled Coarse Aggregate. In *Proceedings of the 2nd International Conference on Design and Manufacturing Aspects for Sustainable Energy, Hyderabad, India, 10–12 July 2020*; E3S Web of Conferences: Les Ulis, France, 2020; Volume 184, p. 01095.
36. Moulya, H.V.; Chandrashekhar, A. Experimental Investigation of Effect of Recycled Coarse Aggregate Properties on the Mechanical and Durability Characteristics of Geopolymer Concrete. *Mater. Today Proc.* **2022**, *59*, 1700–1707. [CrossRef]
37. Nuaklong, P.; Sata, V.; Chindaprasirt, P. Properties of metakaolin-high calcium fly ash geopolymer concrete containing recycled aggregate from crushed concrete specimens. *Constr. Build. Mater.* **2018**, *161*, 365–373. [CrossRef]
38. Muduli, R.; Mukharjee, B.B. Effect of incorporation of metakaolin and recycled coarse aggregate on properties of concrete. *J. Clean. Prod.* **2019**, *209*, 398–414. [CrossRef]
39. Lee, S.; Kim, B.; Seo, J.; Cho, S. Beneficial Use of MIBC in Metakaolin-Based Geopolymers to Improve Flowability and Compressive Strength. *Materials* **2020**, *13*, 3663. [CrossRef]
40. Berhanul, A.; Gebreyouhannes, E.; Zerayohannes, G.; Zeleke, E. The Effect of Metakaolin on the Properties of Recycled Aggregate Concrete, Concrete Structures: New Trends for Eco-Efficiency and Performance. In Proceedings of the fib Symposium 2021, Lisbon, Portugal, 14–16 June 2021.
41. Xu, Z.; Huang, Z.; Liu, C.; Deng, X.; Hui, D.; Deng, S. Research progress on mechanical properties of geopolymer recycled aggregate concrete. *Rev. Adv. Mater. Sci.* **2021**, *60*, 158–172. [CrossRef]
42. Liang, Y.-C.; Ye, Z.-M.; Vernerey, F.; Xi, Y. Development of processing methods to improve strength of concrete with 100% recycled coarse aggregate. *J. Mater. Civ. Eng.* **2013**, *27*, 04014163. [CrossRef]
43. Alqarni, A.S.; Abbas, H.; Al-Shwikh, K.M.; Al-Salloum, Y.A. Treatment of recycled concrete aggregate to enhance concrete performance. *Constr. Build. Mater.* **2021**, *307*, 124960. [CrossRef]
44. Aldemir, A.; Akduman, S.; Kocaer, O.; Aktepe, R.; Sahmaran, M.; Yildirim, G.; Almahmood, H.; Ashour, A. Shear behaviour of reinforced construction and demolition waste based geopolymer concrete beams. *J. Build.* **2022**, *47*, 103861. [CrossRef]
45. Romanazzi, V.; Leone, M.; Aiello, M.; Pecce, M. Bond behavior of geopolymer concrete with steel and GFRP bars. *Compos. Struct.* **2022**, *300*, 116150. [CrossRef]
46. Thangamanibindhu, M.K.; Murthy, D.R. Flexural behavior of reinforced geopolymer concrete beams partially replaced with recycled coarse aggregates. *Int. J. Civ. Eng. Technol.* **2015**, *6*, 13–23.
47. American Society for Testing and Materials (ASTM). *C1161-18; Standard Test Method for Flexural Strength of Advanced Ceramics at Ambient Temperature*. American Society for Testing and Materials (ASTM): West Conshohockenm, PA, USA, 2018.
48. Deepa, R.S.; Jithin, B. Strength and behaviour of recycled aggregate geopolymer concrete beams. *Adv. Concr. Constr.* **2017**, *5*, 145–154. [CrossRef]
49. Srinivas, T.; Abhignya, G. Behaviour of Structural Elements made of Geopolymer Concrete with Recycled Aggregates. In *IOP Conference Series: Materials Science and Engineering, Proceedings of the 3rd International Conference on Inventive Research in Material Science and Technology (ICIRMCT 2021), Coimbatore, India, 22–23 January 2021*; IOP Publishing Ltd.: Bristol, UK, 2021; Volume 1091, p. 1091. [CrossRef]
50. Zhang, H.; Wan, K.; Wu, B.; Hu, Z. Flexural behavior of reinforced geopolymer concrete beams with recycled coarse aggregates. *Adv. Struct. Eng.* **2021**, *24*, 3281–3298. [CrossRef]
51. Raza, A.; Rashedi, A.; Rafique, U.; Hossain, N.; Akinyemi, B.; Naveen, J. On the Structural Performance of Recycled Aggregate Concrete Columns with Glass Fiber-Reinforced Composite Bars and Hoops. *Polymers* **2021**, *13*, 1508. [CrossRef] [PubMed]

Article

Development of Geopolymer-Based Materials with Ceramic Waste for Artistic and Restoration Applications

Laura Ricciotti [1,*], Alessio Occhicone [2], Claudio Ferone [2,3], Raffaele Cioffi [2,3], Oreste Tarallo [4] and Giuseppina Roviello [2,3,*]

1 Department of Architecture and Industrial Design, University of Campania, Luigi Vanvitelli, 81031 Aversa, Italy
2 Department of Engineering, University of Naples 'Parthenope', Centro Direzionale, Isola C4, 80143 Napoli, Italy
3 INSTM Research Group Napoli Parthenope, National Consortium for Science and Technology of Materials, Via G. Giusti, 9, 50121 Firenze, Italy
4 Department of Chemical Sciences, University of Naples Federico II, 80126 Naples, Italy
* Correspondence: laura.ricciotti@unicampania.it (L.R.); giuseppina.roviello@uniparthenope.it (G.R.); Tel.: +39-081-5476799 (L.R.); +39-081-5476781 (G.R.)

Abstract: This contribution presents the preparation and characterization of new geopolymer-based mortars obtained from recycling waste deriving from the production process and the "end-of-life" of porcelain stoneware products. Structural, morphological, and mechanical studies carried out on different kinds of mortars prepared by using several types of by-products (i.e., pressed burnt and extruded ceramic waste, raw pressed and gypsum resulting from exhausted moulds) point out that these systems can be easily cast, also in complex shapes, and show a more consistent microstructure with respect to the geopolymer paste, with a reduced amount of microcracks. Moreover, the excellent adhesion of these materials to common substrates such as pottery and earthenware, even for an elevated concentration of filler, suggests their use in the field of technical-artistic value-added applications, such as restoration, conservation, and/or rehabilitation of historic monuments, or simply as materials for building revetments. For all these reasons, the proposed materials could represent valuable candidates to try to overcome some problems experienced in the cultural heritage sector concerning the selection of environmentally friendly materials that simultaneously meet art and design technical requirements.

Keywords: geopolymer; ceramic wastes; mortars; art and design; recycling

Citation: Ricciotti, L.; Occhicone, A.; Ferone, C.; Cioffi, R.; Tarallo, O.; Roviello, G. Development of Geopolymer-Based Materials with Ceramic Waste for Artistic and Restoration Applications. *Materials* 2022, 15, 8600. https://doi.org/10.3390/ma15238600

Academic Editor: Sujeong Lee

Received: 1 November 2022
Accepted: 29 November 2022
Published: 2 December 2022

Publisher's Note: MDPI stays neutral with regard to jurisdictional claims in published maps and institutional affiliations.

Copyright: © 2022 by the authors. Licensee MDPI, Basel, Switzerland. This article is an open access article distributed under the terms and conditions of the Creative Commons Attribution (CC BY) license (https:// creativecommons.org/licenses/by/ 4.0/).

1. Introduction

Ceramic products embody a fundamental segment of the Italian manufacturing industry: currently, this business, which includes the production of sanitary ceramics, ceramic tiles and slabs, porcelain and tableware, technical ceramics and bricks, and refractory materials, includes 279 companies and over 27,500 employees with a turnover of 6.5 billion euros [1]. In particular, porcelain stoneware represents the main type of production in the Italian ceramic industry: 400 million square metres were manufactured in 2019, accounting for about 80% of the total production [1].

Porcelain stoneware is composed of illitic kaolinitic clays, sodium-potassium feldspars, and feldspar sands as the predominant parts and chromophoric oxides (usually iron and titanium oxides) as the minor part [2]. It is characterized by very low porosity and presents remarkable technological properties such as high mechanical, chemical, abrasion, and stain resistance [3]. These features make it an ideal material for applications in several areas, from building components such as floors to producing artistic objects.

In terms of environmental impacts, the main factors affecting the production of porcelain stoneware are polluting emissions, the extraction of raw materials and their transport (lead, fluorine, boron, powders, CO_2) [4,5], the consistent consumption of energy (mainly consumption of methane gas), water, and the production of solid wastes that usually are destined for landfills [4,5]. Indeed, it has been estimated that about 15 to 30% of this production is considered a by-product and/or waste generally intended for landfills.

Recently, ceramic products and their by-products have been used as fillers in cementing materials for improving their durability and mechanical performances, thanks to the presence of silicoaluminate crystalline components [6]. However, the reuse of ceramic scraps in the building field is rather negligible and in the early stages of its diffusion [7].

The research for alternative sustainable materials has gained great attention due to the green policy introduced by the European Union through the introduction of the Next Generation EU package [8], based on the circular economy and the Italian Ecological Transition Plan, and has created effective routes for the reuse and valorization of a massive amount of industrial wastes.

In this context, while the design of "green" products with a low carbon footprint and a reduced environmental impact according to the principles of the "eco-design" is well structured in sectors such as building and construction [9–14], to date, several issues have been pointed out in the case of the creative and restoring industry, particularly in terms of using of sustainable materials to meet eco-design features [15]. Traditionally, in fact, this sector has only been concerned with providing aesthetic enhancements by using an approach that ignores energy savings and emissions reduction [15]. However, in more recent years the approach has completely shifted toward the so-called eco-design, a new branch of design that aims to integrate environmental aspects during the process of designing products as any other criterion (economic, technological, and so on), to reduce their life cycle impacts. In order to reduce environmental impacts, its directives should be more enforced and routinely integrated into the product development process.

In this regard, geopolymers and alkali-activated materials, which can be made from widely available minerals, are perfect candidates to produce materials with low environmental impacts. In fact, aluminosilicate materials, which are major components of the Earth's crust (65%) [16] as virgin materials or by-products of other industrial processes, are used as raw materials for the synthesis of geopolymer materials [17].

These raw materials can react in alkaline conditions and form amorphous species, characterized by cross-linked networks consisting of Si–O–Al–O bonds [18], that are considered a valid alternative to cement-based materials due to their characteristics such as thermal stability, low shrinkage, freeze-thaw, chemical and fire resistance, recyclability, and long-term durability [19].

One of the most interesting aspects is the possibility of preparing geopolymers by valorizing waste materials from different types of processing. Moreover, it is well known that aluminosilicate materials from bottom ash deriving from blast furnace slag, fly ash from thermoelectric industry residues, red mud, lake residues, and up to rice husk waste can be profitably exploited, giving life to new materials that find applications in various fields, from construction, to the manufacturing of objects, to the development of innovative materials for technological applications, when they are functionalized or produced as hybrids or composites [20–22].

In this regard, the authors of this paper have contributed over the years to this research by developing composites and hybrid materials based on geopolymers able to overcome some limitations of geopolymers that perimeter their extensive application in the construction sector, thus reaching the development of new materials with excellent mechanical performance and good thermal-acoustic insulation properties [23–38].

To the best of our knowledge, the excellent mechanical properties, fast setting time, easy workability, and ability to adhere to different types of substrates of geopolymer-based materials to date were exploited mainly in the construction sector, while no attention was devoted to the application of these materials in the artistic and/or restoration sectors. As a matter of fact, the previously listed properties, combined with the high chemical resistance and good mechanical performances of geopolymer materials, make them excellent candidates for the restoration field and artistic industry.

In this work, geopolymer-based mortars obtained through the valorization of wastes deriving from the production and "end-of-life" of porcelain stoneware products are proposed as eco-friendly materials to be used in the art and design sector. The objectives of this study have been both investigating the potential use of large amounts of ceramic waste in geopolymer-based mortars, since it is an interesting option for ceramic industries and today more and more are interested in recycling and producing sustainable materials, and, at the same time, suggesting the exploitation of these materials in the artistic and/or restoration sectors.

Different kinds of geopolymer mortars were prepared by using all kinds of by-products of the production process of the porcelain stoneware (pressed burnt and extruded ceramic wastes, raw pressed and gypsum resulting from exhausted moulds). These wastes were physically and chemically characterized, and then used as a recycled aggregate.

It is suggested that such materials can be used in the field of the art and design industry since they have excellent rheological properties, are suitable for casting in moulds with complex shapes, and, once consolidated, show good mechanical properties. Moreover, they do not show appreciable shrinkage, have a water absorption capacity similar to unmodified geopolymers, and show an excellent adhesive capacity to various types of substrates such as ceramic, earthenware, tuff, concrete, and marble, even for an elevated concentration of filler, thus suggesting that they could be used also for restoration and consolidation of artistic or archaeological artefacts.

The materials developed in this work potentially offer the possibility of recovering and recycling up to 100% (almost endlessly) of not only waste materials and by-products of the ceramic industries (such as gypsum, raw pressed, and porcelain stoneware waste), but also the geopolymers themselves at the end of their use. In this way, a sensible reduction of the environmental impacts of these materials could be achieved and, in line with the Circular Economy approach, green and economically competitive products could be obtained.

The possibility of extending this approach to valorize and recycle ceramic wastes from different kinds of industries, together with in-depth structural and durability tests and studies on advanced modelling of the architectural composition, could open up new scenarios for making structural and functional elements for sustainable and advanced buildings.

2. Experimental Section

2.1. Materials

Metakaolin MetaMax® (by BASF) was kindly provided by Neuvendis s.p.a. (Milan, Italy) and its composition is reported in Table 1. BASF MetaMax® is a high-purity white mineral admixture that meets or exceeds all the specifications of ASTM C-618 Class N pozzolans. The sodium silicate solution (Table 1) was supplied by Prochin Italia S.r.l (Caserta, Italy), while sodium hydroxide with reagent grade was supplied by Sigma-Aldrich. Ceramic waste and scraps, selected from those produced in greater quantities in the various stages of the porcelain manufacturing process (pressed burnt and extruded ceramic waste, raw pressed and gypsum resulting from exhausted moulds), were kindly supplied by a company producing stoneware and ceramic tiles in the province of Vicenza, Northern Italy.

2.2. Sample Preparation

2.2.1. Geopolymer (MK)

It is well known that the geopolymerization reaction is based on the alkaline activation of an aluminosilicate raw material using a strongly alkaline solution. In this work, NaOH in pellets was dissolved in the sodium silicate solution to prepare the alkaline activating solution.

The solution so prepared was cooled and allowed to equilibrate for 24 h, as reported in refs. [30–38]. The chemical composition of the alkaline activating solution is Na_2O 1.55 SiO_2

12.14 H_2O. BASF Matamax® metakaolin was then incorporated into the activating solution with a liquid-to-solid ratio of 1.4:1 by weight and mixed by a mechanical mixer for 10 min at 800 rpm [25–30]. The composition of the whole geopolymer system can be expressed as Al_2O_3 3.48 SiO_2 1.0 Na_2O 12.14 H_2O, as revealed by EDS analysis carried out on the cured samples. In this paper, the geopolymer sample obtained is indicated as MK.

Table 1. Chemical composition (weight %) of the metakaolin BASF MetaMax® and sodium silicate solution used in this paper.

Compound	Metakaolin	Sodium Silicate
SiO_2	52.2	27.40
Al_2O_3	45.1	-
Na_2O	0.22	8.15
K_2O	0.15	-
TiO_2	1.75	-
Fe_2O_3	0.42	-
CaO	0.04	-
MgO	0.04	-
P_2O_5	0.08	-
H_2O	-	64.45

2.2.2. Geopolymer Mortars

Table 2 shows the composition of the geopolymer-based mortars studied in the present paper. The samples were obtained by adding different percentages by weight of the ceramic wastes (in the range of 6–45 wt.%) to the freshly prepared geopolymer suspension, prepared as described in the previous paragraph, and quickly incorporating by controlled mechanical mixing (5 min at 800 rpm). On the other hand, the mass percentages of the ceramic waste were chosen in order to not change significantly the workability, setting times, and physical-mechanical properties. Ceramic waste was ground before use to obtain a fine powder (particle size in the range of 5–80 μm).

Table 2. Mix design of the geopolymeric samples prepared in this work (MK: neat geopolymer; Gy: gypsum waste; RP: raw pressed ceramic waste; PS: porcelain stoneware waste). In the MIX-MK sample, a mixture of all the used waste was used.

Materials (wt.%)	MK	PS-MK	RP-MK	RP_{dry}-MK [1]	Gy-MK	MIX-MK
Metakaolin	37.5	11	11.8	11.6	15.3	11
NaOH	7.2	5.2	5.6	5.6	7.3	5.2
Sodium silicate	55.3	40	43	42.7	56.1	40
Pressed burnt and extruded ceramic waste	-	43.8	-	-	-	20.9
Raw pressed	-	-	39.6	40.1	-	16.3
Gypsum	-	-	-	-	21.3	6.6

[1] The raw pressed clay was dried in an oven at 750 °C for 5 h before use.

The mixture resulted well workable for several hours (the complete crosslinking and hardening took place in 5–7 h at room temperature, 20 °C). The mortar samples are hereafter indicated as PS-MK, where PS refers to pressed burnt and extruded ceramic waste; RP-MK, where RP refers to raw pressed ceramic waste; RP_{dry}-MK, where RP_{dry} refers to raw pressed ceramic waste annealed at 750 °C for 5 h in air; Gy-MK, where Gy refers to gypsum waste; and MIX-MK, which refers to geopolymer mortars obtained from the addition of all ceramic wastes.

2.2.3. Curing Treatments

As soon as prepared, all the specimens were cast in cubic moulds (50 × 50 × 50 mm^3) and cured in >95% relative humidity conditions at 60 °C for 24 h. Subsequently, the specimens were kept at room temperature for a further 6 days in >95% relative humidity conditions, and then for a further 21 days in air, as reported in refs. [30–38].

2.3. Methods

SEM analysis was carried out using a Phenom Pro X Microscope (Phenom-World B.V., Eindhoven, The Netherlands) on fresh fracture surfaces, after metallization with gold, carried out using a high vacuum sputter-coating technique. The acceleration potential used was between 5 and 15 kV. The EDS analysis was conducted with a BSD detector in full mode.

Hydrostatic weighing for apparent density and open porosity measurements was carried out employing a balance OHAUS-PA213 provided by Pioneer.

To determine the water absorption of mortar specimens, 3 cubes from each series were oven-dried at a temperature of 60 °C for 24 h [39], and their weight was determined as starting weight. The samples were then immersed in water for 24 h and their saturated surface dry weight was recorded as the final weight. Water absorption of specimens was reported as the percentage increase in weight.

X-ray diffraction patterns were obtained with Ni-filtered Cu-K$_\alpha$ radiation (λ = 0.15406 nm) at room temperature (20 °C) with an automatic Rigaku powder diffractometer mod. Miniflex 600, operating in the $\theta/2\theta$ Bragg-Brentano geometry. The phase recognition was carried out by using the PDF-4+ 2021 (International Centre for Diffraction Data®) database and the Rigaku PDXL2 software.

The compressive strength was evaluated according to EN 196-1 and measured by testing cubic concrete specimens (40 × 40 × 40 mm^3) in a Controls MCC8 multipurpose testing machine with a capacity of 100 kN. The tests were performed after 28 days of curing at room temperature, and the values reported are the averages of 5 compression strength values.

The compression tests were performed until the sample was densified and/or ruptured at a constant displacement velocity of 2 kN/s.

3. Results and Discussion

3.1. X-ray Diffraction Characterization

The diffraction patterns of the raw materials used in this work, and the corresponding geopolymer samples and mortars are reported in Figures 1 and 2, respectively. In Table 3, the corresponding degrees of crystallinity are reported.

As far as the raw materials, the porcelain stoneware waste (Figure 1A (PS)) consists of two main crystalline phases: quartz (SiO_2), the main one, and sillimanite (Al_2SiO_5). Considering the range of composition of different PS products [40], this sample can be considered representative of this kind of material. An amorphous halo with a maximum at $2\theta \approx 23°$ is also present (the x_c of the sample is $\approx$87%, Table 3), probably attributable to the melting of part of the alumina present in the system during the stoneware production [41,42]. This amorphous phase is responsible for the reduction of the characteristic porosity of ceramic materials fired at temperatures below 1000 °C, the characteristics of waterproofing properties of the stoneware, and resistance to acids and bases, and an increase in their mechanical properties compared to porous paste ceramics [41,42].

As far as the gypsum waste (Figure 1A (Gy)), it shows, apparently, just a single crystalline phase: gypsum. The diffraction pattern of the metakaolin sample (Figure 1A (metakaolin)) is characterized by a broad amorphous halo centred at 23° with only a crystalline peak at 25.4°, indicating the presence of small amounts of titanium oxide in the form of anatase (x_c = 8%, Table 3).

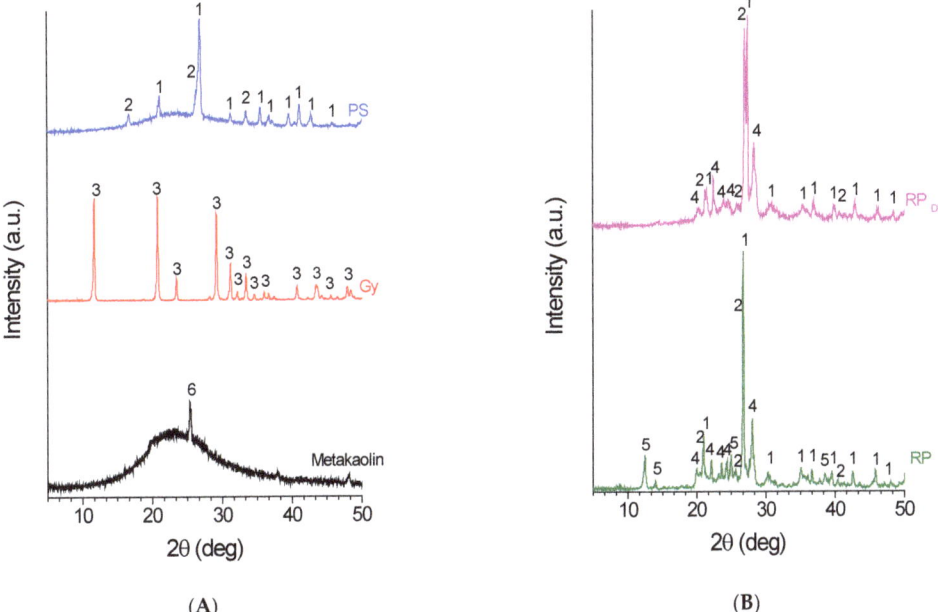

Figure 1. X-ray diffraction pattern of raw materials. Part (**A**): porcelain stoneware (PS), gypsum (Gy), and metakaolin; Part (**B**): Raw pressed (RP) and annealed raw pressed (RP$_{dry}$) waste. The numbers correspond to main diffraction peaks (with relative ICDD PDF-4+ 2021 card numbers) of 1-Quartz (SiO$_2$) (01-083-0539); 2-Sillimanite (Al$_2$SiO$_5$) (01-088-0893); 3-Gypsum (CaSO$_4$·2(H$_2$O)) (00-003-0053); 4-Albite (NaAlSi$_3$O$_8$) (01-089-6427); 5-Kaolinite (Al$_2$Si$_2$O$_5$(OH)$_4$) (01-080-0886); and 6-Anatase (TiO$_2$) (01-070-7348).

RP and RP$_{dry}$ waste (Figure 1B) present appreciably different diffraction patterns, although it is apparent that they are made of different amounts of the same crystalline phases: quartz, albite, and, in the case of the RP sample, kaolinite. The differences are attributable to the thermal treatment at 750 °C of the RP raw to prepare the RP$_{dry}$ sample, which causes the transformation of kaolinite into metakaolin, a fact that is clearly pointed out by the disappearance in the diffraction pattern of the annealed sample at the diffraction peaks of 2θ = 12.5, 13.9, and 21 deg. that are distinctive of kaolinite crystalline structure, and the formation of an appreciable amorphous halo (the degree of crystallinity of the RP waste sample is 97%, while that of RP$_{dry}$ waste is reduced to 78% upon annealing, Table 3) with a maximum intensity at 2θ ≈ 23°, in good agreement with the characterization of the diffraction pattern of MetaMax® metakaolin reported in Figure 1A. As reported in Table 3, the estimated amount of glassy phase in these samples ranges from 92% in the case of metakaolin (as high as expected for this typically amorphous material), to 3% in the case of RP waste.

Finally, the presence of calcium compounds in the samples was excluded by EDX analysis (Table 4 reports the results in the case of PS and RP waste). This analysis was particularly important since, as reported in the literature [43], Ca ions negatively interfere with the geopolymerization reaction, giving the final material poor mechanical properties. In this way, as will be better described below, it was possible to create a geopolymer mortar by loading the mixture with an appreciable quantity of gypsum as filler, without significantly affecting the mechanical properties of the material obtained.

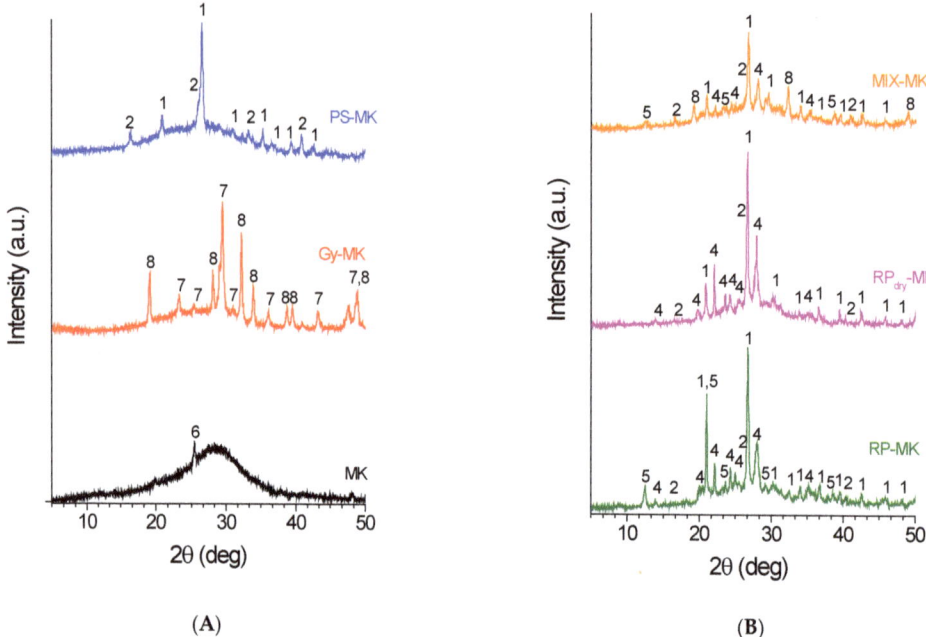

Figure 2. X-ray diffraction pattern of (**A**) geopolymer (MK) and geopolymeric mortars loaded with porcelain stoneware waste (PS-MK), gypsum waste (Gy-MK), and (**B**) geopolymeric mortars loaded with raw pressed (RP-MK) and annealed raw pressed (RP$_{dry}$-MK) waste. Also in (**B**), the X-ray diffraction pattern of the MIX-MK mortar obtained with the aggregate mixtures described in Table 2 is reported. The numbers correspond to the main diffraction peaks of 1-Quartz (SiO$_2$) (01-083-0539); 2-Sillimanite (Al$_2$SiO$_5$) (01-088-0893); 3-Gypsum (CaSO$_4$·2(H$_2$O)) (00-003-0053); 4-Albite (NaAlSi$_3$O$_8$) (01-089-6427); 5-Kaolinite (Al$_2$Si$_2$O$_5$(OH)$_4$) (01-080-0886); 6-Anatase (TiO$_2$) (01-070-7348); 7-Calcite (CaCO$_3$) (01-080-9776); and 8-Thenardite (Na$_2$SO$_4$) (04-010-2457).

Table 3. Degrees of crystallinity (x_c, %) of the starting raw waste and corresponding geopolymeric materials obtained by geopolymerization reaction, whose diffraction patterns are reported in Figures 1 and 2.

Sample Waste	x_c (%)	Geopolymeric Sample Loaded with	x_c (%)
metakaolin	8	geopolymer (MK)	7
porcelain stoneware (PS)	87	porcelain stoneware (PS)	65
gypsum (Gy)	90	gypsum (Gy)	61
raw pressed waste (RP)	97	raw pressed waste (RP)	60
annealed raw pressed waste (RP$_{dry}$)	78	annealed raw pressed waste (RP$_{dry}$)	43
–		MIX waste	42

As far as the geopolymeric materials, the presence in the X-ray diffraction patterns (Figure 2) of an amorphous halo (characterized by a maximum centred at 2θ ≈ 29°) linked to the formation of a disordered network of Si–O–Al bonds shows that all the samples have undergone geopolymeric activation. This amorphous network is generated mainly by the reaction of metakaolin with sodium silicate. The amorphous hump is particularly evident for the geopolymer MK sample (Figure 2A (MK)) obtained from metakaolin only, without the addition of aggregates, in which the crystalline reflection at 2θ ≈ 25.4° is due to the presence in the starting metakaolin of a minor amount of TiO$_2$. Similarly, in the diffraction patterns of all the other samples, crystalline reflections due to unreacted crystalline phases already present in the starting raw materials, or to new crystals formed during the geopolymerization reaction, are still present. It is worth noticing that the presence of these crystalline domains allows the activated material to limit the shrinkage phenomena due to the loss of water that occurs during the thermal curing and the ageing of the samples [43].

Table 4. Chemical composition (weight %) of PS and RP used in this paper as obtained by EXD characterization.

Phase	PS	RP
SiO_2	64.3	65.4
Al_2O_3	29.1	29.1
Na_2O	2.45	2.38
K_2O	3.51	2.19
Other	0.64	0.93

In particular, as far as the PS-MK geopolymer sample, the diffraction pattern (Figure 2A (PS-MK)) shows an amorphous halo more pronounced than that shown by the starting raw material (Figure 1A (PS)), thus suggesting that the geopolymerization reaction was successfully carried out (with reference to Figures 1 and 2 and Table 3, the degree of crystallinity of the geopolymeric mortar is 65%, significantly lower than the 87% of the starting PS waste). It is not possible to observe direct evidence of the amorphous phase involvement of the porcelain stoneware in the geopolymeric reaction during basic activation. This is in line with literature finds [44], stating that crystalline phases do not take part in the geopolymerization reaction and do not significantly influence it.

Meanwhile, the system that shows major variations after the alkaline activation process (Figure 2A) is the Gy-MK. In fact, even if it is well known that gypsum is not involved in geopolymerization, in the used experimental conditions this crystalline phase is transformed into calcite ($CaCO_3$) and thenardite (Na_2SO_4). The first phase is likely to be formed upon reaction, in the basic environment, of calcium ions with atmospheric carbon dioxide. Thenardite, on the other hand, is likely to be formed from the SO_4^{2-} anions (deriving from gypsum) with the Na^+ ions dissolved in the reagent base solution. As will be discussed in the next part, these ionic species greatly contribute to limiting the mechanical, physical, and chemical properties of the formed products, since, for example, the crystalline domain of calcite shows poor mechanical properties and low resistance to acids, bases, and water. Also, in the diffraction pattern of this sample, the presence of a broad hump centred at $2\theta \approx 29°$ is evident, pointing out the success of the geopolymerization.

If we compare the X-ray diffraction patterns of the RP-MK and RP_{dry}-MK geopolymeric materials (Figure 2B (RP-MK) and (RP_{dry}-MK)) with those of the starting raw materials (RP and RP_{dry} patterns of Figure 1B), the geopolymer mortars show the formation of a more pronounced amorphous halo, with a maximum shifted to $2\theta \approx 29°$. In particular, the RP-MK sample shows a smaller decrease in the degree of crystallinity with respect to RP_{dry}-MK ($\Delta x_c = 38\%$ in the case of RP-MK and $\Delta x_c = 45\%$ for RP_{dry}-MK) if compared to the starting raw materials ($x_c = 60\%$ and 43%, respectively, as reported in Table 3). This more pronounced reduction in the degree of crystallinity could be attributed to the fact that in the formation of the RP_{dry}-MK mortar the geopolymerization reaction involved not only the added metakaolin (see Table 2 for the mix design), but also the metakaolin formed during the annealing at 750 °C of the RP sample, as discussed before.

Finally, Figure 2 shows the X-ray diffraction pattern of a geopolymer mortar (Figure 2, MIX-MK) obtained by using the mixture of all the treated waste reported in Table 2. The obtainment of this particular sample was derived from the necessity of valorizing the gypsum waste that, as discussed before in the text, usually has a strong detrimental effect on the properties of the final geopolymeric material since the presence of Ca ions inhibit the geopolymerization reaction [43]. As a matter of fact, the mix design of the MIX-MK sample reported in Table 2 has been tailored to incorporate in the geopolymer mortar the maximum quantity of gypsum without having a significant decrease in the mechanical properties of the new material.

3.2. Microstructural Analysis

To highlight the microstructure of the samples, SEM micrographs at different scale ranges are reported in Figure 3 for waste aggregates, while the SEM micrographs of freshly obtained fracture surfaces of the geopolymeric samples after setting are reported in Figure 4.

Figure 3. Scanning Electron Microscope (SEM) images of PS (**A,B**), RP (**C,D**), RP_{dry} (**E,F**), and Gy (**G,H**) waste at 1000× and 15,000× magnifications.

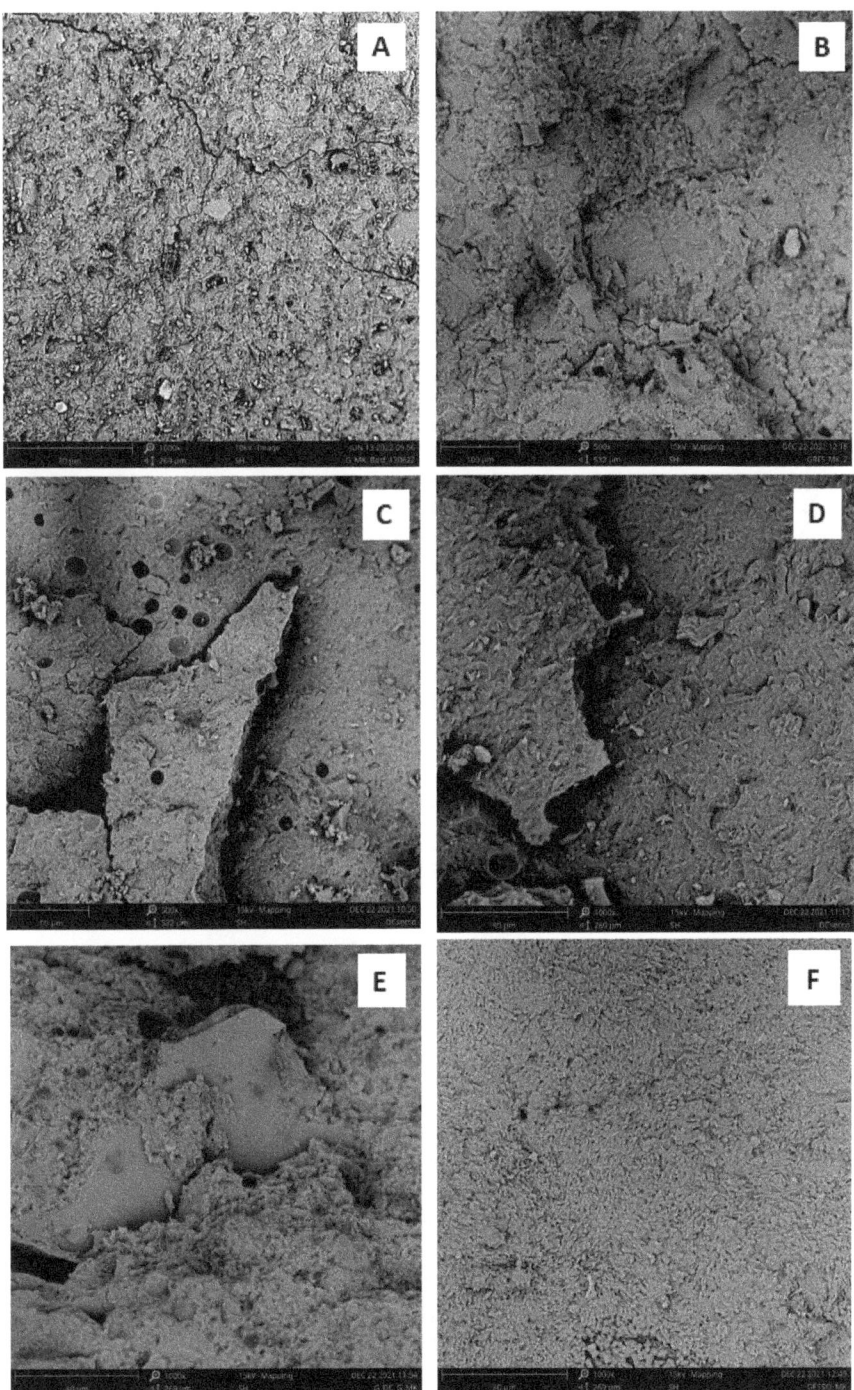

Figure 4. Scanning Electron Microscope (SEM) images of (**A**) neat geopolymer; (**B**) PS-MK; (**C**) RP-MK; (**D**) RP$_{dry}$-MK; (**E**) MIX-MK; and (**F**) Gy-MK mortars at 1000× magnifications.

It is apparent that PS powders (see Figure 3A,B) appear as a collection of particles with irregular shapes and sharp edges, whose dimensions are in the range of 0.2–100 μm. RP and RP_{dry} aggregates (see Figure 3C–F) occur as particles of a few microns (0.5–5 μm) that are organized into larger aggregates. Finally, Gy powders appear as aggregates of elongated crystals of calcium sulfate with dimensions in the range of 1–10 μm.

The neat geopolymer sample (Figure 4A) shows a homogeneous amorphous structure with some cracks (that could have been produced when the sample was fractured to obtain a fresh fracture surface to be analyzed). The largely homogeneous microstructure suggests a good geopolymerization behaviour (as pointed out also by the X-ray diffraction analysis discussed in the previous paragraph). The morphology is very similar to that observed for the neat geopolymer sample and is observable also in the case of the geopolymeric mortar specimens (Figure 4B–F). In the case of the PS-MK sample (Figure 4B), the morphology is characterized by the presence of PS filler particles well-embedded into a homogeneous and compact geopolymer matrix and strongly adhering to it. No apparent fractures are present. Unlike what has been reported in the literature for analogous samples [10,11], the point examination of Figure 4B shows that the PS particles retain their irregular shape with rather sharp and well-defined edges. For this reason, it can be said that there are no clear indications of the possible involvement in the geopolymerization reaction of the PS particles, which therefore behave like a simple filler. As far as the morphology and microstructure of the RP-MK and RP_{dry}-MK samples (Figure 4C,D, respectively), it is very similar and for both of them is not possible to recognize the waste aggregates. Also, in this case, the sample morphology is in line with the diffraction profile, where there is a major amorphous phase in RP_{dry}-MK (the heat treatment at 750 °C for 5 h allowed the kaolinite calcination process that produced a major percentage of amorphous phase to RP-MK). It is worth pointing out that these mortars show a more consistent microstructure with respect to the geopolymer paste, with a reduced amount of microcracks.

A different morphology was observed in the case of the Gy-MK sample (Figure 4F), in which it is possible to identify the presence of crystals of calcium carbonate and sodium sulfate distributed in the geopolymer matrix. Finally, the MIX-MK sample (Figure 4E) also shows a morphology rather homogeneous, where, as in the case of the PS-MK sample, it is possible to recognize aggregate particles of PS with dimensions of about 100 μm strongly interpenetrated in the geopolymer matrix. Concluding, in all cases this homogeneous structure strongly suggests that ceramic waste does not take part, interfere, or inhibit the geopolymerization reaction.

3.3. Physical and Mechanical Properties

Table 5 reports the compressive strength, density, and water absorption values of the geopolymeric samples prepared.

Table 5. Physical and mechanical properties of the geopolymer sample (MK) and geopolymeric mortars obtained with porcelain stoneware waste (PS-MK), raw pressed waste (RP-MK), calcined raw pressed waste (RP_{dry}-MK), and gypsum (Gy-MK) as filler, and of the geopolymeric mortar with the mixed filler whose composition is reported in Table 2 (MIX-MK). The calcined raw pressed clay (RP_{dry}) was obtained by annealing as obtained raw pressed clay (RP) in an oven at 750 °C for 5 h.

Sample Properties	MK	PS-MK	RP-MK	RP_{dry}-MK	Gy-MK	MIX-MK
Density (kg/m^3)	1370	1773 ± 95	1718 ± 80	1687 ± 90	1335 ± 99	1627 ± 77
Water absorption (%)	18 ± 1	16 ± 1	18 ± 1	19 ± 1	>25	>25
Compressive strength (MPa)	25 ± 2	30 ± 1	25 ± 3	41 ± 3	4.0 ± 0.5	38 ± 2

By comparing the samples' densities values (first row of Table 5) it is apparent that, as expected, all the geopolymeric mortars are characterized by values fairly higher than the neat geopolymer. This is due to the presence of a significant amount (see Table 2 for the mix design) of the aggregate that in every case is made of dense materials (the typical density of porcelain stoneware and pressed clays is about 2.6 g/cm^3).

The only exception is the Gy-MK mortar, in which the density is practically the same as the neat geopolymer. This fact is in line with the high water absorption capability of the sample, due to the high affinity of the calcium sulphate with water. This fact decreases the workability of the geopolymer slurry, strongly reducing the set time and thus causing the incorporation of a greater amount of air than the other mortars, resulting in a final product with the lowest density value (1335 kg/m^3).

A similar trend can be pointed out by examining the values of the compressive strength of the different samples (third line of Table 5). As we can see, except for the Gy-MK mortar, all the other samples show a comparable or even higher compressive strength to the neat geopolymer, with an increase of compressive strength values up to 40%, indicating that the addition of the aggregates has a remarkable enhancing effect on the mechanical properties of the materials. This is likely to be caused by the good adhesion and the good dispersions of the filler in the geopolymer matrix, as shown by the SEM images reported in Figure 4, which suggests that the waste particles dispersed within the geopolymer matrix can create a barrier against crack growth, enhancing the mechanical response of the material.

It is worth pointing out that the most evident improvement in the mechanical properties of the material has been recorded for the RP$_{dry}$-MK sample. This important increase of the compressive strength is probably due to the presence in the starting sample of a greater quantity of reactive phase (i.e., the metakaolin added according to the mix design of Table 2 and produced by the calcination of kaolinite at 750 °C). In this way, as also highlighted by the SEM images shown in Figure 2, the geopolymerization reaction led to the obtaining of a geopolymer mortar sample in which the non-reactive crystalline phase is very well-included and dispersed in the geopolymer amorphous matrix.

At variance with this sample, the Gy-MK mortar shows very poor mechanical properties, with a reduction of about 85% in the compressive strength value of the neat geopolymer (MK). Substantially, the presence of gypsum within the paste negatively affects the mechanical properties of the final product because gypsum is a source of Ca ions, which are competitive with the geopolymerization reaction [22,43]. As already discussed, to find a possible strategy to valorize this very abundant waste, an alternative sample composition was developed, in which gypsum was mixed (which hurts the mechanical properties of the final material) with the other wastes, and in particular, RPdry, which significantly improves the mechanical properties of the mortars (as seen in the previous sections). The mix obtained (see Table 2) allowed to include up to ≈7% by weight of gypsum without significantly compromising the mechanical properties of the final product, a result that to our knowledge was never previously obtained using aggregates with high reactive calcium concentration. As a final note, it is worth highlighting that the material obtained presents good mechanical properties despite its high water absorption capacity (>25 wt.%), which is fairly higher than that of the neat geopolymer (18 wt.%).

3.4. Applications in the Field of the Creative Industry and Cultural Heritage

A preliminary investigation of the potentialities of the geopolymer-based materials made from ceramic wastes described so far has shown that such systems could be used for the restoration and conservation of artworks and the creation of products for art and design.

Experimental tests have shown that the geopolymer slurry is characterized by very good thixotropic behaviour and high workability, which makes it easy to spread and model on different substrates that need repair interventions. Moreover, the chemical compatibility of the geopolymer-based materials with different substrates (tuff, cement, ceramic, and porcelain stoneware) ensures good adhesion with them, pointing to the possibility of using this kind of binder as a joining or fixing paste.

To this aim, Figure 5 reports some artefacts made of porcelain stoneware (top of Figure 5) and ceramic (bottom of Figure 5) that were restored and repaired by using the previously described PS-MK and RP$_{dry}$-MK geopolymer-based pastes as fixing materials. Particularly, the two objects were subsequently damaged and then repaired using geopolymer systems as bonding materials. The repair process remained stable even after many months. A very good adhesion in each of the examined cases was observed, even for those materials containing an elevated concentration of filler with respect to others described in the literature [13]. This encouraging result was confirmed by performing SEM characterization of the interface between the geopolymer mortar, used as repairing material, and a pottery fragment, as shown in Figure 6: it is also apparent that at a micrometric level, a very strict adhesion between the geopolymer and ceramic phases can be observed, and, also at micrometric level, the geopolymer material and the ceramic substrate seem to form a continuous phase with no neat interface and no fractures. However, physical properties with a focus on the optimization of adhesion between ceramic artifacts and geopolymer binders should be further studied. For this reason, the incorporation of organic additives in geopolymer formulations developed for this aim is still in progress.

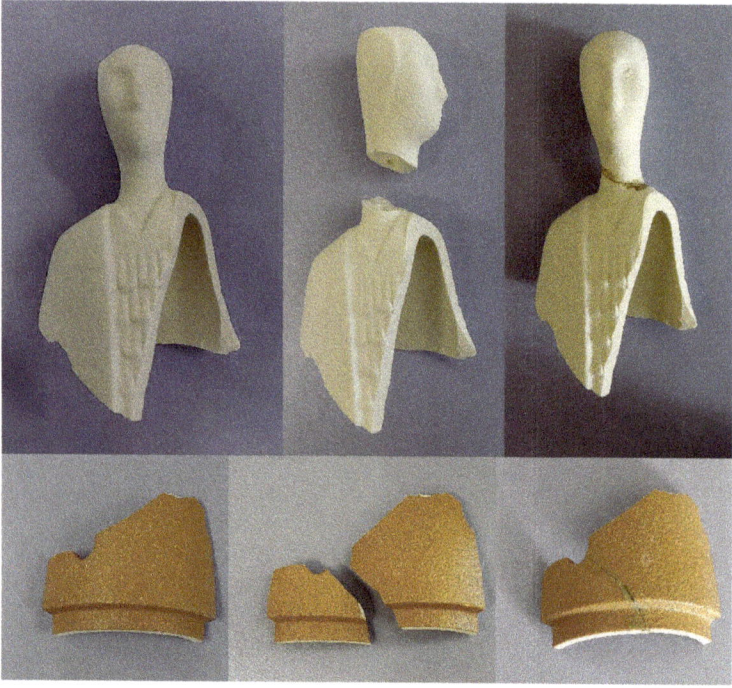

Figure 5. Porcelain stoneware and pottery artworks restored and repaired by using the geopolymer-based materials PS-MK and RP$_{dry}$-MK described in this paper.

These preliminary results suggest that the proposed mortars can be applied as materials to decorate, seal, and repair cracks and fractures in stone, artifacts, tiles, and masonries.

To perform these functions optimally, they must have aesthetic characteristics similar to those of the materials to be repaired, but, as suggested by current restoration practice, still well recognizable: as shown in the lower part of Figure 5, mortars of the same colour as the artifact, but slightly paler, were obtained.

Finally, it is worth noting that the geopolymer products developed and characterized in this work can be easily coloured by simply adding water-based pigments into the slurry and/or through post-panting operations with oil-based paint and/or cold painting. As an example, Figure 7 reports different kinds of artefacts realized with PS-MK mortar, with the addition of water-based pigments. Moreover, the colour of the mortar can be changed by adding rock powders, mimicking the material to be restored.

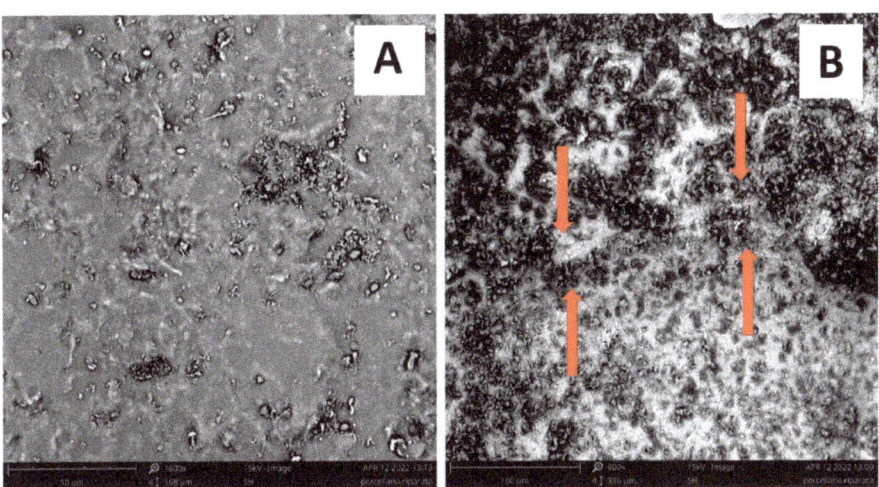

Figure 6. Scanning electron microscope (SEM) micrographs of (**A**) ceramic substrate and (**B**) interface transition zone (indicated by the red arrows) between the ceramic substrate (upper part of the figure) and RP_{dry}-MK geopolymer mortar (low part of the figure).

Figure 7. Examples of coloured artefacts created with the geopolymer-based materials described in this paper. As apparent, even objects with complex shapes can be easily realized by a simple casting procedure.

Future developments will be focused on extending the methodology developed for obtaining building elements with high performance in terms of mechanical strength (rigidity of shape, texture, and geometry), technological integrability (with other materials), low maintenance, substitutability, recoverability at end of life, thermal and acoustic insulation, and architectural and aesthetic characteristics. Moreover, research will be integrated with Life Cycle Assessment (LCA) studies and a deep investigation of the social and economic dimensions, using Life Cycle Costs (LCC) and Social Life Cycle Assessment (S-LCA).

4. Conclusions

Geopolymers are environmental-friendly materials for which their use in many application fields has been proposed. To date, a great effort has been made as far as the structural applications and the replacement of ordinary Portland cement with this class of materials are concerned. However, in the last years, their application in art and design, and more in general, in Cultural Heritage is attracting more and more attention. As a matter of fact, geopolymer-based materials may find interesting applications in these fields thanks to their interesting chemical and physical properties (such as high durability and good flexural and compressive strength). Moreover, these properties can be also tailored in order to guarantee functional and aesthetic compatibility with the original materials on which the restoration action has to be performed.

The present research describes the preparation and characterization of new geopolymer mortars obtained by recycling wastes deriving from the production process and the "end of life" of porcelain stoneware products. In particular, geopolymer mortars were obtained after the consolidation of a slurry prepared by using metakaolin, an activating alkaline solution, and different kinds of by-products of tiles and ceramic production (pressed burnt and extruded ceramic waste, raw pressed and gypsum resulting from exhausted moulds). The obtained materials resulted to be largely composed of an amorphous binding material in which the filler was homogeneously dispersed.

The chemical compatibility of the geopolymer materials with different substrates (tuff, cement, ceramic, and porcelain stoneware) ensures good adhesion with them, pointing to the possibility of using this kind of binder as a joining and fixing material or as sacrificial material for the restoration of stone objects. To this aim, the possible addition of rock powders within the mix design composition could allow obtaining materials that mimic different types of stone, thus reaching good aesthetic compatibility. In particular, ongoing studies on formulations including volcanic rocks and tuffs show promising perspectives in terms of chemical, mineralogical, mechanical, and aesthetic compatibility with the local built heritage since they mimic the traditional materials used in the Vesuvius archaeological area. In this framework, to further increase the adhesion between the substrate and geopolymer binders, the incorporation of organic additives in geopolymer formulations developed for this aim is also in progress. The possibility to develop new materials with the concurrent reduction of their environmental footprint and production cost can create a tangible gain in the sectors of the decorative industry and Cultural Heritage. In fact, the demonstration of real and practical use of sustainable materials for the creative industry and restoration field can improve business around waste collection and conversion. Moreover, continued use of recycled materials in mass production can increase the competitive position of Eco-materials on the market and reduce the overall environmental footprint in the creative industry and Cultural Heritage field.

In this scenario, the authors wish to give a contribution to the research field of sustainable materials to enhance the use of recycled raw materials and also in the field of Art and Design, restoration, conservation, and/or rehabilitation of historic monuments, decorative and architectural intervention.

Author Contributions: Conceptualization, L.R. and G.R.; Methodology, L.R.; Data curation, A.O. and C.F.; Writing—original draft, L.R.; Writing—review & editing, O.T. and R.C.; Supervision, R.C. All authors have read and agreed to the published version of the manuscript.

Funding: This research received no external funding.

Institutional Review Board Statement: Not applicable.

Informed Consent Statement: Not applicable.

Data Availability Statement: Not applicable.

Acknowledgments: The authors thank Neuvendis S.p.A. for the metakaolin supply and Prochin Italia S.r.l. for the silicate solution supply. Giovanni Morieri and Luciana Cimino are warmly acknowledged for their assistance in laboratory activities.

Conflicts of Interest: The authors declare no conflict of interest.

References

1. Confindustria Ceramica. 2019. Available online: http://www.confindustriaceramica.it/site/home/news/documento5751.html (accessed on 21 November 2022).
2. Martín-Márquez, J.; Rincón, J.M.; Romero, M. Effect of firing temperature on sintering of porcelain stoneware tiles. *Ceram. Int.* **2008**, *34*, 1867–1873. [CrossRef]
3. Pérez, J.M.; Romero, M. Microstructure and technological properties of porcelain stoneware tiles moulded at different pressures and thicknesses. *Ceram. Int.* **2014**, *40*, 1365–1377. [CrossRef]
4. Bovea, M.D.; Díaz-Albo, E.; Gallardo, A.; Colomer, F.J.; Serrano, J. Environmental performance of ceramic tiles: Improvement proposals. *Mater. Des.* **2010**, *31*, 35–41. [CrossRef]
5. El-Dieb, A.S.; Taha, M.R.; Kanaan, D.; Aly, S.T. Ceramic waste powder: From landfill to sustainable concretes. *Proc. Inst. Civ. Eng. Constr. Mater.* **2018**, *171*, 109–116. [CrossRef]
6. Pacheco-Torgal, F.; Jalali, S. Retraction Note: Compressive strength and durability properties of ceramic wastes based concrete. *Mater. Struct.* **2021**, *54*, 153. [CrossRef]
7. Silva, T.H.; de Resende, M.C.; de Resende, D.S.; Ribeiro Soares, P.R., Jr.; da Silva Bezerra, A.C. Valorization of ceramic sludge waste as alternative flux: A way to clean production in the sanitary ware industry. *Clean. Eng. Technol.* **2022**, *7*, 100453. [CrossRef]
8. European Commission. *Directorate-General for Budget, The EU's 2021–2027 Long-Term Budget & Next Generation EU: Facts and Figures, Publications Office.* 2021. Available online: https://data.europa.eu/doi/10.2761/808559 (accessed on 21 November 2022).
9. Rambaldi, E.; Esposito, L.; Tucci, A.; Timellini, G. Recycling of polishing porcelain stoneware residues in ceramic tiles. *J. Eur. Ceram. Soc.* **2007**, *27*, 3509–3515. [CrossRef]
10. Sun, Z.; Cui, H.; An, H.; Tao, D.; Xu, Y.; Zhai, J.; Li, Q. Synthesis and thermal behavior of geopolymer-type material from waste ceramic. *Constr. Build. Mater.* **2013**, *49*, 281–287. [CrossRef]
11. Medri, V.; Landi, E. Recycling of porcelain stoneware scraps in alkali bonded ceramic composites. *Ceram. Int.* **2014**, *40*, 307–315. [CrossRef]
12. Reig, L.; Sanz, M.A.; Borrachero, M.V.; Monzó, J.; Soriano, L.; Payá, J. Compressive strength and microstructure of alkali-activated mortars with high ceramic waste content. *Ceram. Int.* **2017**, *43*, 13622–13634. [CrossRef]
13. Luhar, I.; Luhar, S.; Abdullah, M.M.A.B.; Nabiałek, M.; Sandu, A.V.; Szmidla, J.; Jurczyńska, A.; Razak, R.A.; Aziz, I.H.A.; Jamil, N.H.; et al. Assessment of the Suitability of Ceramic Waste in Geopolymer Composites: An Appraisal. *Materials* **2021**, *14*, 3279. [CrossRef] [PubMed]
14. Ramos, G.A.; de Matos, P.R.; Pelisser, F.; Gleize, P.J.P. Effect of porcelain tile polishing residue on eco-efficient geopolymer: Rheological performance of pastes and mortars. *J. Build. Eng.* **2020**, *32*, 101699. [CrossRef]
15. Hayles, C.S. Environmentally sustainable interior design: A snapshot of current supply of and demand for green, sustainable or Fair Trade products for interior design practice. *Int. J. Sustain. Built Environ.* **2015**, *4*, 100–108. [CrossRef]
16. NPCS Board of Consultants & Engineers. *The Complete Technology Book on Steel and Steel Products*; Asia Pacific Business Press Inc.: New Delhi, India, 2008; ISBN 8178330180, 9788178330181.
17. Zeyad, A.M.; Magbool, H.M.; Tayeh, B.A.; de Azevedo AR, G.; Abutaleb, A.; Hussain, Q. Production of geopolymer concrete by utilizing volcanic pumice dust. *Case Stud. Constr. Mater.* **2022**, *16*, e00802. [CrossRef]
18. Provis, J.L. Green concrete or red herring?—Future of alkali-activated materials. *Adv. Appl. Ceram.* **2014**, *113*, 472–477. [CrossRef]
19. Davidovits, J. *Geopolymer Chemistry and Applications*, 5th ed.; Geopolymer Institute: Saint-Quentin, France, 2020; ISBN 9782954453118.
20. Singh, B.; Ishwarya, G.; Gupta, M.; Bhattacharyya, S. Geopolymer concrete: A review of some recent developments. *Constr. Build. Mater.* **2015**, *85*, 78–90. [CrossRef]
21. Duxon, P.; Fernandez-Jiminez, A.; Provis, J.L.; Luckey, G.C.; Palomo, A.; Van Deventure, J.S.J. Geopolymer technology: The current state of the art. *J. Mater. Sci.* **2007**, *42*, 2917–2933. [CrossRef]
22. Provis, J.L.; Van Deventer, J.S.J. (Eds.) *Geopolymers, Structure, Processing, Properties and Application*; Woodhead Publishing Limited: Cambridge, UK, 2009.
23. Davidovits, J. Geopolymers: Inorganic polymeric new materials. *J. Therm. Anal.* **1991**, *37*, 1633–1656. [CrossRef]
24. Nikolov, A.; Rostovsky, I.; Nugteren, H. Geopolymer materials based on natural zeolite. *Case Stud. Constr. Mater.* **2017**, *6*, 198–205. [CrossRef]
25. Nasaeng, P.; Wongsa, A.; Cheerarot, R.; Sata, V.; Chindaprasirt, P. Strength enhancement of pumice-based geopolymer paste by incorporating recycled concrete and calcined oyster shell powders. *Case Stud. Constr. Mater.* **2022**, *17*, e01307. [CrossRef]
26. Occhicone, A.; Vukcevic, M.; Boskovic, I.; Mingione, S.; Ferone, C. Alkali-Activated Red Mud and Construction and Demolition Waste-Based Components: Characterization and Environmental Assessment. *Materials* **2022**, *15*, 1617. [CrossRef] [PubMed]
27. Occhicone, A.; Vukčević, M.; Bosković, I.; Ferone, C. Red mud-blast furnace slag-based alkali-activated materials. *Sustainability* **2021**, *13*, 11298. [CrossRef]

28. Shaikh, F. Review of mechanical properties of short fiber reinforced geopolymer composites. *Constr. Build. Mater.* **2013**, *43*, 37–49. [CrossRef]
29. Sakulich, A. Reinforced geopolymer composites for enhanced material greenness and durability. *Sustain. Cities Soc.* **2011**, *1*, 195–210. [CrossRef]
30. Ricciotti, L.; Molino, A.J.; Roviello, V.; Chianese, E.; Cennamo, P.; Roviello, G. Geopolymer Composites for Potential Applications in Cultural Heritage. *Environments* **2017**, *4*, 91. [CrossRef]
31. Colangelo, F.; Roviello, G.; Ricciotti, L.; Ferrándiz-Mas, V.; Messina, F.; Ferone, C.; Tarallo, O.; Cioffi, R.; Cheeseman, C.R. Mechanical and thermal properties of lightweight geopolymer composites containing recycled expanded polystyrene. *Cem. Concr. Compos.* **2018**, *86*, 266–272. [CrossRef]
32. Roviello, G.; Menna, C.; Tarallo, O.; Ricciotti, L.; Messina, F.; Ferone, C.; Asprone, D.; Cioffi, R. Lightweight geopolymer-based hybrid materials. *Compos. Part B Eng.* **2017**, *128*, 225–237. [CrossRef]
33. Roviello, G.; Menna, C.; Tarallo, O.; Ricciotti, L.; Ferone, C.; Colangelo, F.; Asprone, D.; di Maggio, R.; Cappelletto, E.; Prota, A.; et al. Preparation, structure and properties of hybrid materials based on geopolymers and polysiloxanes. *Mater. Des.* **2015**, *87*, 82–94. [CrossRef]
34. Roviello, G.; Ricciotti, L.; Ferone, C.; Colangelo, F.; Tarallo, O. Fire resistant melamine based organic-geopolymer hybrid composites. *Cement Concrete Compos.* **2015**, *59*, 89–99. [CrossRef]
35. Ferone, C.; Roviello, G.; Colangelo, F.; Cioffi, R.; Tarallo, O. Novel hybrid organic-geopolymer materials. *Appl. Clay Sci.* **2013**, *73*, 42–50. [CrossRef]
36. Roviello, G.; Ricciotti, L.; Ferone, C.; Colangelo, F.; Cioffi, R.; Tarallo, O. Synthesis and Characterization of Novel Epoxy Geopolymer Hybrid Composites. *Materials* **2013**, *6*, 3943–3962. [CrossRef] [PubMed]
37. Colangelo, F.; Roviello, G.; Ricciotti, L.; Ferone, C.; Cioffi, R. Preparation and characterization of new geopolymer-epoxy resin hybrid mortars. *Materials* **2013**, *6*, 2989–3006. [CrossRef] [PubMed]
38. Roviello, G.; Ricciotti, L.; Molino, A.J.; Menna, C.; Ferone, C.; Asprone, D.; Cioffi, R.; Ferrandiz-Mas, V.; Russo, P.; Tarallo, O. Hybrid fly ash-based geopolymeric foams: Microstructural, thermal and mechanical properties. *Materials* **2020**, *13*, 2919. [CrossRef] [PubMed]
39. Sithole, N.T.; Ntuli, F.; Okonta, F. Fixed bed column studies for decontamination of acidic mineral effluent using porous fly ash-basic oxygen furnace slag based geopolymers. *Miner. Eng.* **2020**, *154*, 106397. [CrossRef]
40. Roviello, G.; Chianese, E.; Ferone, C.; Ricciotti, L.; Roviello, V.; Cioffi, R.; Tarallo, O. Hybrid geopolymeric foams for the removal of metallic ions from aqueous waste solutions. *Materials* **2019**, *12*, 4091. [CrossRef] [PubMed]
41. Roviello, G.; Ricciotti, L.; Molino, A.J.; Menna, C.; Ferone, C.; Cioffi, R.; Tarallo, O. Hybrid Geopolymers from Fly Ash and Polysiloxanes. *Molecules* **2019**, *24*, 3510. [CrossRef]
42. Ricciotti, L.; Occhicone, A.; Petrillo, A.; Ferone, C.; Cioffi, R.; Roviello, G. Geopolymer-based hybrid foams: Lightweight materials from a sustainable production process. *J. Clean. Prod.* **2020**, *250*, 119588. [CrossRef]
43. Botti, R.F.; Innocentini, M.D.M.; Faleiros, T.A.; Mello, M.F.; Flumignan, D.L.; Santos, L.K.; Franchin, G.; Colombo, P. Biodiesel Processing Using Sodium and Potassium Geopolymer Powders as Heterogeneous Catalysts. *Molecules* **2020**, *25*, 2839. [CrossRef]
44. Thokchom, S.; Ghosh, P.; Ghosh, S. Eeffect of water absorption, porosity and sorptivity on durability of geopolymer mortars ARPN. *J. Eng. Appl. Sci.* **2009**, *4*, 28–32.

Article

Effects of Na₂CO₃/Na₂SiO₃ Ratio and Curing Temperature on the Structure Formation of Alkali-Activated High-Carbon Biomass Fly Ash Pastes

Chengjie Zhu *, Ina Pundienė, Jolanta Pranckevičienė and Modestas Kligys

Laboratory of Concrete Technology, Institute of Building Materials, Vilnius Gediminas Technical University, Linkmenų Str. 28, LT-08217 Vilnius, Lithuania
* Correspondence: chengjie.zhu@vilniustech.lt

Citation: Zhu, C.; Pundienė, I.; Pranckevičienė, J.; Kligys, M. Effects of Na₂CO₃/Na₂SiO₃ Ratio and Curing Temperature on the Structure Formation of Alkali-Activated High-Carbon Biomass Fly Ash Pastes. *Materials* **2022**, *15*, 8354. https://doi.org/10.3390/ma15238354

Academic Editor: Miguel Ángel Sanjuán

Received: 20 October 2022
Accepted: 22 November 2022
Published: 24 November 2022

Publisher's Note: MDPI stays neutral with regard to jurisdictional claims in published maps and institutional affiliations.

Copyright: © 2022 by the authors. Licensee MDPI, Basel, Switzerland. This article is an open access article distributed under the terms and conditions of the Creative Commons Attribution (CC BY) license (https://creativecommons.org/licenses/by/4.0/).

Abstract: This study explored unprocessed high-carbon biomass fly ash (BFA) in alkali-activated materials (AAM) with less alkaline Na₂CO₃ as the activator. In this paper, the effects of the Na₂CO₃/Na₂SiO₃ (C/S) ratio and curing temperature (40 °C and 20 °C) on the setting time, structure formation, product synthesis, and physical-mechanical properties of alkali-activated BFA pastes were systematically investigated. Regardless of curing temperature, increasing the C/S ratio increased the density and compressive strength of the sample while a decrease in water absorption. The higher the curing temperature, the faster the structure evolution during the BFA-based alkaline activation synthesis process and the higher the sample's compressive strength. According to XRD and TG/DTA analyses, the synthesis of gaylussite and C-S-H were observed in the sample with an increasing C/S ratio. The formation of the mentioned minerals contributes to the compressive strength growth of alkali-activated BFA pastes with higher C/S ratios. The findings of this study contribute to the applicability of difficult-to-recycle waste materials such as BFA and the development of sustainable BFA-based AAM.

Keywords: biomass fly ash; alkali-activated materials; Na₂CO₃/Na₂SiO₃ ratio; curing temperature; structure formation; compressive strength

1. Introduction

Pressures to protect the global environment and provide renewable energy sources have resulted in rising demand for biomass renewable energy sources. Biomass plays a significant role in increasing ash generation, which should be regulated environmentally [1,2]. Biomass ashes, high in carbon due to their biogenic nature, have been one of the most heavily utilized wastes [3]. Primary solid biofuels (plant matter used directly as fuel or processed into solid fuels) represent about 9% of worldwide energy production [4,5]. It is predicted that industrial biomass incineration for combined heat and power will triple by 2035 compared with 2008 levels [6]. This expansion is partly driven by rules that promote biomass as a renewable energy source and burning as a CO₂-neutral process. In most cases, the inorganic elements of biomass do not burn during the incineration process [7–9]. They are left as ashes together with a proportion of unburned carbon, depending on the temperature and efficiency of the combustion process [10,11]. However, biomass ash (except rice husk ash and sugarcane bagasse ash, which are rich in SiO₂) has a low inorganic concentration, so utilizing such residues is difficult. Rice husk ash contains a large amount of SiO₂ [12–21] and is an effective precursor for binder systems (cement and alkali activator-based binders) [22–26].

Aqueous alkaline solutions promote the dissolution of alumina and silica from biomass ash. Residual unburned carbon in biomass ash complicates its usage in cementitious materials, and higher carbon content is associated with increased adsorption of air-entraining admixtures and higher water consumption [27,28]. In many developing countries in Asia and Africa, biomass incineration for energy generation has increased significantly in recent years, resulting in large quantities of biomass ashes [12]. Biomass fly ash (BFA) has traditionally been utilized as a mineral fertilizer in agriculture [29], but commercial utilization of biomass ash is not so widely reported. Due to the expected growth in this waste, it is vital to investigate its potential uses. Most biomass ash produced in thermal power plants is now disposed of in landfills or recycled in agricultural fields or forests with little or no oversight [30]. However, given that the cost of disposing of biomass ashes is increasing and that biomass ash quantities are expanding globally, it is critical to developing a solution for the sustainable management of such waste. In this regard, some researchers have considered using biomass ashes from wood burning as a concrete substitute [28,31].

Research [32–34] proves that biomass ashes can be used in road construction. On the other hand, biomass ashes worsen the rheological behaviour and compressive strength of concrete [30,31,35,36], depending on the pozzolanic nature of biomass ashes and the higher organic content of 20–25% [37]. It has also been pointed out that biomass ashes affect the apparent density, water absorption of concrete [30], and the continuous degradation of compressive strength of specimens when the amount of biomass ash is increased.

One of the more environmentally friendly solutions for recovering biomass ashes is to use them as binders in alkali-activated substances. Due to almost similar binding characteristics, alkali-activated materials (AAM) are a viable alternative to Portland cement [38–40]. AAM has attracted much interest in recent years because of its satisfying early compressive strength, low permeability, and chemical resistance. The use of Na_2CO_3 solutions to activate slags, which have high calcium amounts, has been investigated extensively, while its use in the activation of fly ash is significantly less [39]. Using mixed sodium hydroxide-carbonate activating solutions produces a porous, poorly reacted product, although it is uncertain what role the carbonate component of the activating solution plays in activated fly ash [41]. According to researchers, fly ash (Class F) is less reactive than slags and therefore requires stronger alkalinity than Na_2CO_3 solution.

However, the practical utilization of traditionally used NaOH and Na_2SiO_3 as activators has been hindered by sustainability concerns and handling issues due to the high alkalinity [42,43]. In addition, hydroxides are made from the electrolysis of chloride salts, which consume a lot of energy and emit CO_2. Alkali silicates are made by melting sand with carbonates at 1100–1200 °C and then dissolving them under high pressure at 140–160 °C [44,45]. In addition to its high prices, using NaOH as an activator has the problems of quick setting and drying shrinkage [46–49]. Although much attention has been paid to fly ash/slag mix activated with alkali hydroxides/silicates, alternate activation systems still need to be found to sort out the efflorescence problem associated with NaOH and its handling issue due to the high alkalinity [42,43]. The use of other types of activators, such as $Ca(OH)_2$, Na_2SO_4, Na_2CO_3, CaO, and MgO, offers a promising option for reducing the environmental impact of AAM [50–56].

As an alternative, Na_2CO_3 can be used as an activator to create long-lasting AAM systems in less alkaline environments and with lower drying shrinkage [57–59]. Na_2CO_3 possesses low cost and commercial availability and helps reduce the pH of alkaline activator solution (AAS). Despite these benefits, the use of Na_2CO_3 as an activator in AAS systems has been infrequently reported, owing to low pH limitations resulting in long setting times and limited strength development [60,61]. The reaction kinetics of AAM can be enhanced to some extent by introducing higher concentrations of hydroxides [62–65], sodium silicate [66], or blending with reactive admixtures such as CaO [67]. Nonetheless, all these approaches are ineffective, expensive, or have negative effects. For example, by adding NaOH to the activator, samples can achieve a 1-day compressive strength of about 5–7 MPa, with almost no additional strength development after 7 days of curing [64,68]. A prolonged reaction process was reported in [52,69], whereby demolding took approximately 3–5 days, depending on the composition. The subsequent reaction is cyclic with a buffered alkaline environment regulated by CO_3^{2-} anions and maintained by continuous $CaCO_3$ dissolution [70]. According to the study [52], the precipitation of $CaCO_3$ and consumption of CO_3^{2-} anions produced by the activator are mostly responsible for the delayed reaction process of Na_2CO_3-activated slag. However, on the contrary, another study reported that compositions activated by Na_2CO_3 show satisfactory setting time and compressive strength during 1–3 days of curing, reaching 5–25 MPa [71,72]. The activation potential of activators might be arranged in the following order [73]: Na_2SiO_3 > NaOH > NaOH + Na_2CO_3 > KOH.

It is pointed out that hydroxides accelerate the reaction of Na_2CO_3-activated slag by removing the CO_3^{2-} anions, resulting in a significant rise in pH, and the samples can be hardened within 24 h [74]. The 7-day compressive strength of ground granulated blast-furnace slag (GGBS) pastes activated by $NaOH/Na_2SiO_3$ is three times that of Na_2CO_3-activated GGBS pastes [69,75]. The Ca^{2+} ions released from the dissolved GGBS interact with CO_3^{2-} anions and create carbonates (e.g., calcite and gaylussite). If this reaction is prolonged because of the solution's low pH, the dissolution time of the silicate species is prolonged, as is the setting time [69,76]. The consumption of CO_3^{2-} anions releases hydroxide ions, which raise the pH of the liquid phase, forcing the silicate species to dissolve and C-(A)-S-H to form [77].

In the early stage, the calcium carbonate and sodium-calcium carbonate phases are formed as the Ca^{2+} ions released from slag react with the CO_3^{2-} anions of the dissolved Na_2CO_3 in the pore solution. Meanwhile, the dissolution of the slag releases silicate and aluminate ions, which react with the sodium of the activator to form zeolite NaA [78]. Additionally, synthetic C-S-H seeds [79] or MgO [80–85] are suggested for activation reaction improvement. Such additives provide extra nucleation sites, higher pH, additional formation of hydration products, and early strength gain. Compared with Portland cement-based materials, the application of using biomass ashes to make AAM [30,31,35–37] has excellent potential, and the use of low-cost unprocessed BFA for producing AAM has not been described before.

This research aims to study the characteristics of the used BFA and the development of BFA-based AAM. Therefore, experimental research was carried out to investigate the effects of the Na_2CO_3/Na_2SiO_3 (C/S) ratio and curing temperature (40 °C and ambient temperature) on the fresh-state properties (e.g., setting time), structure formation, product synthesis, and physical-mechanical properties of BFA-based AAM pastes. The use of unprocessed BFA to make AAM can reduce production costs. Moreover, using less alkaline Na_2CO_3 as an activator benefits the development of environmentally friendly products. The findings of this study contribute to the applicability of difficult-to-recycle waste materials such as BFA and the development of sustainable BFA-based AAM. The study results could also encourage the construction industry to use such materials to reduce the negative environmental impact of their storage.

2. Materials and Methods
2.1. Raw Materials

The main precursor in this research is BFA from the biomass power plant. The wooden chips of pine trees were used as biomass. On average, the coniferous wood contains (48–56%) cellulose, (26–30%) lignin, and (23–26%) hemicelluloses containing (10–12%) pentosans and about 13% hexosans. When wood is heated at an increasing temperature, the processes of its drying, pyrolysis and gasification, and accompanying combustion proceed sequentially. At a temperature of 250–350 °C, the wood begins to decompose into components under the effect of high heat. As a result of pyrolysis (thermal destruction) of wood, volatile substances are released. The products of wood pyrolysis are mainly tar, coal, and low molecular weight gases, and large amounts of carbon monoxide and dioxide (CO and CO_2) are also released. Pine burning is characterized by a low combustion temperature of about 610–630 °C and the formation of smoke and soot.

The chemical composition of the BFA is presented in Table 1. As can be observed, the main oxides presented in the BFA are CaO (31.50%) and SiO_2 (22.91%), and carbon occupied 20.40% of the composition. This carbon content is apparently due to the incomplete combustion process of the wood [86]. It is common for BFA to contain such a large amount of carbon because it is difficult to maintain a smooth combustion process in bio-boilers. In the BFA, the alkalis are presented as K_2O (4.003%) and Na_2O (0.259%), and the BFA contains a significant amount of MgO (3.574%). The contents of chlorides and sulfates are not high, while P_2O_5 content is noticeable higher. The presence of heavy metals such as Cr, Cu, Ni, and Zn in the BFA is not high. The most critical components in wood ash are CaO, SiO_2, Al_2O_3, Fe_2O_3, and MgO, as they react in the existence of moistness to create bonding agents.

Table 1. Chemical compositions of BFA.

Element	Concentration (%)	Compound Formula	Concentration (%)
C	20.40	C	20.40
O	33.96	O	5.992
Na	0.192	Na_2O	0.259
Mg	2.155	MgO	3.574
Al	1.390	Al_2O_3	2.627
Si	10.71	SiO_2	22.91
P	1.182	P_2O_5	2.707
S	0.354	SO_3	0.883
Cl	0.098	Cl	0.098
K	3.324	K_2O	4.003
Ca	22.51	CaO	31.50
Ti	0.127	TiO_2	0.212
Cr	0.011	Cr_2O_3	0.016
Mn	1.602	MnO	2.068
Fe	1.624	Fe_2O_3	2.322
Ni	0.010	NiO	0.013
Cu	0.013	CuO	0.017
Zn	0.080	ZnO	0.100
Rb	0.016	Rb_2O	0.017
Sr	0.077	SrO	0.091
Y	0.015	Y_2O_3	0.019
Ba	0.154	BaO	0.172

It is known that fly ash must meet the standard requirements (BS EN 450-1:2012) to be used as a substitute for cement. The chemical composition of the BFA was compared with the requirements for fly ash. The total content of (SiO_2 + Al_2O_3 + Fe_2O_3) in the BFA is almost three times lower, and the LOI is 2.5 times higher than specified for fly ash in the standard. The BFA has a bulk density of 683 kg/m^3, a specific surface area of 493 m^2/kg, and an average particle size of 43.26 μm (Figure 1).

Two components were used as an alkaline activator solution: Na_2CO_3 solution and sodium silicate ($Na_2SiO_3 \cdot nH_2O$, $H_2Na_2O_4Si$) solution. Na_2CO_3 solution (30 wt%) was prepared by dissolving Na_2CO_3 crystalline powder with water. Figure 2 shows the XRD pattern of Na_2CO_3. Na_2SiO_3 solution concentration was 50 wt% with a density of 1.37 kg/m^3, and the molar ratio of SiO_2 to Na_2O ratio was 3.22, and the boiling point of Na_2SiO_3 is 100 °C. Figure 3 shows the XRD pattern of Na_2SiO_3.

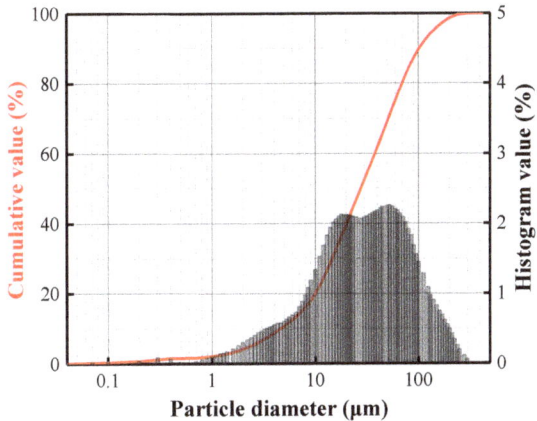

Figure 1. Particle size distribution of the BFA.

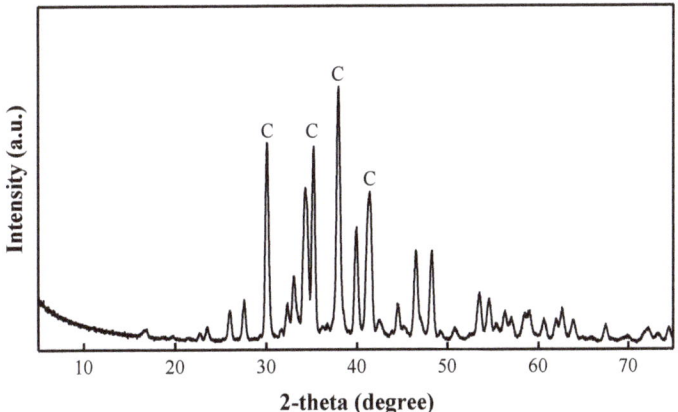

Figure 2. XRD pattern of Na_2CO_3 (C: Na_2CO_3).

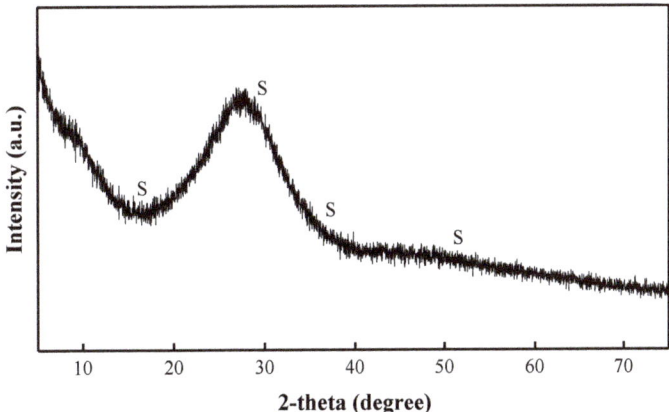

Figure 3. XRD pattern of Na_2SiO_3 (S: Na_2SiO_3).

Polycarboxylate-based superplasticizer (SP) with 27% dry matter content of the solution, with a molecular weight of 51 g/mol, was used to maintain the same consistency in these samples, and the SP was used in a liquid state.

2.2. Paste Design and Sample Preparation

The BFA-based AAM pastes were prepared according to the mix design in Table 2. To study the effect of the C/S ratio on the structure formation of BFA-based AAM pastes, the AAS was designed with different C/S ratios. The Na_2CO_3/Na_2SiO_3 solute ratio varied from 0.40 to 1.20, while the solution ratio varied from 0.67 to 2.00. The AAS and the SP solution were constant at 25 wt% and 5.71 wt% of the BFA, respectively. The water/BFA ratio of these samples was constant at 0.35. The water in the system includes water in the Na_2CO_3-Na_2SiO_3 solution, SP solution, and additional water.

Table 2. Composition of BFA-based AAM pastes (Group 1 and 2).

Group	Sample	Na_2CO_3 (wt% of BFA)	Na_2SiO_3 (wt% of BFA)	Na_2CO_3/Na_2SiO_3 Solute Ratio	Na_2CO_3/Na_2SiO_3 Solution Ratio	Curing Temperature (°C)
1	CS-0.40-40	3.00	7.50	0.40	0.67	40
	CS-0.60-40	3.75	6.25	0.60	1.00	40
	CS-0.75-40	4.17	5.56	0.75	1.25	40
	CS-0.90-40	4.50	5.00	0.90	1.50	40
	CS-1.20-40	5.00	4.15	1.20	2.00	40
2	CS-0.40-20	3.00	7.50	0.40	0.67	20
	CS-0.60-20	3.75	6.25	0.60	1.00	20
	CS-0.75-20	4.17	5.56	0.75	1.25	20
	CS-0.90-20	4.50	5.00	0.90	1.50	20
	CS-1.20-20	5.00	4.15	1.20	2.00	20

After mixing, the fresh AAM pastes were cast in 160 × 40 × 40 mm steel molds and compacted on the vibrating table for 20 s. The BFA-based AAM paste molds were divided into two groups: one group was kept at an ambient temperature of 20 ± 2 °C for 24 h and then demoulded and kept at the same temperature; another group of samples was kept at 40 °C in the oven (Figure 4). After 24 h, samples were demolded and further kept in the oven at 40 °C for all testing times.

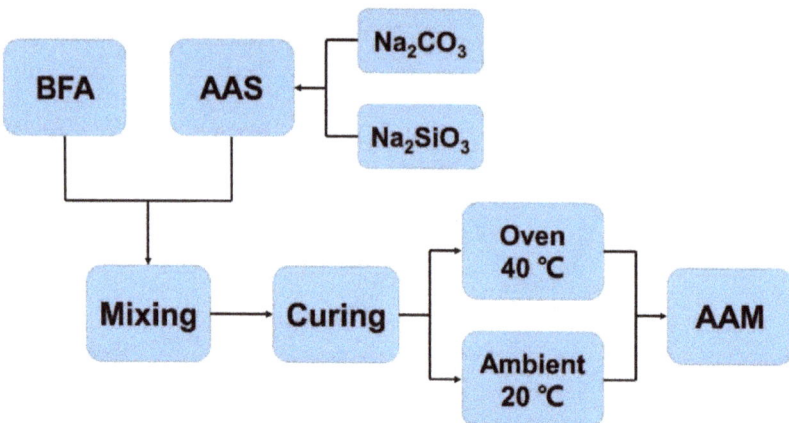

Figure 4. Schematic of BFA-based AAM paste preparation.

2.3. Test Methods

BFA powder analysis was performed using a high multichannel performance sequential Wavelength Dispersive X-ray Fluorescence (WD-XRF) spectrometer (AXIOS-MAX, Panalytical, Eindhoven, Netherlands). The WD-XRF system operated for 1440 s. The quantitative analysis was performed using "Omnian" software, and corresponding standards were used for a standard-less analysis. X-ray diffraction (XRD) patterns of the studied BFA and composites (CS-0.40-40, CS-0.75-40 and CS-1.20-40) were measured using an X-ray diffractometer, SmartLab (Rigaku, Tokyo, Japan), equipped with an X-ray tube with a 9 kW rotating Cu anode. The measurements were performed using Bragg–Brentano geometry with a graphite monochromator on the diffracted beam and a step scan mode with a step size of 0.02° (in 2θ scale) and a counting time of 1 s per step. The measurements were conducted in the 2θ range of 5–75°. Phase identification was performed using the software package PDXL (Ver. 2.8, Rigaku) and the ICDD powder diffraction database PDF4+ (2021 release). Mineral content was estimated using Reference Intensity Ratio (RIR) method.

Thermogravimetry/differential thermal analysis (TG/DTA) curves of the BFA were registered with a Linseis STA PT-1600 thermal analytical instrument up to 1000 °C (with a temperature rise rate of 10 °C/min; the air was used as the heating environment; the weight of the specimens was (50 ± 5) mg). According to standard NF P18-513 and [87], Chapelle's method was used to determine the pozzolanic activity of the BFA. The hydraulic activity of the BFA was calculated based on the standard requirements (BS EN 197-1:2011). The hydraulic index K_3 was calculated based on the chemical composition of the studied BFA and using calculation methods presented in [88] and [89]. The hydraulic index K_3 is defined as the ratio of basic oxides to acidic oxides: (%CaO + %MgO + %Al_2O_3)/(%SiO_2). Values of K_3 above 1.0 indicate good hydraulic properties.

The effect of different concentrations of the prepared Na_2CO_3 solutions and Na_2SiO_3 solution on the electrical conductivity (EC) and pH values were performed with the MPC 227 instrument manufactured by "Mettler-Toledo" (Columbus, OH, USA, pH electrode InLab 410, measuring accuracy 0.01; EC electrode InLab 730, measuring range 0–1000 mS/cm). The EC and pH values of the prepared activator solutions and alkali-activated BFA pastes with different C/S ratios (Table 2) were measured at 20 °C.

The structure formation of the samples cured at 40 °C and 20 °C was evaluated using the ultrasonic pulse velocity (UPV) method using the tester Pundit 7. The sample was placed between two ultrasonic transducers (transmitter and receiver) operating at 54 kHz. The transducers were pressed against the samples at two strictly opposite points, and Vaseline was used to ensure good contact. The ultrasonic pulse velocity (V) was calculated using Equation (1):

$$V = \frac{l}{\tau} \cdot 10^6 \tag{1}$$

where l is the distance between the cylindrical heads and τ is the time of pulse spread.

The compressive strength, bulk density, water absorption, and porosity of the samples cured at 40 °C and 20 °C during 28 days were determined according to standard LST EN ISO 1927-6:2013.

3. Results and Discussion

3.1. Parameters of BFA

According to the X-ray phase analysis (Figure 5), the BFA consists of quartz—52.0%, dicalcium silicate—19.7%, portlandite—15.0%, lime—9.20%, and calcite—3.80%.

The SEM images of the BFA are presented in Figure 6. Most BFA particles have irregular shapes, while other ash particles show approximately spherical shapes. The parts of unburned wood or soot particles are sufficiently noticeable.

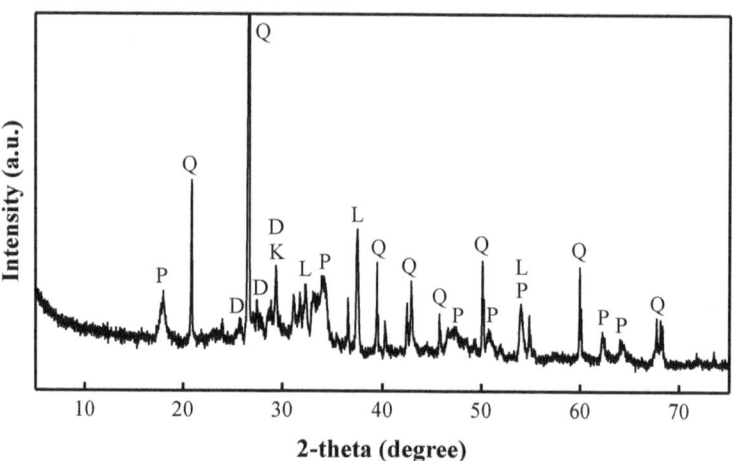

Figure 5. XRD pattern of the BFA (Q: quartz; D: dicalcium silicate; P: portlandite; L: lime; K: calcite).

(a) (b)

Figure 6. SEM images of the BFA. Magnification: (a) 250 and (b) 1200.

It can be observed that the TG/DTA curves of the BFA have one exothermic effect in the temperature interval of 300–520 °C and two endothermic effects in the temperature intervals of 20–200 °C and 680–780 °C (Figure 7). In the temperature interval of 20–200 °C, the unbound water and a part of bound water are removed with a mass loss of approximately 1.8%. There is a clearly visible exothermic peak in the temperature range of 300–520 °C, and the exothermic process can be attributed to the combustion of soot and organic matter (unburned wood aggregate) [90]. During the combustion of wood, some decomposition takes place at temperatures as follows: hemicellulose (200–260 °C); cellulose (240–350 °C); and lignin (280–500 °C) [91,92]. Hemicellulose and cellulose are characterized by fast thermal decomposition, while lignin is more resistant, and during its pyrolysis, carbon layers on the lignin surface are formed. During this burning process, mass loss is the highest. Although portlandite appeared in the XRD test, the decomposition of this mineral is hindered by the exothermal effect of unburned wood aggregate. The endothermal effect in the temperature range of 680–780 °C is attributed to the decomposition of calcite.

The pozzolanic activity of the BFA is 45.7 mg CaO/g. It indicates that the BFA is not rich in amorphous SiO_2, which can react with the $Ca(OH)_2$ to create additional hydration products, such as C-S-H, which improves the physical and mechanical properties of the cementitious system. SiO_2 is usually formed at high temperatures during the incineration of biomass. However, in our research, the BFA was collected from a low-temperature combustion process and is characterized by properties the same as referred to in research [93,94]. In the study [87], after a series of experiments, the following pozzolanic activities of biomass fly ash were determined: sugar cane biomass fly ash—279 mg CaO/g, rice husk biomass fly ash—622 mg CaO/g, wood biomass fly ash—269 mg CaO/g.

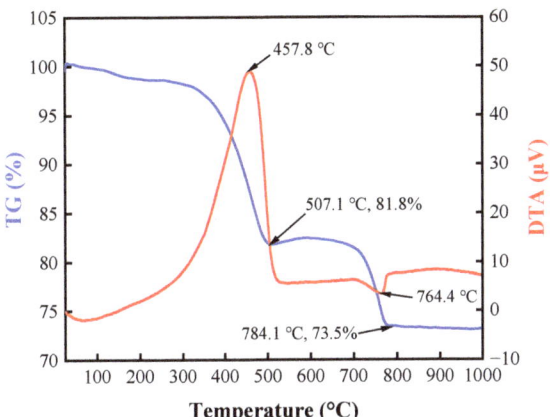

Figure 7. TG/DTA curves of the BFA.

The ability of a material to set and harden when mixed with water and produce hydration products during the hydration process is referred to as hydraulic activity. The contents of SiO_2 and CaO determine the hydraulic activity, and the standard requirement (BS EN 197-1:2011) for hydraulic material is that the mass ratio of CaO/SiO_2 is not less than 2. Based on the data in Table 1, the hydraulic activity of the BFA is 1.375. Hydraulic index K_3 is calculated based on the chemical composition of the studied BFA, defined as the ratio of basic oxides to acidic oxides: $(\%CaO + \%MgO + \%Al_2O_3)/(\%SiO_2)$ [88]. Based on the data in Table 1, the hydraulic index K_3 of the studied BFA is 1.645.

Overall, it can be concluded that the BFA has low hydraulic properties and is reasonable for use as a precursor in alkali-activated materials.

3.2. Activation Solution Characterization

The tests were carried out to describe how Na_2SiO_3 solutions and different concentrations of Na_2CO_3 in solutions affect the values of EC and pH (Table 3). According to the results, the highest pH is Na_2SiO_3 solution; for Na_2CO_3 solutions, the pH varies a little. The concentration of Na_2CO_3 affects the EC value of the solution, as the EC value depends on the number of ions in the solution. The EC value in the 10% Na_2CO_3 solution is 18% lower than in the 20% Na_2CO_3 solution. However, when the concentration of Na_2CO_3 in the solution was increased to 30%, the EC value decreased to 79.2 S/m. This reduction in EC value is assumed to occur due to supersaturation of the solution, as also noted in studies [87,93,94].

Table 3. EC and pH characteristics of Na_2SiO_3 solution and different Na_2CO_3 solutions.

Activator Solution	pH	EC, S/m
Na_2SiO_3	11.90	31.8
10% Na_2CO_3	11.36	71.4
20% Na_2CO_3	11.38	85.9
30% Na_2CO_3	11.38	79.2

The EC values of BFA-based AAM pastes with different C/S ratios are presented in Figure 8. The research results show that the number of ions in the solution increases with the increased amount of Na_2CO_3. The EC value gradually increases by increasing the C/S ratio in the paste. When the C/S ratio triples, the EC value increases by 45.2%. However, the dissolution of BFA minerals and the penetration of ions into the solution were delayed, and the EC values increased slowly in pastes with larger C/S ratios.

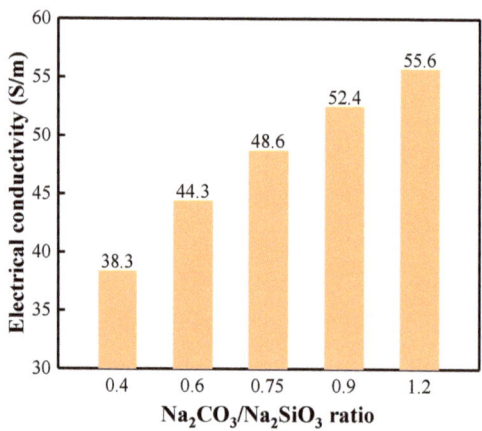

Figure 8. EC of BFA-based AAM pastes with different C/S ratios.

3.3. Initial and Final Setting Times of BFA-Based AAM Pastes

To choose the most appropriate C/S ratio, setting time tests were performed, including the initial and final setting times of prepared BFA-based AAM pastes (Figure 9). Initial setting time results show that CS-0.40 paste, which possesses the lowest EC, is characterized by the longest initial setting time. As the C/S ratio increases, the initial setting time decreases, and the interaction between the AAS with higher Na_2CO_3 content and the BFA with high lime content starts faster. When the Na_2CO_3 content in the BFA-based AAM paste is the highest, the final setting time is one-third of that with the lowest Na_2CO_3 content.

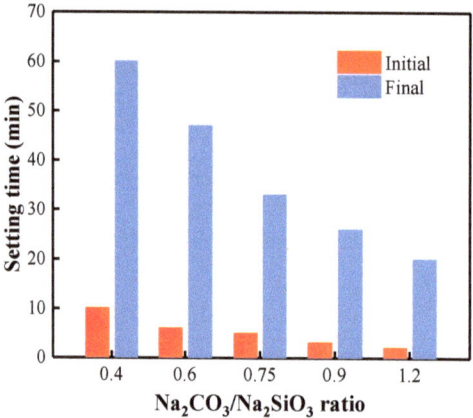

Figure 9. Initial and final setting times of BFA-based AAM pastes with different C/S ratios.

Researchers have studied the workability of mixtures containing different amounts of alkali [62] and found that the initial flow of the manufactured mortar decreased slightly with increasing alkali content. However, it was pointed out that higher doses of Na_2CO_3 lead to faster sintering reaction rates [52,95]. The availability of CO_3^{2-} anions in the pore solution also substantially affects the reaction kinetics [52]. As Na_2CO_3 increases, more Ca^{2+} ions can be dissolved at the early stage and precipitated with CO_3^{2-} anions, thus speeding the reaction and the setting of AAM. Similar results have emerged in the studies [52,69]. It was pointed out that the duration of the induction period of Na_2CO_3 activated slag cement with an alkali concentration of 3 Na_2O% is 1.6 times that with an alkali concentration of 8 Na_2O%. The delayed reaction of slag cement activated by lower concentrations of Na_2CO_3 can be explained by the initial chemical precipitation of $CaCO_3$.

3.4. Density of BFA-Based AAM Pastes

Densities of the samples cured at 40 °C (Figure 10a) decreased rapidly for the first 7 days and then leveled off gradually. A reduction in weight loss of 14.2% to 9.5% after 7 days of curing can be observed as the C/S ratio increases from 0.40 to 1.20 (Figure 10b). After 28 days of curing, the weight loss is in the range of 20.4–13.8%. It can be clearly observed that by increasing the C/S ratio, the weight of samples is reduced less. As the C/S ratio increases visibly, the amount of free water that can be evaporated decreases with its participation in the synthesis reactions. EC studies of the pastes confirm this reasoning (Figure 8), showing that the EC value increases with increasing the C/S ratio. As was pointed out earlier, an increase in CO_3^{2-} anions promote Ca^{2+} ions dissolution and new phase formation, which involve water and prevent evaporation.

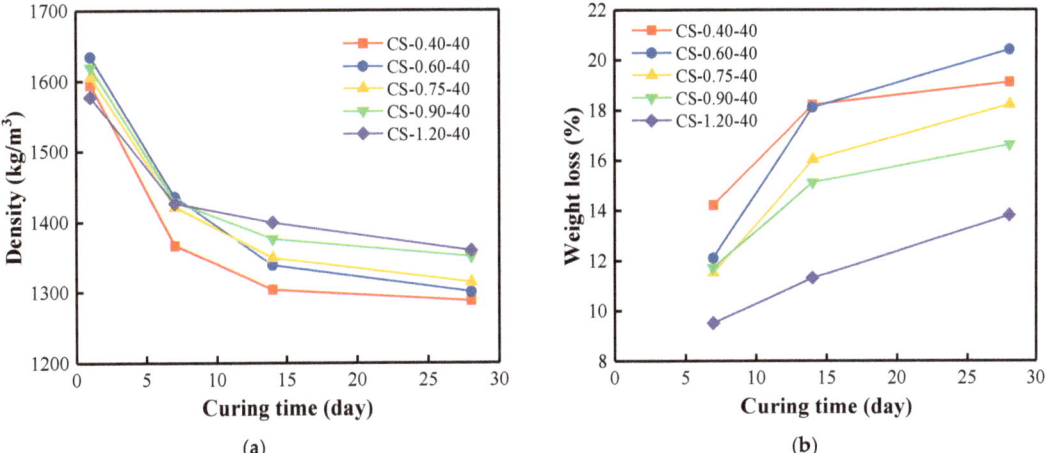

Figure 10. (a) Density of BFA-based AAM pastes cured at 40 °C for 28 days; (b) weight loss of samples in % during 28 days.

The same trend was observed for the samples cured at 20 °C (Figure 11). However, the densities of the samples cured at 20 °C decreased significantly during 14 days, and the weight losses were less than that of the samples cured at 40 °C. The sample with the smallest C/S ratio always showed the smallest density whenever it was at 40 °C or 20 °C. The higher the curing temperature, the faster the structure evolution during the BFA-based alkali activation-synthesis process. The same trend was observed in research for samples cured at ambient temperature [96,97]. The study concluded that the density of the cured samples decreased slightly by about 2% in the first few weeks, but remained almost constant thereafter. In our case, the weight losses were more pronounced because of using the different activators, lower curing temperature, and high carbon content of BFA. The density tests were supplemented by UPV testing of the samples.

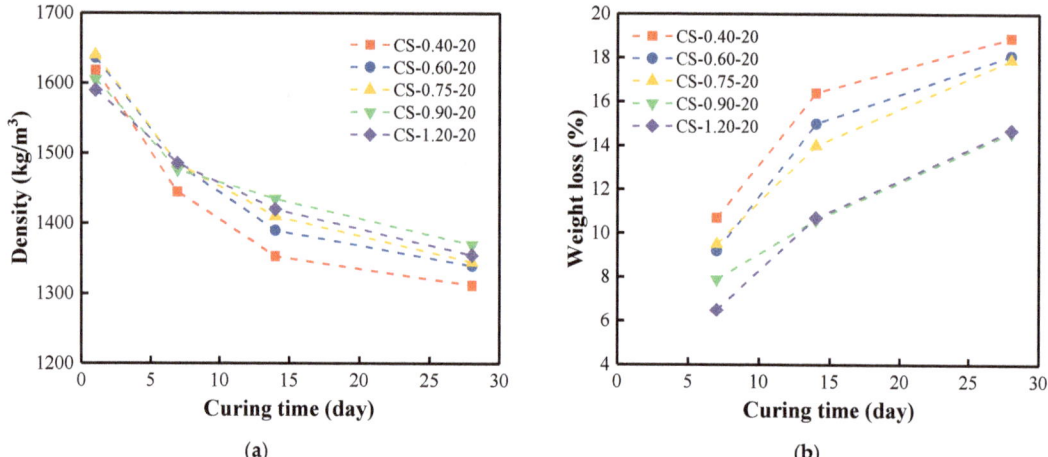

Figure 11. (**a**) Density of BFA-based AAM pastes cured at 20 °C for 28 days; (**b**) weight loss of samples in % during 28 days.

3.5. UPV of BFA-Based AAM Pastes

Density trends are reflected in the structure formation of BFA-based AAM pastes (Figure 12). At the curing temperature of 40 °C, the UPV values of the samples generally showed a downward trend within 14 days, and then the UPV values changed slightly until 28 days. It seems that the structure formation was generally completed in this period. UPV values were consistently lower in the samples with lower C/S ratios. The UPV values of the sample with the smallest C/S ratio decreased the fastest and then leveled off. It is shown that the more Na_2CO_3 in the AAS, the faster the solidification of the AAM paste. At the curing temperature of 20 °C, various increases in UPV values were observed for the structure formation of the samples after 7 days. Further structure formation was slow, and the UPV values decreased slightly. It can be concluded that higher curing temperatures accelerate the AAM synthesis.

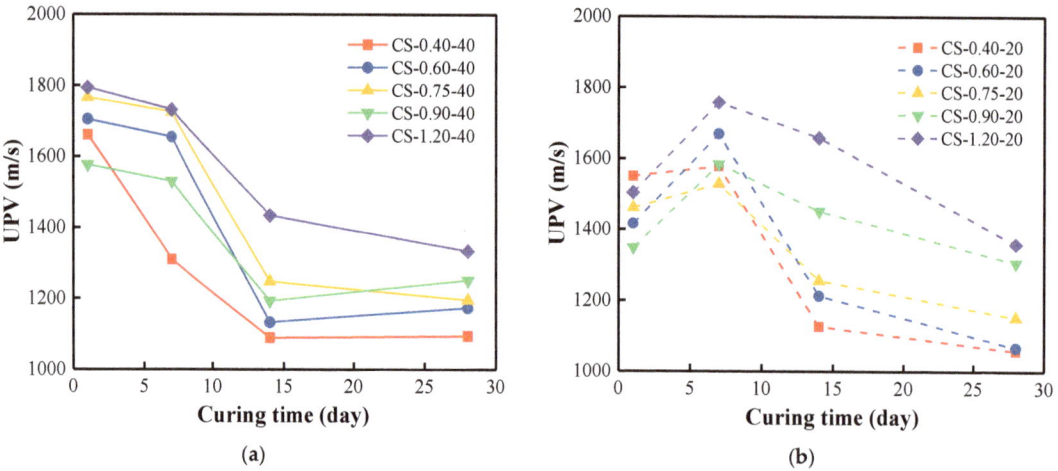

Figure 12. UPV of BFA-based AAM pastes cured at (**a**) 40 °C and (**b**) 20 °C for 28 days.

3.6. Compressive Strength of BFA-Based AAM Pastes

Figure 13a presents the compressive strength of samples with different C/S ratios cured at 40 °C. After curing for 7 days, the compressive strength increased with the increase in the C/S ratio, indicating that the higher the alkali content, the better the mechanical properties [62,98]. Depending on the amount of Na_2CO_3, the compressive strength of the samples at 7-day curing ranged from 3.94 to 4.55 MPa (Figure 13a). However, when the curing time reached 28 days, the strength of the samples showed a decreasing trend, from 2.36 MPa (the lowest C/S ratio) to 3.97 MPa (the highest C/S ratio). This trend is mainly observed when the C/S ratio is lower. From 7 to 28 days, the sample CS-0.40-40 showed the largest strength loss with a 40% decrease. Moreover, the sample with the smallest C/S ratio always showed the smallest density at 40 °C. For the sample CS-1.20-40 with the highest C/S ratio, the compressive strength decreased the least, by only 8.3%.

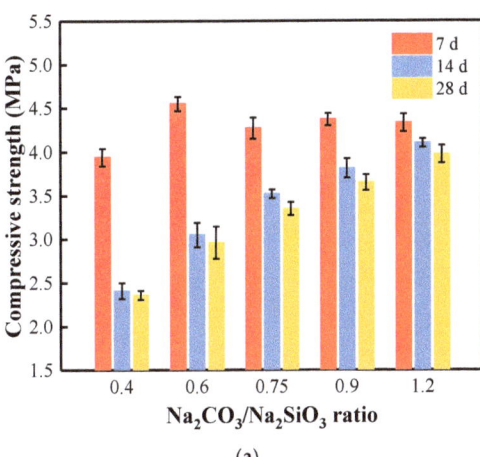
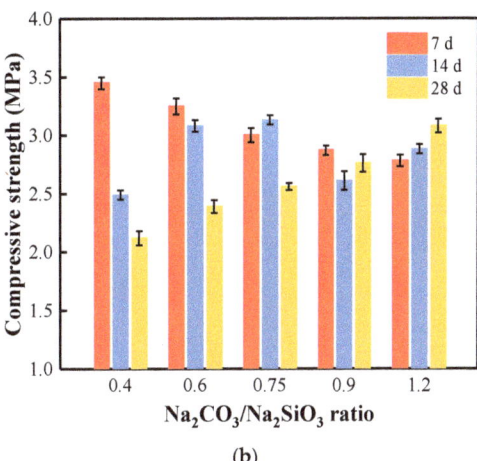

(a)   (b)

Figure 13. Compressive strength of AAM pastes with different C/S ratios cured at (**a**) 40 °C and (**b**) 20 °C.

Overall, the compressive strength of the samples cured at 40 °C was higher than that of the same composition cured at 20 °C. For samples cured at 20 °C, the sample (CS-0.40-20) with the smallest C/S ratio showed a sharp decrease in compressive strength over the curing time (from 3.45 to 2.12 MPa, decreased by 38.6%). However, with increasing the C/S ratio, the compressive strength of CS-1.20-20 increased slightly (from 2.78 to 3.08 MPa, increased by 10.8%) during the curing time (Figure 13b). It can be explained by the fact that dicalcium silicate, one of the key minerals of the BFA, dissolved better in AAS at a higher temperature than in the pastes activated at ambient temperature [99]. Researchers pointed out that amorphous hydrated C-(A)-S-H-like gels with aluminum and sodium ions were generated during structure formation. As a result, these reaction products occupied the cavities and pores, increasing the samples' mechanical strength.

Several research results [100,101] can be presented about using wastes from various organic sources in geopolymers. As pointed out in [100], greener geopolymers have been developed using agricultural and industrial wastes such as rice husk, rice husk ash, metakaolin, ground granulated blast furnace slag, and palm oil fuel ash, activated by different ratios of activators (Na_2SiO_3/NaOH). The results showed that the compressive strength of geopolymers was directly proportional to the ratio of the alkaline solution, while the compressive strength of the samples ranged from 0.8 to 2.8 MPa and up to 4.9 MPa. The same low strength of 4–6 MPa was reported in [96], when only biomass fly ash or fly ash was used in the alkali-activated materials composition. In the synthesized one-part geopolymers, consisting of fly ash, slag, and hydrated lime, the 28-day compressive strength of the fly-ash-based samples cured at ambient temperature reached 5.2 MPa when activated with Na_2SiO_3 [101]. Overall, a higher curing temperature and a higher C/S ratio can improve the strength under certain conditions.

3.7. Water Absorption of BFA-Based AAM Pastes

By comparing the density and water absorption of AAM samples, they both show some significant correlations. Overall, the water absorption of the samples cured at 40 °C is slightly higher than that of the same composition cured at 20 °C (Figure 14). After curing at 40 °C and 20 °C for 28 days, the sample CS-0.40 with the smallest density displayed higher water absorption. When the C/S ratio in the AAM increases from 0.40 to 0.90, the water absorption decreases from 23.32% to 20.78% for samples cured at 40 °C and from 24.23% to 20.33% for samples cured at 20 °C. The presence of BFA in AAM increases the water absorption of the samples [86]. The apparent density of BFA-based composite activated by Na_2CO_3 varies between 1.58–1.6 kg/m^3, and water absorption varies between 20–22%. In our case, this could be due to the high content of unburned carbonized organic matter, which actively adsorbs water due to its highly porous structure [102,103]. The densities of our samples are significantly lower (1.27–1.37 kg/m^3), which results in higher water absorption values.

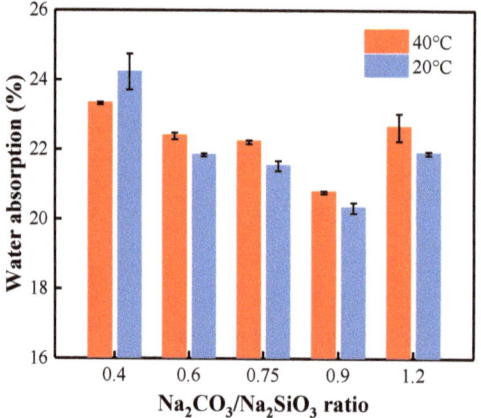

Figure 14. Water absorption of AAM pastes with different C/S ratios cured at 40 °C and 20 °C.

3.8. XRD Patterns of BFA-Based AAM Pastes

Samples cured at the higher temperature of 40 °C show higher compressive strength results than samples cured at 20 °C. Therefore, the reaction products of the samples cured at 40 °C with the C/S ratio of 0.40, 0.75, and 1.20 were characterized by XRD tests (Figure 15). After curing for 28 days, the main mineral in the AAM pastes of all compositions is quartz, which is also the main mineral of the BFA. The mineralogical composition of Na_2SiO_3 showed a diffuse broad hump in the range of 23–35° 2θ (Figure 3) [62]. The most intensive peaks of Na_2CO_3 were observed at 28–46° 2θ (Figure 2). A diffuse broad hump attributed to the amorphous phase has been identified in the diffractograms of all samples. It was assumed that due to the high pH in the AAS, the formation of the crystalline phase from the zeolitic germs formed in the samples was slow, and thus an amorphous (C-S-H/C-A-S-H) phase was produced [41].

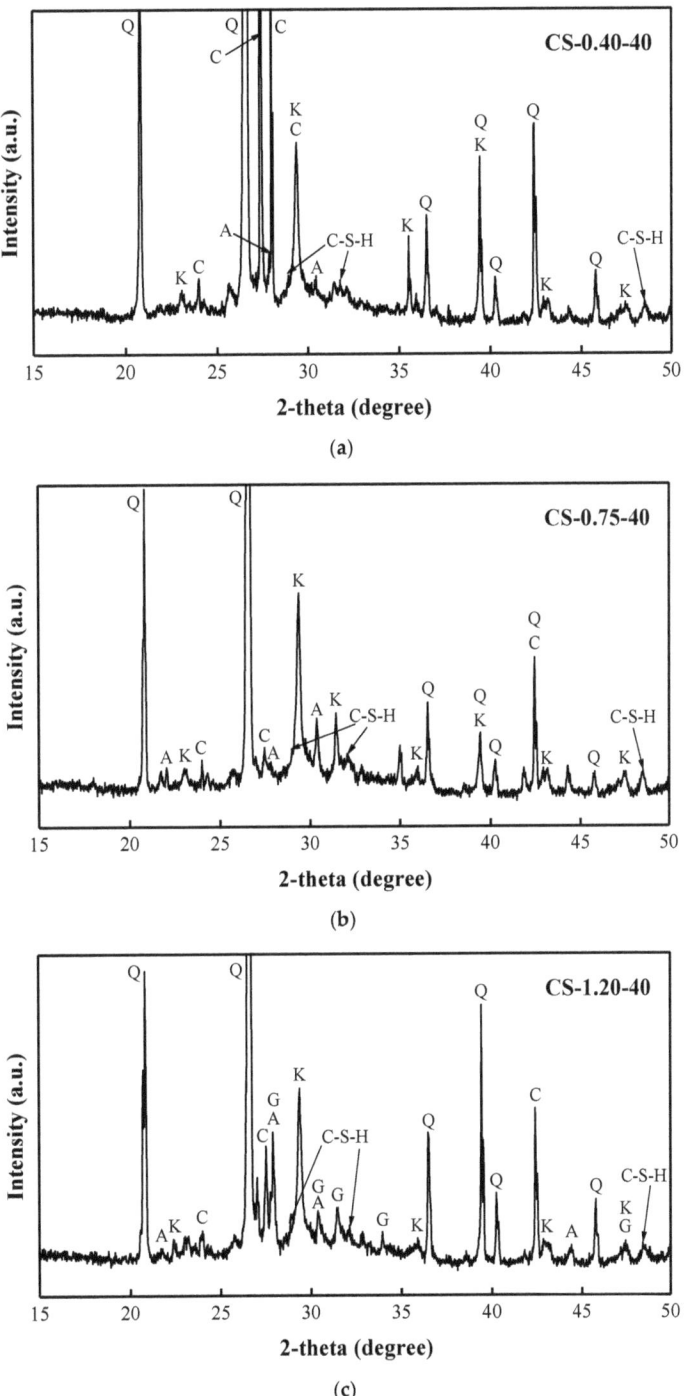

Figure 15. XRD patterns of AAM pastes cured at 40 °C for 28 days: (**a**) CS-0.40-40; (**b**) CS-0.75-40; (**c**) CS-1.20-40 (Q: quartz; K: calcite; G: gaylussite; A: anorthoclase; C: Na_2CO_3; C-S-H: C-S-H gel).

The reaction products of the sample with a C/S ratio of 0.40 are calcite and C-S-H. Additionally, some Na_2CO_3 and anorthoclase have been identified. It should be noted that the most intense peaks of Na_2CO_3 were observed in this composition. As the C/S ratio increases, the intensity of the Na_2CO_3 peak decreases. The reaction products in the sample with a C/S ratio of 0.75 are the same. The growth of gaylussite was observed in the sample with a C/S ratio of 1.20. In addition, the above minerals Na_2CO_3, C-S-H, and anorthoclase have all been identified. These reaction products (calcite, C-S-H, and gaylussite) control the compressive strength growth of the AAM pastes.

Along with the identified reaction products (calcite and gaylussite) in the composition with the same activator, the researchers also identified poorly crystalized C-(A)-S-H gels [62]. According to the researchers, Ca^{2+} ions are available and quickly precipitated with CO_3^{2-} anions generating calcite and gaylussite. The transformation of initially precipitated $CaCO_3$ to other phases is the key to the reaction development [62]. The crystalline phases, such as quartz, in the raw BFA remained unaltered after the activation of BFA pastes. The same observations were reported in the research [104].

3.9. TG/DTA Test of BFA-Based AAM Pastes

The TG/DTA curves of BFA-based AAM pastes (C/S ratios of 0.40, 0.75 and 1.2) cured at 40 °C for 28 days show two endothermic effects in the temperature intervals of 20–250 °C and 680–760 °C, and three exothermic effects in the temperature intervals of 280–320 °C, 330–380 °C, and 400–500 °C (Figure 16). In the temperature interval of 20–200 °C, the unbound and partially bound water was removed from the samples. At the same temperature interval (160–200 °C) from which the decomposition of the C-S-H undergoes [105], dehydration of gaylussite in the temperature interval of 200–250 °C was observed [106–108]. These findings are supported by the XRD results (Figure 15).

The mass loss of the samples in the temperature range of 20–250 °C increases from 5.8% to 7.2% and 8.5% when changing the C/S ratio from 0.4 to 1.2. During the temperature interval (160–200 °C) attributed to C-S-H decomposition, the mass losses in the samples were 0.6%, 0.9%, and 1.1%, respectively. During the temperature interval (200–250 °C) attributed to gaylussite decomposition, the mass losses in the samples were 0.8%, 1.0%, and 1.3%, respectively. It can be found that in the samples with C/S ratios of 0.4 and 0.75, the mass loss was significantly lower than that of 1.2. This could explain why no gaylussite was found in these two samples in the XRD study. Dehydration converted gaylussite to the double carbonate, $Na_2Ca(CO_3)_2$, which generated the exothermic effects above 250 °C in the temperature intervals of 280–320 °C with mass losses of 0.8%, 0.9%, and 1.1%. Between 330 °C and 380 °C, another exothermic peak occurred due to crystal transformation from the low-temperature form of $Na_2Ca(CO_3)_2$ to the high-temperature form with mass losses of 1.6%, 2.0%, and 3.8%, but the generally exothermic peaks in the temperature interval of 300–500 °C (including peaks at 400–500 °C) are due to the decomposition of BFA (combustion of soot and organic matter), and mass losses with increasing C/S ratio increased from 7% to 7.9% and 9.4% in the whole temperature interval. The decomposition of $CaCO_3$ was observed during the endothermic effect (680–780 °C) with mass losses from 0.6% to 1.1%. The decomposition of Na_2CO_3 should be around 850–900 °C, but no clear endothermic peaks were identified, and the decomposition of sodium silicate occurs at temperatures above 1080 °C [108].

The total mass loss of the samples during the TG/DTA test varies significantly, with 17.0% for sample CS-0.40-40, 19.3% for sample CS-0.75-40, and 24.2% for sample CS-1.2-40. This result confirms that C-S-H and gaylussite were formed in the composition. Due to the synthesis of new products, the amount of $CaCO_3$ is low. The overall results prove that the amount of synthetic products increases with increasing the C/S ratio in the composition.

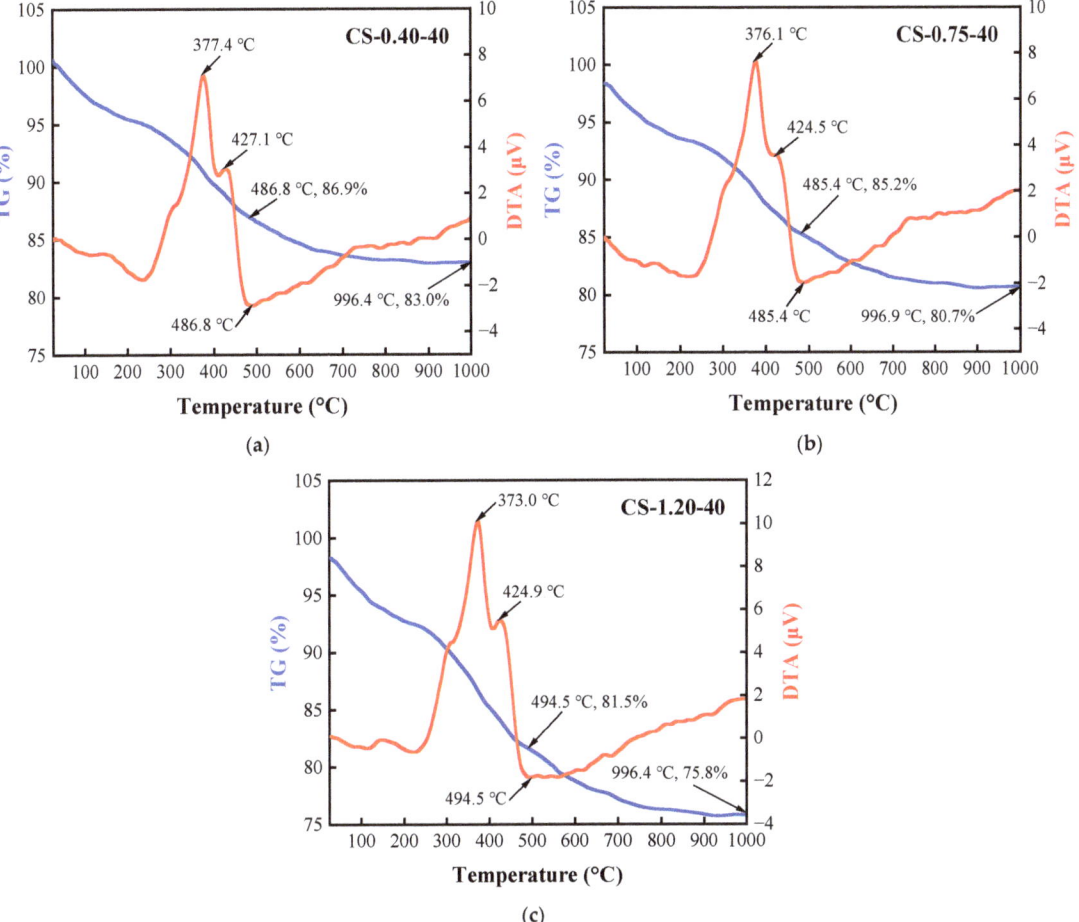

Figure 16. TG/DTA curves of AAM pastes cured at 40 °C for 28 days: (**a**) CS-0.40-40; (**b**) CS-0.75-40; (**c**) CS-1.20-40.

3.10. Microstructure of BFA-Based AAM Pastes

The differences in microstructure of the samples with C/S ratios of 0.40 and 1.20 cured at 40 °C and ambient temperature can be seen in the SEM images (Figures 17–20). The sample with a C/S ratio of 0.40 cured at 20 °C (Figure 17a) has an uneven microstructure. The cracks and cavities are distributed around the unburned organic matter throughout the surface. The samples with a C/S ratio of 1.2 cured at 20 °C (Figure 17b) and 40 °C (Figure 18b) have fewer surface voids and denser structures. The larger visible voids and cracks could be linked with water evaporation from the gels during curing. The samples with smaller C/S ratios seem to have more unbound water in the structure, which significantly decreases the density and compressive strength of samples during evaporation.

Figure 17. Microstructure of the samples cured at 20 °C: (**a**) CS-0.40-20; (**b**) CS-1.20-20 (magnification 500).

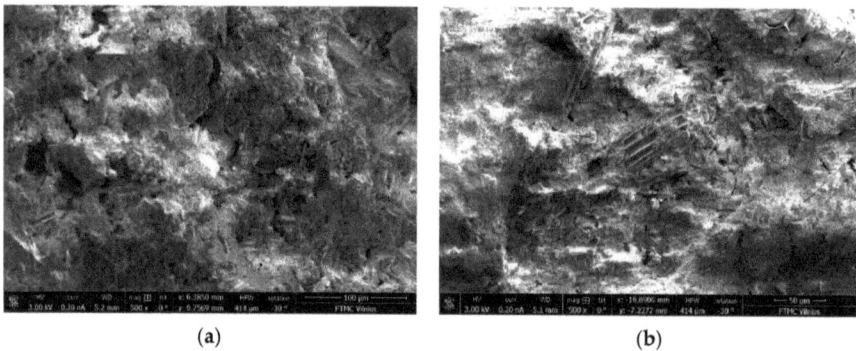

Figure 18. Microstructure of the samples cured at 40 °C: (**a**) CS-0.40-40; (**b**) CS-1.20-40 (magnification 500).

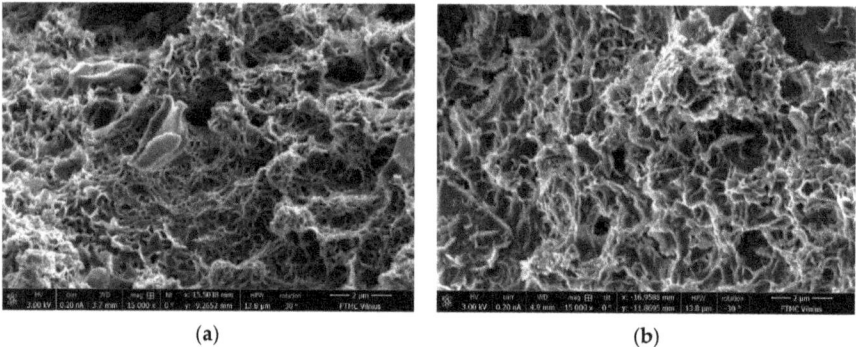

Figure 19. Microstructure of the samples cured at 20 °C: (**a**) CS-0.40-20; (**b**) CS-1.20-20 (magnification 15,000).

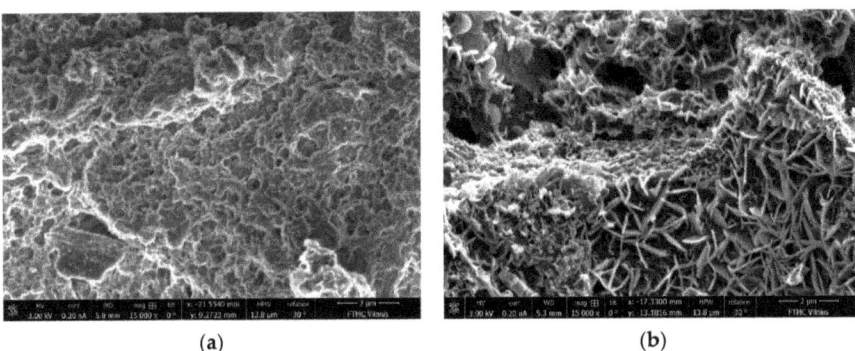

Figure 20. Microstructure of the samples cured at 40 °C: (**a**) CS-0.40-40; (**b**) CS-1.20-40 (magnification 15,000).

Higher magnification shows that the sample cured at 20 °C (Figure 19) shows a highly porous gel structure when the C/S ratio is 0.40. When the C/S ratio is 1.20 in the sample, the gel structure has fewer visible pores. In contrast, the samples cured at 40 °C (Figure 20) exhibits a more compact microstructure with low void content, which could be responsible for their higher strength at 28 days. For samples cured at 40 °C, the gel structure of the sample with a C/S ratio of 0.40 is more even than that cured at 20 °C, and no cracks appeared. On the sample with a C/S ratio is 1.20 surface, the deposition of reaction products is evident. It shows that the reaction products have markedly changed, with calcium released from BFA and participating in activation reactions. The microstructure appeared to be denser than those pastes made with a C/S ratio of 0.4, and this observation supports its higher compressive strength. Overall, it can be underlined that the samples with good homogeneity and dense gel structure corresponded to compositions with higher compressive strength.

4. Conclusions

This study aims to investigate the possibility of utilizing unprocessed carbon-rich BFA in AAM. BFA-based AAM pastes with different Na_2CO_3/Na_2SiO_3 ratios at different curing temperatures (40 °C and 20 °C) were investigated in the study.

It has been established that with the increase in the amount of Na_2CO_3 in the solution, the electrical conductivity of the BFA-based AAM pastes increases and the initial and final setting times decrease. The AAM samples with the smallest C/S ratio are characterized by the lowest density and UPV value and the highest water absorption, whether they were cured at 40 °C or 20 °C. The SEM research confirms the density, water absorption, and UPV results, proving that with an increase in C/S ratio and curing temperature, the structure of the samples became denser and more homogenous, which corresponds with the higher compressive strength of the sample.

Regardless of the curing temperature, the compressive strength of AAM samples after curing for 28 days increased with the increase in the C/S ratio, thus indicating that the higher the alkali content, the better the mechanical properties. According to XRD analysis and TG/DTA tests, with the increase in the C/S ratio in the samples, the synthesis of gaylussite and C-S-H were observed. The formation of the mentioned minerals contributes to the compressive strength growth of the AAM paste with a higher C/S ratio. Overall, the compressive strength of the AAM samples cured at 40 °C was higher than that cured at 20 °C with the same composition. The higher the curing temperature, the faster the structure evolution during the BFA-based alkaline activation synthesis process.

The findings of this study contribute to the applicability of difficult-to-recycle waste materials such as BFA and the development of sustainable BFA-based AAM. The study results could also encourage the construction industry to use such materials to reduce the negative environmental impact of their storage.

Author Contributions: Conceptualization, I.P.; methodology, C.Z. and J.P.; validation, M.K.; formal analysis, C.Z. and I.P.; investigation, C.Z. and J.P.; data curation, C.Z. and M.K.; writing—original draft preparation, C.Z.; writing—review and editing, C.Z., I.P., J.P. and M.K.; visualization, C.Z.; supervision, I.P. All authors have read and agreed to the published version of the manuscript.

Funding: This research received no external funding.

Institutional Review Board Statement: Not applicable.

Informed Consent Statement: Not applicable.

Data Availability Statement: Not applicable.

Conflicts of Interest: The authors declare no conflict of interest.

References

1. Cuenca, J.; Rodríguez, J.; Martín-Morales, M.; Sánchez-Roldán, Z.; Zamorano, M. Effects of Olive Residue Biomass Fly Ash as Filler in Self-Compacting Concrete. *Constr. Build. Mater.* **2013**, *40*, 702–709. [CrossRef]
2. Demirbas, A. Potential Applications of Renewable Energy Sources, Biomass Combustion Problems in Boiler Power Systems and Combustion Related Environmental Issues. *Prog. Energy Combust. Sci.* **2005**, *31*, 171–192. [CrossRef]
3. Chaunsali, P.; Uvegi, H.; Osmundsen, R.; Laracy, M.; Poinot, T.; Ochsendorf, J.; Olivetti, E. Mineralogical and Microstructural Characterization of Biomass Ash Binder. *Cem. Concr. Compos.* **2018**, *89*, 41–51. [CrossRef]
4. IEA. *Key World Energy Statistics 2016*; IEA: Paris, France, 2016. [CrossRef]
5. Scurlock, J.M.O.; Hall, D.O. The Contribution of Biomass to Global Energy Use (1987). *Biomass* **1990**, *21*, 75–81. [CrossRef]
6. IEA. *World Energy Outlook 2010*; OECD Publishing: Paris, France, 2010. [CrossRef]
7. U.S. EIA. *Assumptions to the Annual Energy Outlook 2015*; U.S. EIA: Washington, DC, USA, 2015.
8. Hughes, E. Biomass Cofiring: Economics, Policy and Opportunities. *Biomass Bioenergy* **2000**, *19*, 457–465. [CrossRef]
9. IRENA. *Renewable Power Generation Costs in 2014*; IRENA: Bonn, Germany, 2015.
10. Vassilev, S.V.; Baxter, D.; Andersen, L.K.; Vassileva, C.G. An Overview of the Chemical Composition of Biomass. *Fuel* **2010**, *89*, 913–933. [CrossRef]
11. Vassilev, S.V.; Baxter, D.; Vassileva, C.G. An Overview of the Behaviour of Biomass during Combustion: Part II. Ash Fusion and Ash Formation Mechanisms of Biomass Types. *Fuel* **2014**, *117*, 152–183. [CrossRef]
12. Chaunsali, P.; Uvegi, H.; Traynor, B.; Olivetti, E. Leaching Characteristics of Biomass Ash-Based Binder in Neutral and Acidic Media. *Cem. Concr. Compos.* **2019**, *100*, 92–98. [CrossRef]
13. World Energy Council. *World Energy Resources 2016*; World Energy Council: London, UK, 2016.
14. Vassilev, S.V.; Baxter, D.; Andersen, L.K.; Vassileva, C.G. An Overview of the Composition and Application of Biomass Ash. Part 1. Phase-Mineral and Chemical Composition and Classification. *Fuel* **2013**, *105*, 40–76. [CrossRef]
15. Bernal, S.A.; Rodríguez, E.D.; Mejia De Gutiérrez, R.; Provis, J.L.; Delvasto, S. Activation of Metakaolin/Slag Blends Using Alkaline Solutions Based on Chemically Modified Silica Fume and Rice Husk Ash. *Waste Biomass Valoriz.* **2012**, *3*, 99–108. [CrossRef]
16. Billong, N.; Melo, U.C.; Kamseu, E.; Kinuthia, J.M.; Njopwouo, D. Improving Hydraulic Properties of Lime-Rice Husk Ash (RHA) Binders with Metakaolin (MK). *Constr. Build. Mater.* **2011**, *25*, 2157–2161. [CrossRef]
17. Duxson, P.; Fernández-Jiménez, A.; Provis, J.L.; Lukey, G.C.; Palomo, A.; Van Deventer, J.S.J. Geopolymer Technology: The Current State of the Art. *J. Mater. Sci.* **2007**, *42*, 2917–2933. [CrossRef]
18. Martirena Hernández, J.F.; Middendorf, B.; Gehrke, M.; Budelmannt, H. Use of Wastes of the Sugar Industry as Pozzolana in Lime-Pozzolana Binders: Study of the Reaction. *Cem. Concr. Res.* **1998**, *28*, 1525–1536. [CrossRef]
19. Nair, D.G.; Fraaij, A.; Klaassen, A.A.K.; Kentgens, A.P.M. A Structural Investigation Relating to the Pozzolanic Activity of Rice Husk Ashes. *Cem. Concr. Res.* **2008**, *38*, 861–869. [CrossRef]
20. Yeboah, N.N.N.; Shearer, C.R.; Burns, S.E.; Kurtis, K.E. Characterization of Biomass and High Carbon Content Coal Ash for Productive Reuse Applications. *Fuel* **2014**, *116*, 438–447. [CrossRef]
21. Cordeiro, G.C.; Toledo Filho, R.D.; De Moraes Rego Fairbairn, E. Use of Ultrafine Rice Husk Ash with High-Carbon Content as Pozzolan in High Performance Concrete. *Mater. Struct. Constr.* **2009**, *42*, 983–992. [CrossRef]
22. Medina, J.M.; Sáez del Bosque, I.F.; Frías, M.; Sánchez de Rojas, M.I.; Medina, C. Characterisation and Valorisation of Biomass Waste as a Possible Addition in Eco-Cement Design. *Mater. Struct. Constr.* **2017**, *50*, 207. [CrossRef]
23. Mehta, P.K.; Pitt, N. Energy and Industrial Materials from Crop Residues. *Resour. Recover. Conserv.* **1976**, *2*, 23–38. [CrossRef]
24. Detphan, S.; Chindaprasirt, P. Preparation of Fly Ash and Rice Husk Ash Geopolymer. *Int. J. Miner. Metall. Mater.* **2009**, *16*, 720–726. [CrossRef]

25. Castaldelli, V.N.; Akasaki, J.L.; Melges, J.L.P.; Tashima, M.M.; Soriano, L.; Borrachero, M.V.; Monzó, J.; Payá, J. Use of Slag/Sugar Cane Bagasse Ash (SCBA) Blends in the Production of Alkali-Activated Materials. *Materials* **2013**, *6*, 3108–3127. [CrossRef]
26. Ali, S.H.; Emran, M.Y.; Gomaa, H. Rice Husk-Derived Nanomaterials for Potential Applications. In *Waste Recycling Technologies for Nanomaterials Manufacturing*; Topics in Mining, Metallurgy and Materials Engineering; Springer: Cham, Switzerland, 2021; pp. 541–588. [CrossRef]
27. Freeman, E.; Gao, Y.M.; Hurt, R.; Suuberg, E. Interactions of Carbon-Containing Fly Ash with Commercial Air-Entraining Admixtures for Concrete. *Fuel* **1997**, *76*, 761–765. [CrossRef]
28. Wang, S.; Miller, A.; Llamazos, E.; Fonseca, F.; Baxter, L. Biomass Fly Ash in Concrete: Mixture Proportioning and Mechanical Properties. *Fuel* **2008**, *87*, 365–371. [CrossRef]
29. Basu, M.; Pande, M.; Bhadoria, P.B.S.; Mahapatra, S.C. Potential Fly-Ash Utilization in Agriculture: A Global Review. *Prog. Nat. Sci.* **2009**, *19*, 1173–1186. [CrossRef]
30. Rajamma, R.; Senff, L.; Ribeiro, M.J.; Labrincha, J.A.; Ball, R.J.; Allen, G.C.; Ferreira, V.M. Biomass Fly Ash Effect on Fresh and Hardened State Properties of Cement Based Materials. *Compos. Part B Eng.* **2015**, *77*, 1–9. [CrossRef]
31. Maschio, S.; Tonello, G.; Piani, L.; Furlani, E. Fly and Bottom Ashes from Biomass Combustion as Cement Replacing Components in Mortars Production: Rheological Behaviour of the Pastes and Materials Compression Strength. *Chemosphere* **2011**, *85*, 666–671. [CrossRef]
32. Sherwood, P.T. *Alternative Materials in Road Construction: A Guide to the Use of Recycled and Secondary Aggregates*, 2nd ed.; T. Telford: London, UK, 2001.
33. Forteza, R.; Far, M.; Seguí, C.; Cerdá, V. Characterization of Bottom Ash in Municipal Solid Waste Incinerators for Its Use in Road Base. *Waste Manag.* **2004**, *24*, 899–909. [CrossRef]
34. Kumar, S.; Patil, C.B. Estimation of Resource Savings Due to Fly Ash Utilization in Road Construction. *Resour. Conserv. Recycl.* **2006**, *48*, 125–140. [CrossRef]
35. Elinwa, A.U.; Mahmood, Y.A. Ash from Timber Waste as Cement Replacement Material. *Cem. Concr. Compos.* **2002**, *24*, 219–222. [CrossRef]
36. Siddique, R. Utilization of Wood Ash in Concrete Manufacturing. *Resour. Conserv. Recycl.* **2012**, *67*, 27–33. [CrossRef]
37. Martirena, F.; Middendorf, B.; Day, R.L.; Gehrke, M.; Roque, P.; Martínez, L.; Betancourt, S. Rudimentary, Low Tech Incinerators as a Means to Produce Reactive Pozzolan out of Sugar Cane Straw. *Cem. Concr. Res.* **2006**, *36*, 1056–1061. [CrossRef]
38. Duan, P.; Yan, C.; Zhou, W.; Luo, W. Fresh Properties, Mechanical Strength and Microstructure of Fly Ash Geopolymer Paste Reinforced with Sawdust. *Constr. Build. Mater.* **2016**, *111*, 600–610. [CrossRef]
39. Shi, C.; Jiménez, A.F.; Palomo, A. New Cements for the 21st Century: The Pursuit of an Alternative to Portland Cement. *Cem. Concr. Res.* **2011**, *41*, 750–763. [CrossRef]
40. Pacheco-Torgal, F.; Castro-Gomes, J.; Jalali, S. Alkali-Activated Binders: A Review. Part 2. About Materials and Binders Manufacture. *Constr. Build. Mater.* **2008**, *22*, 1315–1322. [CrossRef]
41. Fernández-JIMÉNEZ, A.; Palomo, A.; Criado, M. Activación Alcalina de Cenizas Volantes. Estudio Comparativo Entre Activadores Sódicos y Potásicos. *Mater. Constr.* **2006**, *56*, 51–65. [CrossRef]
42. Zhang, Z.; Provis, J.L.; Reid, A.; Wang, H. Fly Ash-Based Geopolymers: The Relationship between Composition, Pore Structure and Efflorescence. *Cem. Concr. Res.* **2014**, *64*, 30–41. [CrossRef]
43. Xue, X.; Liu, Y.L.; Dai, J.G.; Poon, C.S.; Zhang, W.D.; Zhang, P. Inhibiting Efflorescence Formation on Fly Ash–Based Geopolymer via Silane Surface Modification. *Cem. Concr. Compos.* **2018**, *94*, 43–52. [CrossRef]
44. Dung, N.T.; Hooper, T.J.N.; Unluer, C. Accelerating the Reaction Kinetics and Improving the Performance of Na_2CO_3 Activated GGBS Mixes. *Cem. Concr. Res.* **2019**, *126*, 105927. [CrossRef]
45. Provis, J.L.; Bernal, S.A. Geopolymers and Related Alkali-Activated Materials. *Annu. Rev. Mater. Res.* **2014**, *44*, 299–327. [CrossRef]
46. Brough, A.R.; Atkinson, A. Sodium Silicate-Based, Alkali-Activated Slag Mortars—Part I. Strength, Hydration and Microstructure. *Cem. Concr. Res.* **2002**, *32*, 865–879. [CrossRef]
47. Brough, A.R.; Holloway, M.; Sykes, J.; Atkinson, A. Sodium Silicate-Based Alkali-Activated Slag Mortars. Part II. The Retarding Effect of Additions of Sodium Chloride or Malic Acid. *Cem. Concr. Res.* **2000**, *30*, 1375–1379. [CrossRef]
48. Rashad, A.M. A Comprehensive Overview about the Influence of Different Additives on the Properties of Alkali-Activated Slag—A Guide for Civil Engineer. *Constr. Build. Mater.* **2013**, *47*, 29–55. [CrossRef]
49. Živica, V. Effects of Type and Dosage of Alkaline Activator and Temperature on the Properties of Alkali-Activated Slag Mixtures. *Constr. Build. Mater.* **2007**, *21*, 1463–1469. [CrossRef]
50. Bian, Z.; Jin, G.; Ji, T. Effect of Combined Activator of $Ca(OH)_2$ and Na_2CO_3 on Workability and Compressive Strength of Alkali-Activated Ferronickel Slag System. *Cem. Concr. Compos.* **2021**, *123*, 104179. [CrossRef]
51. Mobasher, N.; Bernal, S.A.; Provis, J.L. Structural Evolution of an Alkali Sulfate Activated Slag Cement. *J. Nucl. Mater.* **2016**, *468*, 97–104. [CrossRef]
52. Bernal, S.A.; Provis, J.L.; Myers, R.J.; San Nicolas, R.; van Deventer, J.S.J. Role of Carbonates in the Chemical Evolution of Sodium Carbonate-Activated Slag Binders. *Mater. Struct. Constr.* **2014**, *48*, 517–529. [CrossRef]
53. Yuan, B.; Straub, C.; Segers, S.; Yu, Q.L.; Brouwers, H.J.H. Sodium Carbonate Activated Slag as Cement Replacement in Autoclaved Aerated Concrete. *Ceram. Int.* **2017**, *43*, 6039–6047. [CrossRef]

54. Kim, M.S.; Jun, Y.; Lee, C.; Oh, J.E. Use of CaO as an Activator for Producing a Price-Competitive Non-Cement Structural Binder Using Ground Granulated Blast Furnace Slag. *Cem. Concr. Res.* **2013**, *54*, 208–214. [CrossRef]
55. Jin, F.; Al-Tabbaa, A. Strength and Drying Shrinkage of Slag Paste Activated by Sodium Carbonate and Reactive MgO. *Constr. Build. Mater.* **2015**, *81*, 58–65. [CrossRef]
56. Yang, K.H.; Cho, A.R.; Song, J.K.; Nam, S.H. Hydration Products and Strength Development of Calcium Hydroxide-Based Alkali-Activated Slag Mortars. *Constr. Build. Mater.* **2012**, *29*, 410–419. [CrossRef]
57. Ishwarya, G.; Singh, B.; Deshwal, S.; Bhattacharyya, S.K. Effect of Sodium Carbonate/Sodium Silicate Activator on the Rheology, Geopolymerization and Strength of Fly Ash/Slag Geopolymer Pastes. *Cem. Concr. Compos.* **2019**, *97*, 226–238. [CrossRef]
58. Noushini, A.; Babaee, M.; Castel, A. Suitability of Heat-Cured Low-Calcium Fly Ash-Based Geopolymer Concrete for Precast Applications. *Mag. Concr. Res.* **2015**, *68*, 163–177. [CrossRef]
59. Singh, B.; Rahman, M.R.; Paswan, R.; Bhattacharyya, S.K. Effect of Activator Concentration on the Strength, ITZ and Drying Shrinkage of Fly Ash/Slag Geopolymer Concrete. *Constr. Build. Mater.* **2016**, *118*, 171–179. [CrossRef]
60. Xu, H.; Gong, W.; Syltebo, L.; Izzo, K.; Lutze, W.; Pegg, I.L. Effect of Blast Furnace Slag Grades on Fly Ash Based Geopolymer Waste Forms. *Fuel* **2014**, *133*, 332–340. [CrossRef]
61. Chithiraputhiran, S.; Neithalath, N. Isothermal Reaction Kinetics and Temperature Dependence of Alkali Activation of Slag, Fly Ash and Their Blends. *Constr. Build. Mater.* **2013**, *45*, 233–242. [CrossRef]
62. Yuan, B.; Yu, Q.L.; Brouwers, H.J.H. Evaluation of Slag Characteristics on the Reaction Kinetics and Mechanical Properties of Na_2CO_3 Activated Slag. *Constr. Build. Mater.* **2017**, *131*, 334–346. [CrossRef]
63. Duran Atiş, C.; Bilim, C.; Çelik, Ö.; Karahan, O. Influence of Activator on the Strength and Drying Shrinkage of Alkali-Activated Slag Mortar. *Constr. Build. Mater.* **2009**, *23*, 548–555. [CrossRef]
64. Bakharev, T.; Sanjayan, J.G.; Cheng, Y.B. Effect of Admixtures on Properties of Alkali-Activated Slag Concrete. *Cem. Concr. Res.* **2000**, *30*, 1367–1374. [CrossRef]
65. Puertas, F.; Torres-Carrasco, M. Use of Glass Waste as an Activator in the Preparation of Alkali-Activated Slag. Mechanical Strength and Paste Characterisation. *Cem. Concr. Res.* **2014**, *57*, 95–104. [CrossRef]
66. Yuan, B.; Yu, Q.L.; Brouwers, H.J.H. Reaction Kinetics, Reaction Products and Compressive Strength of Ternary Activators Activated Slag Designed by Taguchi Method. *Mater. Des.* **2015**, *86*, 878–886. [CrossRef]
67. Kovtun, M.; Kearsley, E.P.; Shekhovtsova, J. Chemical Acceleration of a Neutral Granulated Blast-Furnace Slag Activated by Sodium Carbonate. *Cem. Concr. Res.* **2015**, *72*, 1–9. [CrossRef]
68. Collins, F.; Sanjayan, J.G. Early Age Strength and Workability of Slag Pastes Activated by NaOH and Na_2CO_3. *Cem. Concr. Res.* **1998**, *28*, 655–664. [CrossRef]
69. Fernández-Jiménez, A.; Puertas, F. Setting of Alkali-Activated Slag Cement. Influence of Activator Nature. *Adv. Cem. Res.* **2001**, *13*, 115–121. [CrossRef]
70. Xu, H.; Provis, J.L.; Van Deventer, J.S.J.; Krivenko, P.V. Characterization of Aged Slag Concretes. *ACI Mater. J.* **2008**, *105*, 131–139. [CrossRef]
71. Fernández-Jiménez, A.; Palomo, J.G.; Puertas, F. Alkali-Activated Slag Mortars: Mechanical Strength Behaviour. *Cem. Concr. Res.* **1999**, *29*, 1313–1321. [CrossRef]
72. Jin, F.; Gu, K.; Al-Tabbaa, A. Strength and Drying Shrinkage of Reactive MgO Modified Alkali-Activated Slag Paste. *Constr. Build. Mater.* **2014**, *51*, 395–404. [CrossRef]
73. Silva, G.J.B.; Santana, V.P.; Wójcik, M. Investigation on Mechanical and Microstructural Properties of Alkali-Activated Materials Made of Wood Biomass Ash and Glass Powder. *Powder Technol.* **2021**, *377*, 900–912. [CrossRef]
74. Shi, C.; Day, R.L. A Calorimetric Study of Early Hydration of Alkali-Slag Cements. *Cem. Concr. Res.* **1995**, *25*, 1333–1346. [CrossRef]
75. Fernández-Jiménez, A.; Puertas, F. Effect of Activator Mix on the Hydration and Strength Behaviour of Alkali-Activated Slag Cements. *Adv. Cem. Res.* **2003**, *15*, 129–136. [CrossRef]
76. Shi, C.; Day, R.L. Some Factors Affecting Early Hydration of Alkali-Slag Cements. *Cem. Concr. Res.* **1996**, *26*, 439–447. [CrossRef]
77. Ke, X.; Bernal, S.A.; Provis, J.L. Controlling the Reaction Kinetics of Sodium Carbonate-Activated Slag Cements Using Calcined Layered Double Hydroxides. *Cem. Concr. Res.* **2016**, *81*, 24–37. [CrossRef]
78. Gao, X.; Yu, Q.L.; Brouwers, H.J.H. Reaction Kinetics, Gel Character and Strength of Ambient Temperature Cured Alkali Activated Slag-Fly Ash Blends. *Constr. Build. Mater.* **2015**, *80*, 105–115. [CrossRef]
79. Hubler, M.H.; Thomas, J.J.; Jennings, H.M. Influence of Nucleation Seeding on the Hydration Kinetics and Compressive Strength of Alkali Activated Slag Paste. *Cem. Concr. Res.* **2011**, *41*, 842–846. [CrossRef]
80. Burciaga-Díaz, O.; Betancourt-Castillo, I. Characterization of Novel Blast-Furnace Slag Cement Pastes and Mortars Activated with a Reactive Mixture of MgO-NaOH. *Cem. Concr. Res.* **2018**, *105*, 54–63. [CrossRef]
81. Jin, F.; Gu, K.; Al-Tabbaa, A. Strength and Hydration Properties of Reactive MgO-Activated Ground Granulated Blastfurnace Slag Paste. *Cem. Concr. Compos.* **2015**, *57*, 8–16. [CrossRef]
82. Jin, F.; Gu, K.; Abdollahzadeh, A.; Al-Tabbaa, A. Effects of Different Reactive MgOs on the Hydration of MgO-Activated GGBS Paste. *J. Mater. Civ. Eng.* **2015**, *27*, B4014001. [CrossRef]
83. Dung, N.T.; Unluer, C. Influence of Nucleation Seeding on the Performance of Carbonated MgO Formulations. *Cem. Concr. Compos.* **2017**, *83*, 1–9. [CrossRef]

84. Dung, N.T.; Unluer, C. Development of MgO Concrete with Enhanced Hydration and Carbonation Mechanisms. *Cem. Concr. Res.* **2018**, *103*, 160–169. [CrossRef]
85. Dung, N.T.; Unluer, C. Performance of Reactive MgO Concrete under Increased CO_2 Dissolution. *Cem. Concr. Res.* **2019**, *118*, 92–101. [CrossRef]
86. Carrillo-Beltran, R.; Corpas-Iglesias, F.A.; Terrones-Saeta, J.M.; Bertoya-Sol, M. New Geopolymers from Industrial By-Products: Olive Biomass Fly Ash and Chamotte as Raw Materials. *Constr. Build. Mater.* **2021**, *272*, 121924. [CrossRef]
87. Quarcioni, V.A.; Chotoli, F.F.; Coelho, A.C.V.; Cincotto, M.A. Indirect and Direct Chapelle's Methods for the Determination of Lime Consumption in Pozzolanic Materials. *Rev. IBRACON Estrut. Mater.* **2015**, *8*, 1–7. [CrossRef]
88. Berra, M.; Mangialardi, T.; Paolini, A.E. Reuse of Woody Biomass Fly Ash in Cement-Based Materials. *Constr. Build. Mater.* **2015**, *76*, 286–296. [CrossRef]
89. Shi, C.; Roy, D.; Krivenko, P. *Alkali-Activated Cements and Concretes*, 1st ed.; Taylor & Francis: New York, NY, USA, 2006. [CrossRef]
90. Straka, P.; Havelcová, M. Polycyclic Aromatic Hydrocarbons and Other Organic Compounds in Ashes from Biomass Combustion. *Acta Geodyn. Geomater.* **2012**, *9*, 481–490.
91. Roberts, A.F. A Review of Kinetics Data for the Pyrolysis of Wood and Related Substances. *Combust. Flame* **1970**, *14*, 261–272. [CrossRef]
92. Sharypov, V.I.; Grishechko, L.I.; Tarasova, L.S.; Baryshnikov, S.V.; Celzard, A.; Kuznetsov, B.N. Investigation of Thermal Decomposition of Lignin Samples Isolated from Aspen Wood by Various Methods. *J. Sib. Fed. Univ. Chem.* **2011**, *3*, 221–232.
93. Cabrera, M.; Galvin, A.P.; Agrela, F.; Carvajal, M.D.; Ayuso, J. Characterisation and Technical Feasibility of Using Biomass Bottom Ash for Civil Infrastructures. *Constr. Build. Mater.* **2014**, *58*, 234–244. [CrossRef]
94. Maschowski, C.; Kruspan, P.; Garra, P.; Talib Arif, A.; Trouvé, G.; Gieré, R. Physicochemical and Mineralogical Characterization of Biomass Ash from Different Power Plants in the Upper Rhine Region. *Fuel* **2019**, *258*, 116020. [CrossRef]
95. Abdalqader, A.F.; Jin, F.; Al-Tabbaa, A. Characterisation of Reactive Magnesia and Sodium Carbonate-Activated Fly Ash/Slag Paste Blends. *Constr. Build. Mater.* **2015**, *93*, 506–513. [CrossRef]
96. Islam, A.; Alengaram, U.J.; Jumaat, M.Z.; Bashar, I.I. The Development of Compressive Strength of Ground Granulated Blast Furnace Slag-Palm Oil Fuel Ash-Fly Ash Based Geopolymer Mortar. *Mater. Des.* **2014**, *56*, 833–841. [CrossRef]
97. Wallah, S.E.; Rangan, B.V. Low-Calcium Fly Ash-Based Geopolymer Concrete: Long-Term Properties. 2006. Available online: http://hdl.handle.net/20.500.11937/34322 (accessed on 6 May 2022).
98. Askarian, M.; Tao, Z.; Samali, B.; Adam, G.; Shuaibu, R. Mix Composition and Characterisation of One-Part Geopolymers with Different Activators. *Constr. Build. Mater.* **2019**, *225*, 526–537. [CrossRef]
99. Puertas, F.; Martínez-Ramírez, S.; Alonso, S.; Vázquez, T. Alkali-Activated Fly Ash/Slag Cements. Strength Behaviour and Hydration Products. *Cem. Concr. Res.* **2000**, *30*, 1625–1632. [CrossRef]
100. Emdadi, Z.; Asim, N.; Amin, M.H.; Yarmo, M.A.; Maleki, A.; Azizi, M.; Sopian, K. Development of Green Geopolymer Using Agricultural and Industrialwaste Materials with High Water Absorbency. *Appl. Sci.* **2017**, *7*, 514. [CrossRef]
101. Nematollahi, B.; Sanjayan, J.; Shaikh, F.U.A. Synthesis of Heat and Ambient Cured One-Part Geopolymer Mixes with Different Grades of Sodium Silicate. *Ceram. Int.* **2015**, *41*, 5696–5704. [CrossRef]
102. Perná, I.; Šupová, M.; Hanzlíček, T.; Špaldoňová, A. The Synthesis and Characterization of Geopolymers Based on Metakaolin and High LOI Straw Ash. *Constr. Build. Mater.* **2019**, *228*, 116765. [CrossRef]
103. Košnář, Z.; Mercl, F.; Perná, I.; Tlustoš, P. Investigation of Polycyclic Aromatic Hydrocarbon Content in Fly Ash and Bottom Ash of Biomass Incineration Plants in Relation to the Operating Temperature and Unburned Carbon Content. *Sci. Total Environ.* **2016**, *563–564*, 53–61. [CrossRef] [PubMed]
104. Oh, J.E.; Monteiro, P.J.M.; Jun, S.S.; Choi, S.; Clark, S.M. The Evolution of Strength and Crystalline Phases for Alkali-Activated Ground Blast Furnace Slag and Fly Ash-Based Geopolymers. *Cem. Concr. Res.* **2010**, *40*, 189–196. [CrossRef]
105. Troëdec, M.L.; Dalmay, P.; Patapy, C.; Peyratout, C.; Smith, A.; Chotard, T. Mechanical Properties of Hemp-lime Reinforced Mortars: Influence of the Chemical Treatment of Fibers. *J. Compos. Mater.* **2011**, *45*, 2347–2357. [CrossRef]
106. Johnson, D.B.; Robb, W.A. Gaylussite: Thermal Properties by Simultaneous Thermal Analysis. *Am. Mineral.* **1973**, *58*, 778–784.
107. Siriwardane, R.V.; Poston, J.A., Jr.; Robinson, C.; Simonyi, T. Effect of Additives on Decomposition of Sodium Carbonate: Precombustion CO_2 Capture Sorbent Regeneration. *Energy Fuels* **2011**, *25*, 1284–1293. [CrossRef]
108. Kim, J.W.; Lee, H.G. Thermal and Carbothermic Decomposition of Na_2CO_3 and Li_2CO_3. *Metall. Mater. Trans. B* **2001**, *32*, 17–24. [CrossRef]

Review

A Review on Geopolymer Technology for Lunar Base Construction

Sujeong Lee [1,2] and Arie van Riessen [3,*]

[1] Resources Utilization Division, Korea Institute of Geoscience and Mineral Resources, Daejeon 34132, Korea; crystal2@kigam.re.kr
[2] Resources Recycling, University of Science and Technology, Daejeon 34113, Korea
[3] John de Laeter Centre, Curtin University, Perth, WA 6845, Australia
* Correspondence: a.vanriessen@curtin.edu.au; Tel.: +61-4011-03-352

Abstract: Geopolymer is a synthetic amorphous aluminosilicate material that can be used as an inorganic binder to replace ordinary Portland cement. Geopolymer is produced by mixing aluminosilicate source materials with alkali activators and curing the mixture either at ambient or low temperatures. Geopolymer research for lunar-based construction is actively underway to enable astronauts to stay on the moon for long periods. This research has been spurred on by earnest discussions of in situ resource utilization (ISRU). Recent research shows that the lunar regolith simulant-based geopolymers have high application potential to protect astronauts from the harsh moon environment. However, not all the simulants perfectly reproduce the lunar regolith, and the characteristics of the lunar regolith vary depending on the site. Issues remain regarding the applicability of geopolymer technology to contribute to ISRU through an elaborate and systematic plan of experiments. In this paper, the potential of geopolymers is assessed as a lunar-based construction material with the latest research results. Future work to develop the lunar regolith-based geopolymer technology is also proposed.

Keywords: geopolymer; lunar base construction; lunar regolith; future work; ISRU

1. Introduction

Geopolymer is an amorphous aluminosilicate inorganic binder that can potentially replace ordinary Portland cement in concrete. The representative source materials for geopolymers are metakaolin and coal-fired fly ash. Geopolymers are produced by mixing an optimum amount of alkali activators with source materials and curing the mixture at ambient to elevated temperatures. In situ resource utilization (ISRU) on the moon is now being discussed in earnest, and research exploring the possibility of utilizing geopolymers as a building material for a lunar base for long-term stay astronauts is actively underway [1–6]. The catalyst for this research arises because the main component of the lunar regolith is aluminosilicate containing glassy phases ideal for reacting with alkali activators. Geopolymer technology has much to offer with advantages such as rapid strength gain, impressive fire resistance and notable thermal insulation performance. In most of the studies, lunar regolith simulants are used as the scarcity of actual lunar regolith prevents it from being used for experimentation. The extreme temperature of the lunar surface, with low gravity and near vacuum, raises considerable technological problems for the manufacturing of geopolymer. While the atmospheric environment in which charged dust particles are suspended is also significantly different from those on Earth and likely to create new technological problems such as electrical sparking and vacuum welding. Therefore, it is necessary to carefully examine whether it is possible to build a lunar base using geopolymer made from lunar regolith.

The lunar regolith is a mixture of unconsolidated material and rock debris covering the lunar surface [7]. In space, large and small meteoroids and charged particles from other planets and the sun constantly collide with the moon's surface. The lunar regolith is formed by space weathering through sputtering and melting caused by solar wind and cosmic radiation. More than a third of the moon's regolith is thought to contain glassy materials, which have the potential to be geopolymer precursors [8]. Due to the limited availability of actual lunar regolith, various simulants have been developed from rocks or volcanic ash on the Earth's surface. Although these simulants have been designed to reproduce the mineral composition, particle size distribution and geotechnical properties of the lunar regolith, no simulant is identical to the lunar regolith. In addition, the lunar simulants are developed to address specific research purposes, such as resource extraction or to mimic local lunar regolith.

This review focuses on the assessment of the feasibility of geopolymer technologies for the construction of lunar bases by taking a general view of the mechanical properties, durability and cosmic radiation shielding performance of geopolymers manufactured from lunar regolith simulants based on the latest research results. Furthermore, the direction of future research and development, as well as the limitations of the research conducted to date, including technological uncertainties, are covered.

2. Applicability of Geopolymers as a Lunar Base Construction Material

2.1. Source Materials for Geopolymers

The most representative source material for geopolymers is metakaolin. When crystalline kaolin is heated to 700–800 °C for a few hours, dehydroxylation occurs, creating metakaolin ($Al_2Si_2O_7$), an amorphous aluminosilicate. Since metakaolin is amorphous, it can be assumed that all of it is available as a precursor to producing geopolymer. When metakaolin, NaOH, silica fume and water are mixed in the ratio $Na_2O:Al_2O_3:4SiO_2:11H_2O$ [9], geopolymers with high compressive strength can be obtained. The Si/Al ratio of the geopolymer produced at this ratio is theoretically 2.0. In order to develop high strength in metakaolin-based geopolymers, the mixing ratio should be derived to achieve the Si/Al ratio in the range of 2 to 2.5 [10]. It is important to appreciate that even in ideal laboratory conditions, not all of the aluminosilicate precursors react, leaving a microstructure with geopolymer, unreacted metakaolin and pores [11].

Low Ca Class F fly ash is the second most used source material for geopolymers. Class F fly ash emitted by pulverized coal combustion (PCC) typically contains mullite that is crystallized at high temperatures plus quartz and iron oxides. However, the most important component of fly ash for making geopolymer is the amorphous material, with the amount and ratio of Al_2O_3 and SiO_2 being paramount for activation by alkali. As every fly ash has a different level of amorphous content with varying Al_2O_3 and SiO_2 ratios, it is essential to accurately determine these values to enable a geopolymer formulation to be calculated. Quantitative X-ray diffraction (QXRD) is the most direct and robust technique for determining the amount of amorphous material and its composition [11]. With the correct formulation and proper curing process, the compressive strength of fly ash-based geopolymers can reach values ≥ 100 MPa [11–15].

A third geopolymer precursor is a volcanic ash made up of rock and glassy substances ejected when a volcano erupts. Volcanic ash is largely divided into pumice and scoria according to its chemical composition and particle shape. Scoria contains crystalline minerals because it is formed from basaltic magma by slow cooling. Its SiO_2 content is around 50%, and the color is dark. In many cases, scoria is used as source material for lunar regolith simulant because the basaltic chemical composition of scoria and the crystalline–glassy mixture properties are similar to lunar regolith (Table 1). Both pumice and scoria can be used as source materials for geopolymers, as they contain large amounts of reactive Al_2O_3 and SiO_2, such as fly ash from thermal power plants.

Table 1. Lunar regolith simulants used in the studies of manufacturing geopolymers for lunar base station. PSD = particle size distribution.

Simulant	Chemistry (wt.%)			Source	Note	Ref.
	SiO_2	Al_2O_3	CaO			
JSC-1A	46.67	15.79	9.90	Basalt cinders from Merriam Crater	Similar to low-titanium lunar mare terrain, formulated to be close to JSC-1	[6,16]
BP-1	47.2	16.7	9.2	San Francisco Volcanic Field	Lack of chemical similarity to Apollo samples	[17]
LHS-1	48.1	25.8	18.4	Not sourced from any particular terrestrial source	High similarity to the highlands soil in terms of chemical composition and PSD	[1]
GVS	43.3	16.5	8.8	Volcanic scoria cones	Same origin of CAS-1 and NEU-1	[3]
LN	44.83	14.18	8.93	Volcanic scoria cones	Similar mineralogy to Apollo samples	[18]
BH-1	43.3	16.5	8.8	Volcanic scoria cones	Mineralogical and chemical analog to Apollo 16 samples	[19]
BH-2	43.3	16.5	8.8	Volcanic scoria cones	Upgraded to have the same gradation to Apollo 17 samples	[20]
DNA-1	47.79	19.16	8.28	Dini Engineering srl for Monolite UK Ltd.	Glass content of 25 vol%	[21]
LMS-1	42.81	14.13	5.94	Exolith Lab.	Lunar mare simulant	LMS-1 Fact Sheet, Exolith Lab, FL

2.2. Advantages of Geopolymers as Lunar Base Construction Materials

The potential for geopolymers as a lunar base construction material is best demonstrated by comparing them with ordinary Portland cement (OPC) (Table 2). If a properly formulated geopolymer mixture is cured below 100 °C, the ultimate strength of a low-Ca fly ash geopolymer can be obtained in 24 h. Ambient curing of geopolymer is usually achieved by adding low amounts of Ca with ultimate strength gained over 28 days. For OPC-based concrete compressive strength at 28 days is used as the design reference for ambient cured samples.

Table 2. Comparison of geopolymer with OPC on Earth.

	Geopolymer	OPC
Advantages	• Rapid strength gain • Higher chloride resistance • Acid and sulfate resistance [24,25] • Excellent fire resistance [13,26] • Impressive heat insulation • Superior acid resistance [24] • Frost resistance • Little or no alkali–silica reaction [27] • Strong ITZ [23]	• Shorter setting time • Faster hardening • Ambient curing • Vasts amounts of available resource
Disadvantages	• Lower workability • May need thermal curing • Safety issues re: working with highly alkaline solutions	• Higher drying shrinkage and cracking • Lower durability • High CO_2 emission • Alkali–silica reaction • Weaker ITZ [22]

Unlike the interfacial transition zone (ITZ) between cement and aggregate, which is the weakest part of ordinary Portland cement concrete [22], the ITZ in geopolymers is generally stronger [23]. For this reason, when a geopolymer concrete fractures, failure may occur through the aggregate rather than along the boundary between the geopolymer and the aggregate (Figure 1).

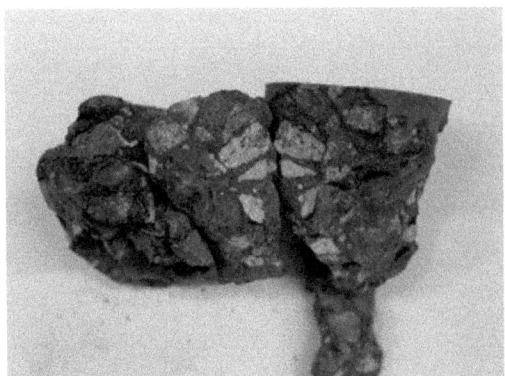

Figure 1. Fracture surface showing cracking through aggregate particles in geopolymer concrete.

In addition, geopolymers are resistant to acids [24], sulfates and chlorides, which cause cement concrete deterioration. Moreover, ASR (alkali–silica reaction), which causes cracking and deterioration of cement concrete, is less severe in geopolymer [27]. Above all, geopolymer is a material with excellent fire resistance that can withstand temperatures up to 900~1000 °C without spalling. Depending on the formulation, the compressive strength of some geopolymers was observed to increase at high temperatures [13,26]. The superior physical and chemical properties of geopolymer compared to OPC makes it a promising prospect to be manufactured from lunar regolith and is thus an ideal construction material for the purposes of ISRU.

A potential drawback of geopolymerization is the high viscosity of the alkalis and the subsequent paste, creating workability issues. NaOH-activated geopolymer paste is highly viscous, while KOH-activated geopolymer paste is much less viscous, albeit more expensive [27]. In order to improve the workability of viscous geopolymers, plasticizers used for the purpose of reducing the amount of mixing water required for OPC are also used in geopolymers but are not very effective [28,29]. In the case of metakaolin-based geopolymers, the use of methyl isobutyl carbinol (MIBC) has the effect of simultaneously improving flowability and strength [30].

3. Utilization of Lunar Regolith as a Raw Material for Geopolymers

3.1. Composition of Lunar Regolith and Its Simulants

The reason lunar regolith has excellent potential as a geopolymer precursor is its mineral composition. The major constituent minerals of the lunar regolith are olivine, pyroxene, plagioclase, ilmenite and silica (Table 3) [31–33]. The regolith thus consists of aluminates and/or silicates except for the ilmenite. The bulk composition of the lunar regolith is similar to the composition of the Earth's crust, which is 40–50% SiO_2 and 10–20% Al_2O_3. Collins et al. (2022) conducted a thorough characterization of a range of lunar simulants [34] and provided a summary of the amorphous content of regolith from the Apollo mission flights. The average amorphous content was found to be approximately 33 wt.%. In addition, more than one-third of the lunar regolith is made up of agglutinates and vitreous micro-spherules [31,35]. Because agglutinate can account for 60–70% of the lunar regolith [35], it is a highly promising source of reactive aluminosilicate source material for geopolymers.

Table 3. Mineralogical properties of major and minor minerals present on the moon.

Mineral	Formula	Specific Gravity	Mohs Scale	Impurities
		Major minerals		
olivine	$(Mg, Fe)_2SiO_4$	3.2–4.5	6.5–7	Mn, Ni
pyroxene	$(Ca, Mg, Fe)_2Si_2O_6$	3.2–3.3 (enstatite)	5–6 (enstatite)	Mn, Li, Na, Al, Sc, Na, Ti, Co
plagioclase	$Ca_2Al_2Si_2O_8$	2.76 (anorthite)	6–6.5	
ilmenite	$FeTiO_3$	4.7–4.8	5–6	Mn, Mg
silica	SiO_2	2.2–2.6	7 (quartz)	Ti, Fe, Mn (quartz)
		Minor minerals		
apatite	$Ca_5(PO_4)(F, Cl)$	3.2	5	REE [#]
baddeleyite	ZrO_2	5.5–6	6.5	Hf
chromite-ulvöspinel	$FeCr_2O_4$-Fe_2TiO_4	4.8–5	5.5–6	Al, V, Mn, Mg, Ca
iron	Fe(Ni, Co)	7.9		Ni, Co
merrillite [*]	$(Ca_3)(PO_4)_2$	3.1		Mg, Na
pleonaste	$(Fe, Mg)(Al, Cr)_2O_4$	3.6–3.9	7.5–8	Mn
rutile	TiO_2	4.2	6–6.5	Nb, Ta
feldspar	$(Ca, Na, K)AlSi_3O_8$	2.6	6–6.5	Rb, Ba
troilite [#]	FeS	4.7–4.8		
zircon	$ZrSiO_4$	4.6–4.7	7.5	
zirkelite-zirconolite	$(Ca, Fe)(Zr, Y, Ti)_2O_7$	4.7	5.5	Th, U, Ce, Nb
dysanalyte	$(Ca, Fe)(Ti, REE)O_3$	4–4.3 (perovskite)	5–5.5 (perovskite)	
thorite	$ThSiO_4$	6.6–7.2	4.5–5	U
titanite	$CaTiSiO_5$	3.5–3.6	5–5.5	Fe, Al, REE, Th
tranquillityite [*]	$Fe_8(Zr, Y)Ti_3Si_3O_{24}$	4.7		Y, Al, Mn, Cr, Nb, REE
yittrobetafite [*]	$(Ca, Y, U, Th, Pb, REE)_2(Ti, Nb)_2O_7$			

[*] Extraterrestrial only; [#] REE = rare earth element.

Since the lunar regolith simulants reflect the chemical composition of the lunar regolith, the SiO_2 and Al_2O_3 content are suitable for the production of geopolymer (Table 1). In addition, with the CaO content in the range of 6~10%, it is also suitable for geopolymerization (Table 1). Simulant JSC-1A is manufactured from the basalt of Merriam Crater and is similar to the regolith of the Mare area. Simulant BP-1 has a higher TiO_2 content than the chemical composition of the lunar regolith since it was developed for geotechnical purposes. LHS-1 is a simulant that reproduces the chemical composition and particle size distribution of the regolith in the Highland area and has higher Al_2O_3 and CaO content compared to other stimulants. Among the simulants manufactured from Chinese volcanic ash (GVS, LN, BH-1, BH-2), BH-1 reproduces the particle size distribution of the Apollo 16 sample. DNA-1 is a simulant developed by the European Space Agency to reproduce the regolith of the lunar Mare area, consisting of 75% crystalline and 25% glassy particles.

3.2. Recycling of the Mixing Water

Water is not a component of geopolymer, with its role being to transport ions in the geopolymer mixture and enable the mixing of the ingredients. Essentially, water is necessary for the production of the geopolymer, but ideally, the mixing water can be recovered after the geopolymer has set and hardened. In 2018, NASA announced that a significant amount of ice water was present in craters on the lunar poles. This discovery brought a positive shift to the concept of ISRU on the moon. However, it would be very difficult to retrieve water from the polar craters due to the cryogenic temperatures experienced in this region of the moon. In 2020, observations from SOFIA, a joint observatory between NASA and the German Space Agency, were presented with clear evidence of water, for the first time, on the sunlit surface of the moon [36]. This positive discovery suggests that mixing water can be secured for the production of geopolymers in several areas on the moon.

As mentioned above, an advantage of geopolymers as a lunar base construction material is that the mixing water can be recycled. Wang et al. (2016) presented a sustainable model in which most of the water used is recovered and reused after the geopolymer is manufactured from tektites, a round gravel-sized material that has been melted by meteorite impact, ejected up into the atmosphere and then fallen back to Earth [5]. The main components of tektite used by Wang et al. (2016) are SiO_2 69.84% and Al_2O_3 12.16%. In this study, the residual moisture content was found to be 0.8~1.77% in tektite-based geopolymers cured at 60 °C for one day and then heated in a vacuum at 120 °C for 8 h. In geopolymers, there is physically bound vaporizable moisture and chemically bound residual moisture. Although there is controversy about the form of chemical moisture, as present in Barbosa's model [37], residual moisture is likely to exist in the Si-Al tetrahedral framework in the form of OH^- bound to cations or in hydrated Na ion clusters. The strength of tektite-based geopolymer was maintained or decreased by only about 10% even after 30 cycles of 30 min at -196 °C and 30 min at room temperature. The sustainable production of geopolymers from lunar regolith accomplished by recycling most of the mixing water, as proposed in Wang et al. (2016), is likely to be realized when the following two requirements are satisfied [5]. First, the lunar regolith as source material for geopolymers must be an aluminosilicate with the low calcium content. Many studies reported that hydrates such as C-A-S-H or C-S-H are formed in the reaction product in the presence of calcium [38–40]. The presence of hydrates, depending on their content, weakens the advantages of the geopolymer and reduces the amount of evaporable moisture that can be recovered. The second is to seal the curing space to prevent moisture from escaping during the initial curing and mixing of the geopolymer in the near-vacuum atmosphere of the moon. This would not be technically easy in the moon's atmosphere, which is made up of a thick layer of suspended charged dust particles.

4. High Potential of Lunar Regolith Simulant-Based Geopolymers

4.1. Selection of Alkali Activator

In general, geopolymers are prepared by activating aluminosilicate source materials with high pH alkali hydroxides or silicates. The most commonly used activators are alkali hydroxides, such as caustic soda (NaOH) or potassium hydroxide (KOH), with the former more widely used as it is less expensive. In most studies of lunar regolith simulant-based geopolymer, caustic soda solution was solely used (Table 4). Activators made from a combination of soda solution and sodium silicate or potassium silicate were also used. The advantages of NaOH are that some crystalline aluminosilicate minerals are generally more soluble in NaOH than in KOH [41], in addition to the more readily dissolvable amorphous material. The greater the degree of Si and Al dissolution, the greater the potential strength of the geopolymer. However, NaOH's high viscosity and solubility decrease rapidly as the temperature reduces [42], suggesting that mixing and curing times need to be carefully selected. In the case of KOH, the viscosity is lower than that of NaOH, and geopolymer prepared with KOH may be preferred in that it exhibits lower thermal expansion [43]. However, since K is larger than Na, it is more exothermic when dissolved in water [44], which needs to be managed when processing geopolymer. Sodium silicate solution, or water glass, is not always used alone as an activator but may be combined with caustic soda so that targeted Si/Al and Na/Al values can be achieved in the geopolymer.

One-part geopolymer, instead of a conventional two-part geopolymer design, can be produced by adding free water to a mixture of powdered activator and geopolymer source material [45,46]. Solid sodium silicate, caustic soda powder, CaO, MgO, red mud, etc., may all be used as activators [43]. The mechanical strength of one-part geopolymers was found to be comparable to that of two-part geopolymers prepared with liquid activators [46]. On the moon, a one-part geopolymer manufacturing method would be more appropriate. It may be more suitable to use KOH rather than NaOH in that it can exhibit higher flowability during mixing and better heat resistance of the final geopolymer. Depending on the amount of reactive silica and alumina in the lunar regolith, optimization of mechanical properties, heat resistance and durability of the geopolymer will be possible by changing the targeted Si/Al ratio by selecting either sodium silicate or sodium aluminate powder as required.

Table 4. Comparison of properties reported for lunar regolish simulant-based geopolymers.

Source	Simulant	Activator	Curing Temperature	Compressive Strength (MPa)	Note
[1]	BP-1 JSC-1A LHS-1	SS	20 °C at 1 atm and vacuum for 7 d, followed by −80 to 600 °C curing	5–10 (20 °C at 1 atm) 18–35 (20 °C (1 d)→600 °C (1 h)) 1–4 (20 °C at 1 atm (7 d)→vacuum at 20 °C) Unconsolidated (20 °C (4 d)→−80 °C (3 d)	Reduced CS for GPs cured under vacuum and exposed to sub-zero temperatures, Positive effect of high amorphous Al-Si content and high proportion of fines
[2]	BH-1	NaOH	30.7→99.6→33.5 °C (discontinuous) for 24 and 72 h, followed the temperature variation cycle ranging from −178.9 °C to 99.6 °C	16–38 at different temperature regimes, 15–18% decrease in CS after the cycle, 49–70% decrease in FS after the cycle	Durability test (lunar surface high and cryogenic temperature variation cycle at 30° latitude). Noticeable degradation after the cryogenic attack with increased porosity
[3]	GVS	NaOH+SS	20, 40, 60, 80 °C at 1 atm	19 (20 °C), 42 (40 °C), 69 (60 °C), 76 (80 °C)—28 d	Curing temperature—the most significant factor influencing CS
[47]	JSC-1A	SS (s)	Mixing simulant with SS followed by calcining at 260 °C for 1 h and 127 °C in air and vacuum for 1 h	Rockwell Hardness of 75 (RH 80 for annealed titanium)	Adequate space radiation shielding of 'Regishell' (simulant + 10% SS binder) (by Monte Carlo simulations)
[18]	LN	NaOH+SS	60 °C for 7 d	59 (7 d) 50 at 120 °C 80 at −30 °C	Increased CS after 40 cycles of thermal shock (−196 °C for 1 h to 25 °C for 1 h)
[48]	DNA-1	NaOH	80 °C for 3 h, followed by a lunar day-and-night cycle at −80 to 114	1 (at 1 atm), 13 (after lunar cycle at 1 atm), 4 (after lunar cycle at vacuum)	Beneficial use of urea 3% Increased CS after LDN cycle Reduced CS by vacuum dehydration
[20]	BH-2	NaOH	30.7–99.6 °C at 1 atm and at vacuum for 0–72 h	19 (24 h), 38 (72 h) at vacuum 20 (24 h), 33 (72 h) at 1 atm	Cured under lunar surface T variation Higher CS under vacuum curing
[21]	DNA-1	NaOH	80 °C for 6 h, followed by freeze-thaw cycles at −80 to 80	16 (0 cycles), 25 (2 cycles), 24 (4 cycles), 32 (8 cycles)	Beneficial use of urea 3% for 3D printing, highest CS for pure GPs
[49]	JSC-1A	NaOH, NaOH+ K_2SiO_3	at RT for 28 d	2 (2 M NaOH)-18 (8 M NaOH)	Less reduction in flexural strength with respect to CS beneficial use of urea.
[6]	JSC-1A	NaOH+SS	26 °C at 1 atm 26 °C at vacuum 106 °C at 1 atm	10–12 (7 d) 11–12 (7 d), 9–10 (28 d) 9–13 (7), 10–20 (28 d)	Compression molding, 106 °C = average lunar daytime heat
[4]	JSC-1A	NaOH+SS	106 °C at vacuum 23 °C for 7 d 60 °C for 3 d (pouring and compression molding)	17 (3 d, conventional pouring) 38 (3 d, compression molding) 33 (7 d, compression molding)	Adequate radiation shielding and thermal insulation of 'Lunamer' (by FLUKA simulations)

Code: SS = sodium silicate; CS = compressive strength; GP = geopolymer; FS = flexual strength; RT = room temperature.

4.2. Compressive Strength and Durability of Lunar Regolith Simulant-Based Geopolymers

It is advantageous to use a combination of sodium silicate solution and NaOH to achieve higher compressive strength, especially when targeting specific Si/Al values [50]. Since the gravitational force on the moon is only 1/6 of that of the Earth, approximately 6 MPa is sufficient compressive strength for lunar base construction, compared with

35 MPa required for a one-story structure on the Earth [4]. Based on this value, it is clear that sufficient compressive strength of lunar regolith simulant-based geopolymers can be obtained when cured at ambient in the air (Table 4) [1,3,4,6,49].

The compressive strength of lunar regolith simulant-based geopolymers was found to decrease by about 50% when cured in a vacuum compared to ambient curing [1]. On the other hand, geopolymers cured at slightly higher temperatures (30.7–99.6 °C) for 72 h were found to have higher strength when cured in a vacuum [3]. In addition, it was stated that curing in a vacuum is advantageous for strength development by preventing efflorescence [3]. Generally, rapid drying of geopolymer during curing causes microcracking, leading to strength loss. Pilehvar et al. (2021) reported that curing in a vacuum increases porosity and reduces strength, although the pressure (vacuum) used in Pilehvar's experiment was many orders of magnitude greater than that experienced on the moon [51]. From these results, it was inferred that careful selection of the curing scheme is essential to maximize the strength of geopolymers manufactured on the moon, and importantly, the processing must also be practical on the lunar surface.

Pilehvar et al. (2020, 2021) found that when conducting durability tests, the compressive strength of geopolymers increased when repeatedly exposed to a low–high-temperature cycle similar to that experienced on the lunar surface [21,51]. The compressive strength of geopolymer was found to increase by a factor of 6 or more when exposed to a thermal cycle of $-80\sim114$ °C in air and vacuum. In another study by Xiong et al. (2022), geopolymer exposed to temperature cycles of $-190\sim25$ °C was found to have gained compressive strength while the compressive strength of samples exposed to $0\sim-30$ °C cycles increased from 60 MPa to 80 MPa [18]. Xiong et al.'s explanation for the strength increase was that the moisture trapped in the pores was frozen, and since the strength of the ice was greater at lower temperatures, the ice in the pores contributed to the increase in the strength of the geopolymer [18]. Zhang et al. (2022), on the other hand, reported opposite results in that the compressive strength of geopolymers exposed to the lunar surface temperature range of $-179.8\sim99.6$ °C decreased [3]. According to mercury intrusion porosimetry (MIP) measurement, after exposing samples to lunar surface temperature cycles, the strength of geopolymers decreased by 15~18% due to the increase in porosity and cracking [3]. Even after the reduction in compressive strength, the residual strength was at least 13 MPa, which is still sufficient for the construction of a lunar base. The flexural strength, however, showed a larger decrease of 49~70%, compared with a mild decrease in compressive strength [3]. Although the results of changes in the compressive strength of geopolymers exposed to lunar surface temperature cycles are inconsistent, the research results overall are generally positive.

The cosmic radiation shielding effect has only been studied for geopolymers made from JSC-1A simulant (Table 4). By using a simulation program, geopolymer was found to have shielding performance adequate for a 12-month stay by humans on the moon, assuming no extreme solar flare events. The equivalent dose would be equal to 5 cSv, which is the annual whole-body radiation worker limit on Earth [4]. In another study by Ferrone et al. (2022), a regolith-binder thickness of 1 m would be required to reduce the galactic cosmic radiation (GCR) dose by half [47]. Ferrone et al. (2022) based their simulations on a 14-day lunar stay [47]. In these two studies, the geopolymers were prepared with a combined activator of caustic soda and water glass, but the Si/Al ratio of the geopolymers prepared by each team was not specified. As the most important factor influencing the properties of a geopolymer is the chemical composition of the geopolymer, it is necessary to evaluate the shielding properties of geopolymer with different Si/Al ratios. Much of the above discussion is based on geopolymer made from a simulant. Once successful in achieving adequate geopolymer binder strength, the potential of adding filler or aggregate using other lunar regolith minerals can be explored. Two things are achieved by doing this: first, a low binder:aggregate ratio means less binder is needed, and secondly, aggregate with high cosmic radiation absorption properties can be included.

In order for more extensive dissolution during the early stages of geopolymerization, it is important to preserve moisture in the geopolymer mixture (Figure 2), as this is when water is consumed. The lunar surface temperature varies depending on the location, and it can reach up to 125 °C [52]. Temperatures above 50 °C last for about 9 days at 30° latitude on the moon. Compared to conventional thermal curing for 24 h on Earth, ambient curing provides sufficient strength on the moon. However, evaporation of moisture inhibits the initial geopolymer reaction and leads to a decrease in compressive strength. Therefore it is technically crucial to mix the source materials and seal the geopolymer mixture during curing.

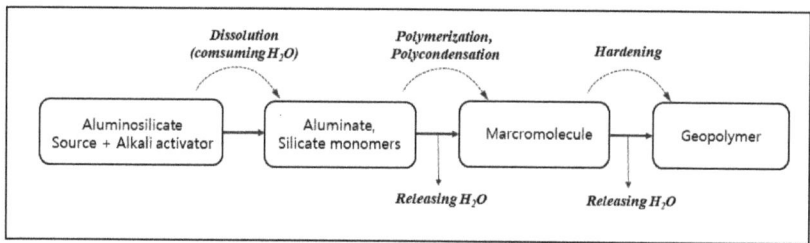

Figure 2. Geopolymerization reaction based on the conceptual model of Duxson et al. (2007) [53].

5. Conclusions

Research results based on geopolymers made from lunar topsoil simulants are very positive and support the proposition that geopolymers could be used as a lunar base construction material. The geopolymer structures would protect long-stay astronauts from extreme temperature and cosmic radiation while achieving ISRU targets of the utilization of lunar regolith. However, given that any lunar regolith simulant does not perfectly reproduce the lunar regolith, and the characteristics of the lunar regolith vary depending on the landing site, issues remain to be solved for this applicability to be realized in the ISRU scope.

Ideally, in ISRU, it would be the use of a one-part geopolymer that is produced by adding only water rather than the conventional two-part geopolymers. For this to be achieved, the type and amount of the activator in powder form that is most suitable in terms of cost and performance of the geopolymer should be thoroughly evaluated. Above all, since most of the lunar regolith particles have an angular shape, ways to increase the workability of geopolymers on the moon and how to promote curing in a vacuum over a wide temperature range of −171 °C to 120 °C or higher will need to be carefully addressed in the future. Recycling the blended water is absolutely necessary, and the durability of geopolymers is expected to be improved if moisture is recovered. The most difficult problem is to develop a quantitative mixing method tailored to the characteristics of lunar regoliths, such as a method for calculating the mix proportions of fly ash-based geopolymers with high compressive strength.

Author Contributions: Conceptualization, S.L.; formal analysis, A.v.R.; resources, S.L.; writing—original draft preparation, S.L.; writing—review and editing, A.v.R.; project administration, S.L.; funding acquisition, S.L. All authors have read and agreed to the published version of the manuscript.

Funding: This research was funded by Hanwha Aerospace; project title: Planning study on technology development for space in situ resource utilization by Hanwha-GRIs cooperation.

Institutional Review Board Statement: Not applicable.

Informed Consent Statement: Not applicable.

Data Availability Statement: Not applicable.

Conflicts of Interest: The authors declare no conflict of interest.

References

1. Mills, J.N.; Katzarova, M.; Wagner, N.J. Comparison of lunar and Martian regolith simulant-based geopolymer cements formed by alkali-activation for in-situ resource utilization. *Adv. Space Res.* **2022**, *69*, 761–777. [CrossRef]
2. Zhang, R.; Zhou, S.; Li, F. Preparation of geopolymer based on lunar regolith simulant at in-situ lunar temperature and its durability under lunar high and cryogenic temperature. *Constr. Build. Mater.* **2022**, *318*, 126033. [CrossRef]
3. Zhou, S.; Zhu, X.; Lu, C.; Li, F. Synthesis and characterization of geopolymer from lunar regolith simulant based on natural volcanic scoria. *Chin. J. Aeronaut.* **2022**, *35*, 144–159. [CrossRef]
4. Montes, C.; Broussard, K.; Gongre, M.; Simicevic, N.; Mejia, J.; Tham, J.; Allouche, E.; Davis, G. Evaluation of lunar regolith geopolymer binder as a radioactive shielding material for space exploration applications. *Adv. Space Res.* **2015**, *56*, 1212–1221. [CrossRef]
5. Wang, K.; Tang, Q.; Cui, X.; He, Y.; Liu, L. Development of near-zero water consumption cement materials via the geopolymerization of tektites and its implication for lunar construction. *Sci. Rep.* **2016**, *6*, 29659. [CrossRef]
6. Davis, G.; Montes, C.; Eklund, S. Preparation of lunar regolith based geopolymer cement under heat and vacuum. *Adv. Space Res.* **2017**, *59*, 1872–1885. [CrossRef]
7. Papike, J.J.; Ryder, G.; Shearer, C.K. Lunar Materials. In *Planetary Materials, Reviews in Mineralogy*; Pakike, J.J., Ed.; Mineralogical Society of America: Washington, DC, USA, 1998; Volume 36, pp. 5.1–5.23.
8. Isachenkov, M.; Chugunov, S.; Akhatov, I.; Shishkovsky, I. Regolith-based additive manufacturing for sustainable development of lunar infrastructure—An overview. *Acta Astronaut.* **2021**, *180*, 650–678. [CrossRef]
9. Kriven, W.M. Inorganic polysialates or 'geopolymers'. *Am. Ceram. Soc. Bull.* **2010**, *89*, 31–34.
10. Rowles, M.R.; Hanna, J.V.; Pike, K.J.; Smith, M.E.; O'Connor, B.H. ^{29}Si, ^{27}Al, ^{1}H and ^{23}Na MAS NMR study of the bonding character in aluminosilicate inorganic polymers. *Appl. Magn. Reson.* **2007**, *32*, 663–689. [CrossRef]
11. Williams, R.; Hart, R.; van Riessen, A. Quantification of the extent of reaction of metakaolin based geopolymers using XRD, SEM and EDS. *J. Am. Ceram. Soc.* **2011**, *94*, 2663–2670. [CrossRef]
12. Williams, R.; van Riessen, A. Determination of the reactive component of fly ashes for geopolymer production using XRF and XRD. *Fuel* **2010**, *89*, 3683–3692. [CrossRef]
13. Rickard, D.A.W.; Williams, R.; Temuujin, J.; van Riessen, A. Assessing the suitability of three Australian fly ashes as an aluminosilicate source for geopolymers in high temperature applications. *Mater. Sci. Eng. A* **2011**, *528*, 3390–3397. [CrossRef]
14. Van Riessen, A.; Chen-Tan, N. Beneficiation of Collie fly ash for synthesis of geopolymer. Part 2—Geopolymers. *Fuel* **2013**, *106*, 569–575. [CrossRef]
15. Lee, S.; van Riessen, A.; Chon, C.-M. Benefits of sealed-curing on compressive strength of fly ash-based geopolymers. *Materials* **2016**, *9*, 598. [CrossRef]
16. Gustafson, R. JSC-1A Lunar regolith simulant production summary and lessons learned. In Proceedings of the Lunar Regolith Simulant Workshop, Huntsville, AL, USA, 17–20 March 2009.
17. Stoeser, D.B.; Wilson, S.; Rickman, D.L. *Preliminary Geological Findings on the BP-1 Simulant*; NASA Marshall Space Flight Center: Huntsville, AL, USA, 2010; pp. 1–24.4.
18. Xiong, G.; Guo, X.; Yuan, S.; Xia, M.; Wang, Z. The mechanical and structural properties of lunar regolith simulant based geopolymer under extreme temperature environment on the moon through experimental and simulation methods. *Constr. Build. Mater.* **2022**, *325*, 126679. [CrossRef]
19. Zhou, S.; Lu, C.; Zhu, X.; Li, F. Preparation and characterization of high-strength geopolymer based on BH-1 lunar soild simulant with low alkali content. *Engineering* **2021**, *7*, 1631–1645. [CrossRef]
20. Zhou, S.; Yang, Z.; Zhang, R.; Zhu, X.; Li, F. Preparation and evaluation of geopolymer based on BH-2 lunar regolith simulant under lunar surface temperature and vacuum condition. *Acta Astronaut.* **2021**, *189*, 90–98. [CrossRef]
21. Pilehvar, S.; Arnhof, M.; Pamies, R.; Valentini, L.; Kjøniksen, A.-L. Utilization of urea as an accessible superplasticizer on the moon for lunar geopolymer mixtures. *J. Clean. Prod.* **2020**, *247*, 119177. [CrossRef]
22. Lilliu, G.; van Mier, J.G.M. 3D lattice type fracture model for concrete. *Eng. Fract. Mech.* **2003**, *70*, 927–941. [CrossRef]
23. Luo, Z.; Li, W.; Wang, K.; Castel, A.; Shah, S. Comparison on the properties of ITZ in fly ash-based geopolymer and Portland cement concretes with equivalent flowability. *Constr. Build. Mater.* **2021**, *143*, 106392. [CrossRef]
24. Gourley, J.T. Geopolymers in Australia. *J. Aust. Ceram. Soc.* **2014**, *50*, 102–110.
25. Yang, M.; Paudel, S.; Asa, E. Comparison of pore structure in alkali activated fly ash geopolymer and ordinary concrete due to alkali-silica reaction using micro-computed tomography. *Constr. Buildi. Mater.* **2020**, *236*, 117524. [CrossRef]
26. An, E.-M.; Cho, Y.-H.; Chon, C.-M.; Lee, D.-G.; Lee, S. Synthesizing and assessing fire-resistant geopolymer from rejected fly ash. *J. Korean Ceram. Soc.* **2013**, *52*, 253–263. [CrossRef]
27. Provis, J.L. Chapter 4 & 9 Activating solution chemistry for geopolymers. In *Geopolymers: Structure, Processing and Industrial Applications*; Provis, J.L., van Deventer, J.S.J., Eds.; Woodhead Publishing Limited: Cambridge, UK; Oxford, UK, 2009; pp. 50–71.
28. Nematollahi, B.; Sanjayan, J. Efficacy of available superplasticizers on geopolymers. *Res. J. Appl. Sci. Eng. Technol.* **2014**, *7*, 1278–1282. [CrossRef]
29. Carabba, L.; Manzi, S.; Bignozzi, M.C. Superplasticizer addition to carbon fly ash geopolymers activated at room temperature. *Materials* **2016**, *9*, 586. [CrossRef]
30. Lee, S.; Kim, B.; Seo, J.; Cho, S. Beneficial use of MIBC in metakaolin-based geopolymers to improve flowability and compressive strength. *Materials* **2020**, *13*, 3663. [CrossRef]

31. Agosto, W.N. Electrostatic concentration of lunar soil minerals. In *Lunar Bases and Space Activities of the 21st Century*; Mendell, W.W., Ed.; Lunar and Planetary Institute: Houston, TX, USA, 1985; pp. 453–464.
32. Papike, J.J.; Simon, S.B.; Laul, J.C. The lunar regolith: Chemistry, mineralogy and petrology. *Rev. Geophys.* **1982**, *20*, 761–826. [CrossRef]
33. Smith, J.V.; Steele, I.M. Lunar mineralogy: A heavenly detective story. Part II. *Am. Miner.* **1976**, *61*, 1059–1116.
34. Collins, P.J.; Edmunson, J.; Fiske, M.; Radlinska, A. Materials characterization of various lunar regolith simulants for use in geopolymer lunar concrete. *Adv. Space Res.* **2022**, *69*, 3941–3951. [CrossRef]
35. Noble, S. The lunar regolith. In Proceedings of the Lunar Regolith Simulant Workshop, Huntsville, AL, USA, 17–20 March 2009.
36. Honniball, C.I.; Lucey, P.G.; Shenoy, S.; Orlando, T.M.;; Hibbitts, C.A.; Hurley, D.M.; Farrell, W.M. Molecular water detected on the sunlit Moon by SOFIA. *Nat. Astron.* **2020**, *5*, 121–127. [CrossRef]
37. Barbosa, V.F.F.; MacKenzie, K.J.D.; Thaumaturgo, C. Synthesis and characterization of materials based on inorganic polymers of alumina and silica: Polysialate polymers. *Int. J. Inorg. Mater.* **2000**, *2*, 309–317. [CrossRef]
38. Alventosa, K.M.; White, C.E. The effects of calcium hydroxide and activator chemistry on alkali-activated metakaolin pastes. *Cem. Concr. Res.* **2021**, *145*, 106453. [CrossRef]
39. Sankar, K.; Sutrisno, A.; Kriven, W.M. Slag-fly ash and slag-metakaolin binders: Part II—Properties of precursors and NMR study of poorly ordered phases. *J. Am. Ceram. Soc.* **2019**, *102*, 3204–3227. [CrossRef]
40. Kim, B.; Lee, S.; Chon, C.-M.; Cho, S. Setting behavior and phase evolution on heat treatment of metakaolin-based geopolymers containing calcium hydroxide. *Materials* **2022**, *15*, 194. [CrossRef]
41. Xu, H.; van Deventer, J.S.J. The geopolymerisation of alumino-silicate minerals. *Int. J. Miner. Process.* **2000**, *59*, 247–266. [CrossRef]
42. Gurvich, L.V.; Bergman, G.A.; Gorokhov, L.N.; Ioris, V.S.; Leonidov, V.Y.; Yungman, V.S. Thermodynamic properties of alkali metal hydroxides. Part 1. lithium and sodium hydroxides. *J. Phys. Chem. Ref. Data* **1996**, *25*, 1211–1276. [CrossRef]
43. Duxson, P.; Lukey, G.C.; van Deventer, J.S.J. The thermal evolution of metakaolin geopolymers: Part 2—Phase stability and structural development. *J. Non-Cryst.* **2007**, *353*, 2186–2200. [CrossRef]
44. Gurvich, L.V.; Bergman, G.A.; Gorokhov, L.N.; Ioris, V.S.; Leonidov, V.Y.; Yungman, V.S. Thermodynamic properties of alkali metal hydroxides. Part II. potassium, rubidium, and desium hydroxides. *J. Phys. Chem. Ref. Data* **1997**, *26*, 1031–1110. [CrossRef]
45. Hajimohammadi, A.; van Deventer, J.S.J. Characterisation of one-part geopolymer binders made from fly ash. *Waste Biomass Valorization* **2017**, *8*, 225–233. [CrossRef]
46. Luukkonen, T.; Abdollahnejad, Z.; Yliniemi, J.; Kinnunen, P.; Illikainen, M. One-part alkali-activated materials: A review. *Cem. Concr. Res.* **2018**, *103*, 21–34. [CrossRef]
47. Ferrone, K.L.; Taylor, A.B.; Helvajian, H. In situ resource utilization of structural material from planetary regolith. *Adv. Space Res.* **2022**, *69*, 2268–2282. [CrossRef]
48. Momi, J.; Lewis, T.; Alberini, F.; Meyer, M.E.; Alexiadis, A. Study of the rheology of lunar regolith simulant and water slurries for geopolymer applications on the Moon. *Adv. Space Res.* **2021**, *68*, 4496–4504. [CrossRef]
49. Alexiadis, A.; Alberini, F.; Meyer, M.E. Geopolymers from lunar and Martian soil simulants. *Adv. Space Res.* **2017**, *59*, 490–495. [CrossRef]
50. Castillo, H.; Collado, H.; Droguett, T.; Sánchez, S.; Vesely, M.; Garrido, P.; Palma, S. Factors affecting the compressive strength of geopolymers: A review. *Materials* **2021**, *11*, 1317. [CrossRef]
51. Pilehvar, S.; Arnhof, M.; Erichsen, A.; Valentini, L.; Kjøniksen, A.-L. Investigation of severe lunar environmental conditions on the physical and mechanical properties of lunar regolith geopolymers. *J. Mater. Res. Technol.* **2021**, *11*, 1506–1516. [CrossRef]
52. Casanova, I. Feasibility study and applications of sulfur concrete for lunar base development: A preliminary study. In Proceedings of the 28th Annual Lunar and Planetary Science Conference, Houston, TX, USA, 17–21 March 1997; p. 209.
53. Duxson, P.; Fernandez-Jimenez, A.; Provis, J.L.; Lukey, G.C.; Palomo, A.; van Deventer, J.S.J. Geopolymer technology: The current state of the art. *J. Mater. Sci.* **2007**, *42*, 2917–2933. [CrossRef]

Article

Prediction of Mechanical Properties of Fly-Ash/Slag-Based Geopolymer Concrete Using Ensemble and Non-Ensemble Machine-Learning Techniques

Muhammad Nasir Amin [1,*], Kaffayatullah Khan [1], Muhammad Faisal Javed [2], Fahid Aslam [3], Muhammad Ghulam Qadir [4] and Muhammad Iftikhar Faraz [5]

1. Department of Civil and Environmental Engineering, College of Engineering, King Faisal University, P.O. Box 380, Al-Ahsa 31982, Saudi Arabia; kkhan@kfu.edu.sa
2. Department of Civil Engineering, Abbottabad Campus, COMSATS University Islamabad, Abbottabad 22060, Pakistan; arbabfaisal@cuiatd.edu.pk
3. Department of Civil Engineering, College of Engineering in Al-Kharj, Prince Sattam Bin Abdulaziz University, Al-Kharj 11942, Saudi Arabia; f.aslam@psau.edu.sa
4. Department of Environmental Sciences, Abbottabad Campus, COMSATS University Islamabad, Abbottabad 22060, Pakistan; hashir785@gmail.com
5. Department of Mechanical Engineering, College of Engineering, King Faisal University, Al-Ahsa 31982, Saudi Arabia; mfaraz@kfu.edu.sa
* Correspondence: mgadir@kfu.edu.sa; Tel.: +966-13-589-5431; Fax: +966-13-581-7068

Abstract: The emission of greenhouse gases and natural-resource depletion caused by the production of ordinary Portland cement (OPC) have a detrimental effect on the environment. Thus, an alternative means is required to produce eco-friendly concrete such as geopolymer concrete (GPC). However, GPC has a complex cementitious matrix and an ambiguous mix design. Aside from that, the composition and proportions of materials utilized may have an impact on the compressive strength. Similarly, the use of robust and efficient machine-learning (ML) approaches is now required to forecast the strength of such a composite cementitious matrix. As a result, this study anticipated the compressive strength of GPC with waste resources using ensemble and non-ensemble ML algorithms. This was accomplished through the use of Anaconda (Python). To build a strong ensemble learner by integrating weak learners, adaptive boosting, random forest (RF), and ensemble learner bagging were employed. Furthermore, ensemble learners were utilized on non-ensemble or weak learners, such as decision trees (DT) and support vector machines (SVM) via regression. The data encompassed 156 statistical samples in which nine variables, namely superplasticizer (kg/m^3), fly ash (kg/m^3), ground granulated blast-furnace slag (GGBS), temperature (°C), coarse and fine aggregate (kg/m^3), sodium silicate (Na$_2$SiO$_3$), and sodium hydroxide (NaOH), were chosen to anticipate the results. Exploring it in depth, twenty sub-models with ensemble boosting and bagging approaches were trained, and tuning was performed to achieve the highest possible coefficient of determination (R^2). Moreover, cross K-Fold validation analysis and statistical checks were performed via indicators for the evaluation of the models. The result revealed that ensemble approaches yielded robust performance compared to non-ensemble algorithms. Generally, an ensemble learner with the RF and bagging approach on a DT yielded robust performance by achieving a better R^2 as 0.93, and with the lowest statistical errors. The communal model in artificial-intelligence analysis, on average, improved the accuracy of the model.

Keywords: fly ash; slag; machine-learning; validation; parametric analysis; ensemble approaches

1. Introduction

The emissions of greenhouse gas (GHG) in the environment have caused the melting of glacier reservoirs, which tremendously contributes to major threats to the globe [1]. The concrete sector is believed to be the most significant source of greenhouse-gas emissions, contributing up to 50% of world emissions [2]. Thus, Portland cement (PC), an essential component of concrete, significantly contributes to GHG emissions [3]. The production of PC contributes around 7% to the atmosphere and the environment. Furthermore, the calcination of calcium oxide (CaO) during the cement-manufacturing process accounts for 50% of CO_2 emissions [4]. Currently, 4000 million tons of PC are produced annually, with an anticipated of about 6000 million tons by 2060 [5]. These figures show the need for alternative measures to meet the rising demand for concrete while using fewer resources and effectively emitting less CO_2 [6,7]. Therefore, the utilization of leftover recycled and waste substances in concrete is one of the proposed scientific and realistic remedies for reducing its high demand [8–11]. This will not only meet the growing need for concrete, but it will also reduce the risk to the environment [9]. In this regard, fly ash (FA) and GGBS as natural pozzolanic materials can be effectively used as supplemental cementitious materials in the construction sector [12–15]. Thus, their use in the building sector could reduce the environmental consequences associated with the manufacturing and usage of cement in the building industry. Moreover, the addition of these materials with alkaline solvents such as Sodium silicate (Na_2SiO_3) and sodium hydroxide (NaOH) produces viable and eco-friendly environmental concrete such as geopolymer concrete (GPC) [16–19]. The amorphous gel form of GPC possesses many outstanding and attractive characteristics, including resistance to sulfate attack, acid resistance, enhanced durability, fire resistance, and an undoubtedly greater compressive strength than conventional concrete [12,20–23]. Likewise, their use in the construction industry can extensively lessen CO_2 emissions in the atmosphere [24]. Moreover, the difference between ordinary Portland cement (OPC) and GPC is illustrated in Table 1. Studies have revealed that the chemical and physical properties of the matrix have a major influence on the strength of GPC. Thus, the fly-ash-to-NaOH ratio, Na_2SiO_3-to-NaOH ratio, workability, fly-ash-to-sand ratio, molarity, and alkaline ratio affect the strength of concrete [25–27]. Ukritnukun et al. [28] observed that the blast-furnace slag concentration, curing temperature, and silicate modulus all had a beneficial effect. Additionally, Asghar et al. [29] determined the ideal molar ratios of Ca/Si (calcium oxide/silica) and (Na + K)/Si ((sodium + potassium)/silica), as well as the ideal volume ratio $(H_2O/solid)_{vol}$ for increasing the strength properties of GPC. Songpiriyakij et al. [30] found that a Si-to-Al ratio of 15.9 resulted in the formation of GPC with the relatively high compressive strength of 73 MPa. Puertas et al. [31] examined the strength and growth characteristics of fly-ash/slag-paste-hydration products. After 28 days of curing at 25 °C, they reported that the mechanical properties of the mix with a fly-ash/slag ratio of 1.0 that was cured at 25 °C and stimulated with a 10 M NaOH solution exceeded 50 MPa. Moreover, according to Rai et al. [31], the cumulative effect of NaOH molarity, curing temperature, and activator-to-binder ratio directly impacts the initial compressive strength, while the $NaOH/Na_2SiO_3$ ratio is not statically important, and the target strength can be attained more quickly at high temperatures than at room temperatures.

Table 1. Comparison of GPC with OPC.

Attributes	GPC	OPC	Summary	References
Tensile strength	Greater	Lower	GPC has higher strength due to presence of aluminosilicate, activators and types of activators that enhance the strength at early age.	[32]
Acid attack	More resistance	Less resistance	Presence of aluminosilicate, activators and types of activators show enhanced resistance to acidic attack	[33]
Durability	More resistance	Less resistance	Presence of aluminosilicate, activators and types of activators show enhanced resistance to acidic attack	[34]

Table 1. Cont.

Attributes	GPC	OPC	Summary	References
Compressive strength	Higher	Lower	Same factors as tensile strength	[35]
Porosity	Significantly less	Moderate	Internal geopolymeric structure and presence of aluminosilicate, activators and types of activators influence GPC porosity.	[36]
Fire resistance	Significantly higher	Limited	GPC concrete is more resistant to deterioration caused by high temperatures as compared to OPC.	[37]
CO_2 emission	Lower	Higher	Utilization of waste materials shows lesser CO_2 as compared to OPC	[38]

To make GPC, pozzolanic materials with binding properties are polymerized at high temperatures in an alkaline medium [39]. As a result, a crystalline and amorphous compound is formed, which can be used to achieve the desired mechanical properties [39]. However, the high demand for heat curing in the production of a geopolymerization compound is not recommended for in-field application. Due to the high heat demand of curing, this will limit the use of FA-GPC in the construction domain [40]. Thus, heat demand can be reduced by using a slag blend with a high concentration of calcium, silica, and alumina. The use of the GGBS slag blend in conjunction with FA gives a dense microstructure with hydrated and polymerization products that significantly improve the early age strength of GPC. Yazdi et al. [41] examined the outcome of GPC by varying the dosage of FA with GGBS. The author showed that replacing FA with GGBS results in a significant increase in compressive and flexural strength of 100 MPa and 10 MPa, respectively. Furthermore, Fang et al. [40] studied the varying dosage of slag content on the flexural and split tensile strength of FA-GPC. The author revealed a higher strength due to the formation of C-A-S-H gel and N-A-S-H. This ultimately speed up the reaction process of GPC [40]. The compressive strength of concrete is typically evaluated by conducting physical tests. In general, concrete specimens that are cubical and cylindrical in shape are produced by using precise mixture ratios and curing with water for approximately 28 days to yield the hydrated products [42]. Afterwards, the compressive strength is determined using a compression-testing machine. This approach is common in the execution of work in the field and laboratory, yet it is inefficient and time-consuming. Rather than using standard experimental procedures to determine the compressive strength of concrete, empirical regression methodologies are preferable for estimating the strength of concrete [43]. On the other hand, the literature reveals that the chemical composition and physical proportions of variables have a significant impact on the GPC [44]. Moreover, heterogeneity exists in the production of GPC as a result of the variety of parameters involved. While various algorithms and methods based on statistical approaches are capable of evaluating the compressive nature of GPC, the relationship between factors and mechanical strength is not well understood. Thus, machine-learning (ML) approaches may now be used to predict the compressive strength of concrete, thanks to recent advances in artificial-intelligence algorithms [45–52]. The evolution of the advanced prediction algorithms could be used for a variety of purposes, such as regression, classification, and clustering of data [53]. Estimating the compressive loading capacity of concrete is just one application of the ML regression function. The ML methodology, in contrast to prior regression methods, delivers very precise results [54,55]. The discovery of artificial-intelligence algorithms such as genetic engineering programming (GEP), support vector machine (SVM), artificial neural network (ANN), and ensemble approaches has enabled researchers to address tough problems [56–61].

This research will investigate the effect of network- and tree-based models for prediction by employing boosting, AdaBoost (bagging), and utilizing modified bagging random forest (RF). Unlike previous research, this study does not exclusively depend on ensemble techniques, but also discusses the tree- and network-based studies on ensemble learning. Second, this study is based on modeling of ensembles over individual models in order to anticipate the compressive behavior of GPC using secondary raw materials. To the authors' knowledge, no work similar to ensemble ML models for GPC has been employed. Furthermore, this modeling was carried out in Anaconda navigator version 1.9.12 with Python version 3.7.

2. Database Presentation Using Python

For the representation of the database, the Anaconda-based Python programming (version 3.7) was utilized from the published literature (Table S1) [62–72]. The data were gathered from the accessible literature and comprise nine parameters, namely as fly ash (kg/m^3), alkaline activator (kg/m^3), aggregate (kg/m^3), GGBS (kg/m^3), NaOH molarity, SP dosage (kg/m^3), curing temperature (°C) and an output parameter of compressive strength as illustrated in Figure 1. Every parameter that was chosen had a significant impact on the strength qualities of fly-ash-slag-based concrete. Moreover, the Python programming language was used to find the link between these variables and concrete compressive strength. Additionally, the influential variables in forecasting the mechanical strength were evaluated through the use of permutation features. Furthermore, Table 2 illustrates the variable range values with maxima and minima based on the 156 data points, while Table 3 displays the results of the statistical-analysis check, which includes the mean, the count, and the standard deviation. The parameters used in making the models have a substantial influence on the model's robustness. Seaborn, a command in Python, is used to employ machine learning (ML) and to depict the correlation between two variables.

Table 2. Contribution of parameters with ranges.

Variables Used	Acronym	Minima	Maxima
Input variables			
Fly ash	FA	0	400
Fine aggregate	FIA	547	810.6
Ground granulated blast furnace slag	GGBS	0	409
Coarse aggregate	CAA	966	1293
Sodium hydroxide	NaOH	9	143.3
Sodium silicate	Na$_2$SiO$_3$	54	192.9
Super plasticizer	SP	0	180
Temperature	T °C	0	60
Output			
Compressive strength	f_c'	10.5	89.6

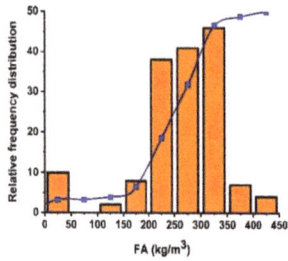
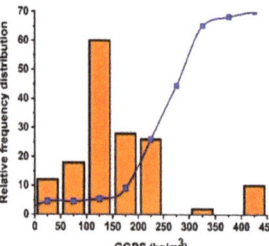
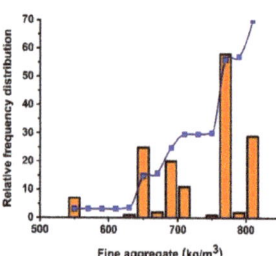

Figure 1. *Cont.*

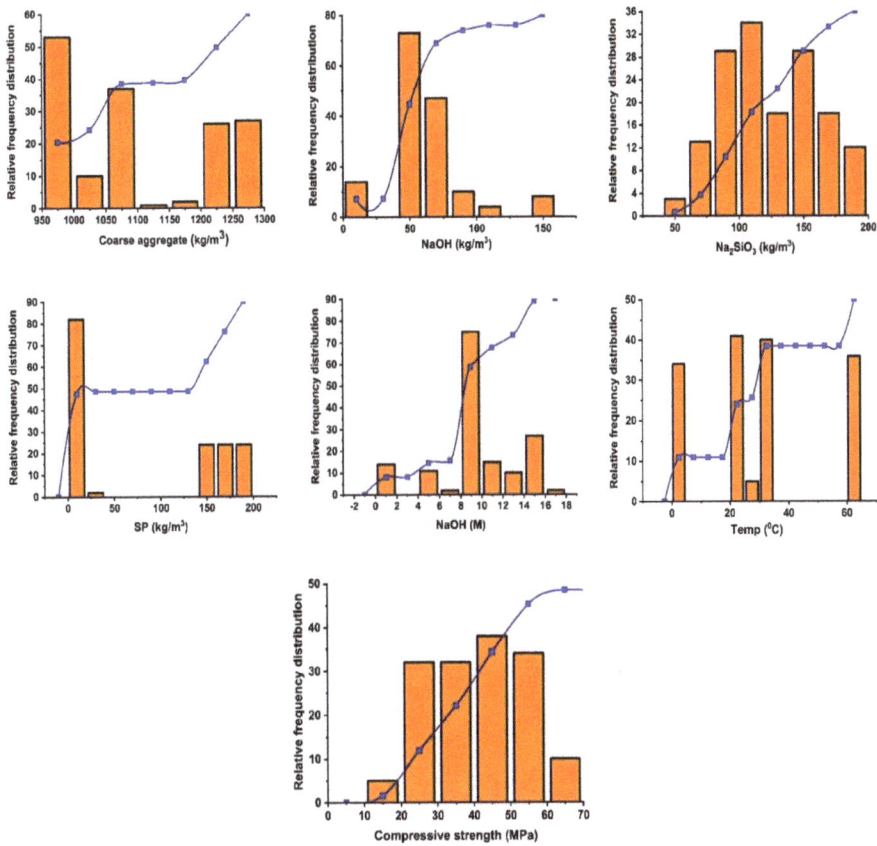

Figure 1. Frequency distribution of input and output parameters.

Table 3. Descriptive data of parameters.

Statistical Description	FA	GGBS	Fine	Coarse	NaOH	Na$_2$SiO$_3$	SP	NaOH	Temp.
Mean	252.5	151.4	729.8	1096.0	60.5	123.0	77.6	8.6	28.1
Standard Error	6.9	6.9	5.4	9.4	2.1	2.9	6.5	0.3	1.6
Median	270.0	135.0	760.5	1090.8	57.1	115.7	7.9	8.0	25.0
Mode	303.8	101.3	774.0	1090.8	81.0	81.0	0.0	8.0	30.0
Standard Deviation	86.3	86.7	68.0	117.9	26.8	35.7	81.0	3.9	20.6
Sample Variance	7442.7	7522.7	4620.5	13,889.3	720.4	1275.1	6558.3	15.2	422.4
Kurtosis	2.5	2.2	0.0	−1.5	3.0	−0.9	−1.9	0.2	−0.9
Skewness	−1.4	1.3	−0.8	0.3	1.2	0.1	0.2	−0.5	0.3
Range	400.0	409.0	263.6	327.0	134.3	138.9	180.0	16.0	60.0
Minimum	0.0	0.0	547.0	966.0	9.0	54.0	0.0	0.0	0.0
Maximum	400.0	409.0	810.6	1293.0	143.3	192.9	180.0	16.0	60.0
Sum	39,384.5	23,624.5	113,849.2	170,980.4	9432.3	19,185.4	12,100.6	1336.0	4380.0
Count	156.0	156.0	156.0	156.0	156.0	156.0	156.0	156.0	156.0

3. Methods

ML technologies are now being used in a wide range of industries to anticipate and understand the nature of various constituents. In this study, ML-based methods such as SVM, the decision tree (DT), RF, and multiple linear regressions (MLR), were utilized to estimate the compressive strength of GPC. These methods were chosen for their popularity, robustness in predicting outcomes, and were recognized as the top evaluated algorithms. Furthermore, the ensemble model with weak learners was utilized to model the strength of GPC utilizing AdaBoost and bagging. Moreover, Figure 2 depicts the entire systematic diagram of the individual and ensemble learning approach.

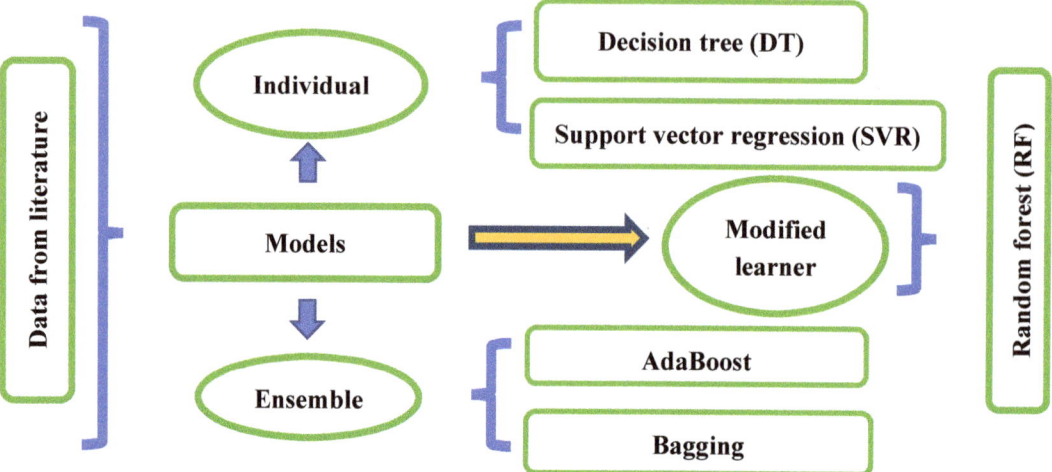

Figure 2. Flow diagram of models used in this research.

3.1. Decision Tree

This is a supervised ML approach that creates a tree-like model from training data using the DT. It is similar to a schematic flow in that each of the vertices reflects a test of a characteristic and that each route reflects the result of the test o that feature. It is referred to as a DT due to the fact that its form is comparable to that of a tree. This is accomplished through the use of partitions in predictors, which allows the target variables to be based primarily on divisions between the input parameters. Due to the fact that the regression tree automatically picks values, the educated regression tree presents parameters that are much more relevant to anticipate target variables from the preceding tree node than variables, which are less important to predict target variables. Because the specified dataset has no classifications, a regression model is fitted to the target variable using the independent variables. Every variable has several sites of division. The technique compares the predicted and actual numbers for each division point. The split point errors for all variables are summed, and the variable with the fitness function's fewest values is chosen as the split point. This process is repeated.

3.2. Random Forest

The RF approach is both a regression and a cataloging approach, and it has been the subject of the majority of the research work. Breiman invented RF regression in 2001, and it is widely regarded as an improvement over traditional classification regression methods. It is reported that the key advantages of RF are its flexibility and speed in building input–output relationships. The main difference between DT and RF is that DT only builds one tree whereas RF builds a forest of trees where dissimilar data are randomly picked and given to each tree. The data are organized into rows and columns for each model tree, with different sizes of columns and rows being used for different trees. Moreover, the development of every tree is carried out in the sequence of phases shown below.

1. Approximately two-thirds of the entire dataset is picked at random for each forest and is symbolized by the data frame, a process known as bagging. In order to discover the optimum node-splitting technique, the predictor parameters are chosen at random.
2. Out-of-bag error is assessed for all of the trees based on the data that are available. Then, the mistakes from each tree are added together in order to yield the final output for each tree.
3. Each tree provides a statistical analysis based on regression, and the algorithm chooses the forest that receives the greatest number of votes. The votes might be 0 or 1. The fraction of 1 s is a prediction probability.

3.3. Support Vector Regression

Vapnik is considered to be the originator of SVM, which was initially utilized in the year 1995. It is now frequently used for classification, prediction, and regression. Because SVMs can effectively handle nonlinear regression problems, they are commonly utilized in input–output analysis. This is accomplished by applying a static diagraming strategy to the SVM analysis data in order to map them into n-dimensional function space. After that, the nonlinear activation operations are employed to match the substantially high-dimensioned space in which the information on the input parameters is more distinct from the original data, leading to a much more precise match. The linear function in space is denoted by the symbol $f(x, w)$, which may be written as follows:

$$f(x,w) = \sum_{j=1}^{n} w_j g_j(x) + b \qquad (1)$$

where, 'b', '$g_j(x)$', and 'w' denote the nonlinear bias term, input space, and weight vector transformations determined by enhancing the normalized risk function, respectively. Assessment quality is also calculated by a loss function L_ε, where L_ε can be given as follows.

$$L_\varepsilon = L_\varepsilon(y, f(x,w)) = \begin{cases} 0 & if\ |y - f(x,w)| \leq \varepsilon \\ |y - f(x,w)| & otherwise \end{cases} \qquad (2)$$

SVM regression is unique in that it uses an ε-insensitive loss function to compute a linear regression function for the additional higher-dimensional space while minimizing model complexity $||w||^2$. This job is proven by non-negative slack variables $\xi_i + \xi_i^*$, where $I = 1, \ldots, n$ is used to find models from the π-insensitive field. Thus, the SVM regression can be built by streamlining the function as follows:

$$\min \frac{1}{2}||w||^2 + C \sum_{i=1}^{n} (\xi_i + \xi_i^*) \qquad (3)$$

$$\text{subject to} \begin{cases} y_i - f(x_i, w) \leq \varepsilon + \xi_i^* \\ f(x_i, w) - y_i \leq \varepsilon + \xi_i^* \\ \xi_i, \xi_i^* \geq 0,\ i = 1, \ldots, n \end{cases} \qquad (4)$$

This optimization issue may be turned into a dual situation that can be resolved by

$$f(x) = \sum_{i=1}^{nsv} (\alpha_i + \alpha_i^*) K(x, x_i) \text{ subject to } 0 \leq \alpha_i^* \leq C,\ 0 \leq \alpha_i \leq C \qquad (5)$$

where n_{SV} is the quantity of provision vectors. The kernel function is

$$K(x, x_i) = \sum_{i=1}^{m} (g_i(x) + g_i(x_i)) \qquad (6)$$

During the training process, selected SVM kernel functions such as the linear, radial basis, polynomial, and sigmoid functions are used to determine support vectors along the function surface of the function surface. The kernel settings are influenced by the type of kernel used and the software that is implemented.

3.4. Boosting and Bagging Ensemble Approaches

Ensemble techniques are used to improve ML recognition and prediction accuracy. By integrating and aggregating numerous weaker prediction models, these methods generally assist in alleviating over-fitting issues (component sub-models). It is possible to make a smarter learner by intelligently altering training data and constructing several sub-models (A, B, ... , N). Furthermore, the ideal model may be made by merging prominent sub-models using voting and averaging combination measures to reach the best possible result, as illustrated in Figure 3. Bagging is among the most widely used ensemble modeling techniques, which uses the bootstrap resampling method to gather data and calculate benefits. During the bagging procedure, the first training set substitutes partial models from the actual model. A few data samples can appear in multiple models, whilst some do not appear at all in any product models. The final model outcome is then calculated by taking an average of the outputs from all of the component models.

The boosting process, like the bagging technique, generates a cumulative model that results in the construction of a number of components that are more precise than non-ensemble models. Additionally, boosting is the process of using weighted averages by relying on sub-models to determine where it should be included in the finalized model. Based on individual learners such as SVM, DT, and RT, this study predicts the strength of GPC using boosting and bagging techniques.

There are two types of tuning parameters utilized in communal (ensemble) algorithms: (i) parameters that are connected with the perfect amount of model learners, and (ii) learning rates. The boosting and bagging algorithms with twenty ensemble models were made from the individual base learner and the best model constructs were picked based on strong correlation coefficient values, as shown in Figure 3 and Table 4. It can be seen that the DT with AdaBoost and bagging with N = 5 and 9 yields an R^2 of 0.92. Moreover, support vector regression (SVR) shows a similar trend with an estimator of 4 and 12 yielding a strong correlation of about 0.90 and 0.93, respectively.

Table 4. N-estimator response of models.

Technique Used	Ensemble Approaches	Machine-Learning Methods	Ensemble Models	Optimum Estimator	R^2-Value
Individual	-	DT	-	-	0.7623
	-	SVR	-	-	0.7923
Ensemble	Bagging	DT - Bagging	(10,20,30 ... .200)	09	0.9206
		SVR - Bagging	(10,20,30 ... .200)	12	0.9300
Ensemble	Boosting	DT - AdaBoost	(10,20,30 ... .200)	05	0.9257
		SVR - AdaBoost	(10,20,30 ... .200)	04	0.9005
Modified learner		RF	(10,20,30 ... .200)	10	0.9388

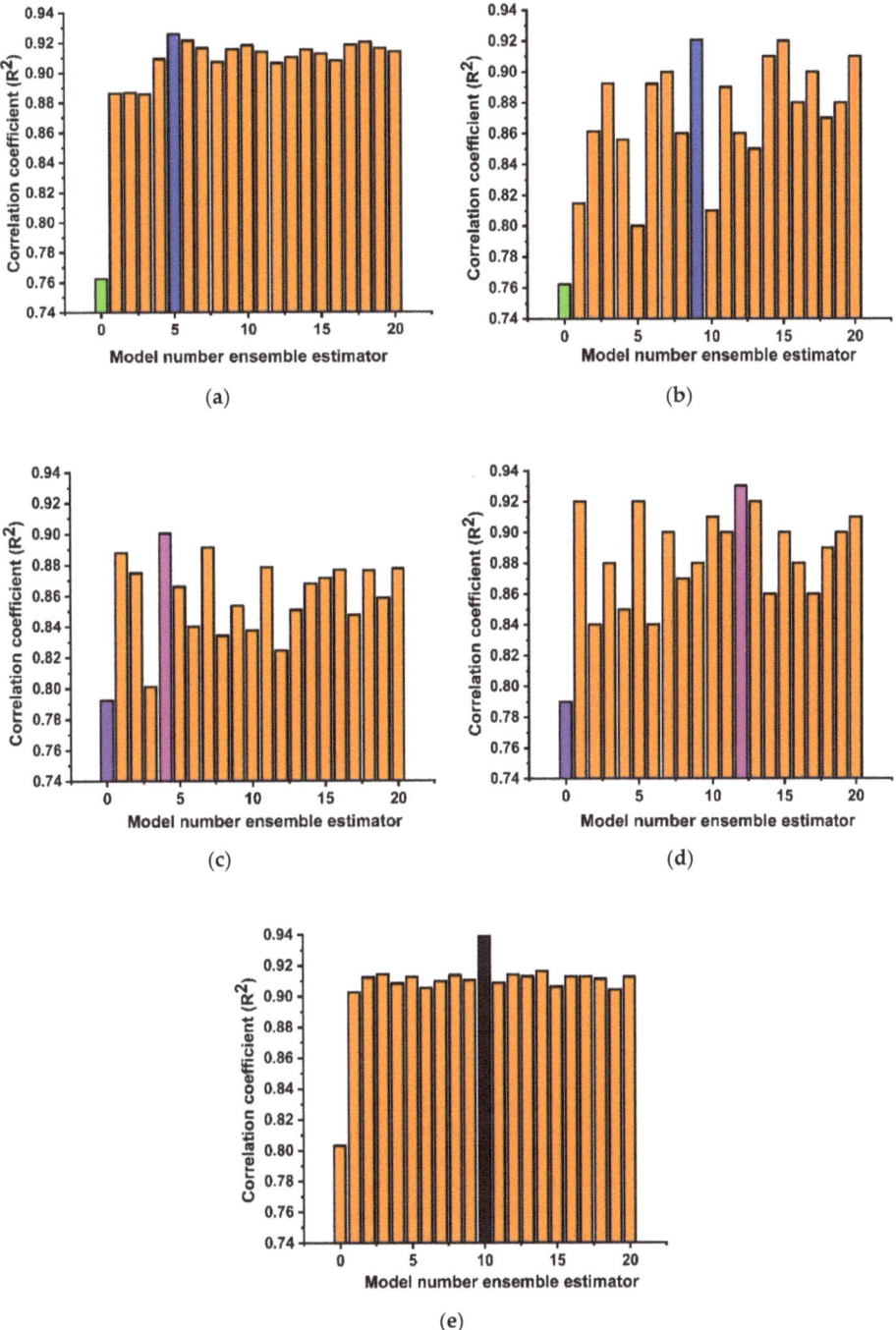

Figure 3. Ensemble modeling (**a**) DT with AdaBoost algorithm; (**b**) DT with bagging algorithm; (**c**) SVR with AdaBoost algorithm; (**d**) SVR with bagging algorithm; (**e**) RF ensembling.

4. Model Assessment Using Statistical Measures

The robustness of the model is evaluated by statistical checks in the form of error measures for individual and ensemble models are presented from Equations (7) and (8)

$$\text{MAE} = \frac{1}{n}\sum_{i=1}^{n}|x_i - x| \qquad (7)$$

$$\text{RMSE} = \sqrt{\sum \frac{\left(y_{pred} - y_{ref}\right)^2}{N}} \qquad (8)$$

5. Result

A linear regression model for predicting GPC with variable influences is illustrated in Figure 4. It should be noted that the Python-based approach has a strong correlation in the prediction of strength, as demonstrated in Figure 4a. However, this approach shows a lesser correlation in prediction with $R^2 = 0.637$. In addition, the difference between the prediction and target in terms of its absolute-error distribution is illustrated in Figure 4b, showing that the majority of the predicted outcomes depict greater error with 17.87 MPa (maximum), 0.29 MPa (minimum), and 7.69 MPa (average) absolute error, specifying that the data set of the model is biased. It shows that linear regression may be used to anticipate non-linear analysis results to a limited extent. Although, the MLR model cannot be used for non-linear analysis outcomes that have the strongest correlation to their outcome.

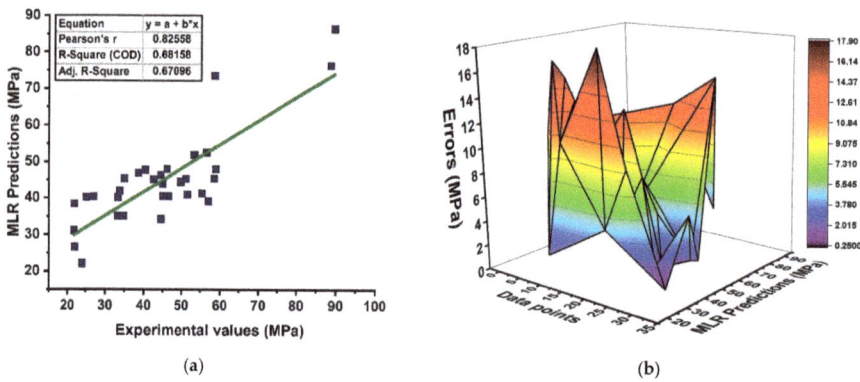

Figure 4. (a) Linear regression modeling; (b) distribution of errors via regression.

5.1. Decision Tree

The supervised and nonlinear regression model with a DT provided a soundly favorable prediction outcome, as depicted in Figure 5. In addition, the DT was modeled using several ensemble methods, such as bagging and boosting, as depicted in Figure 5. It can be seen in Figure 5a that the DT as an individual algorithm produces a good relationship with $R^2 = 0.76$. Moreover, the performance of the model can also be assessed by its absolute error, as demonstrated in Figure 5b. However, the model accuracy and outcome prediction can also be evolved by using ensemble approaches due to its performances and robustness. In addition, adding a boosting regressor to the weak or individual learner shows a positive correlation with $R^2 = 0.92$, as depicted in Figure 5c, with its reduced error distribution in Figure 5d. The bagging model illustrates a good $R^2 = 0.92$ with average errors of 15.78 MPa (lesser maximum), 0.26 MPa (minimum), and 3.22 MPa compared to MLR, as shown in Figure 5e,f. Although, the same individual model was modeled with AdaBoost regressor, showing a clear significant enhancement of the model. Moreover, the efficiency of the model can also be judged by its absolute errors, as depicted in Figure 5g. Its shows that the model performance is significantly enhanced as compared to the MLR model.

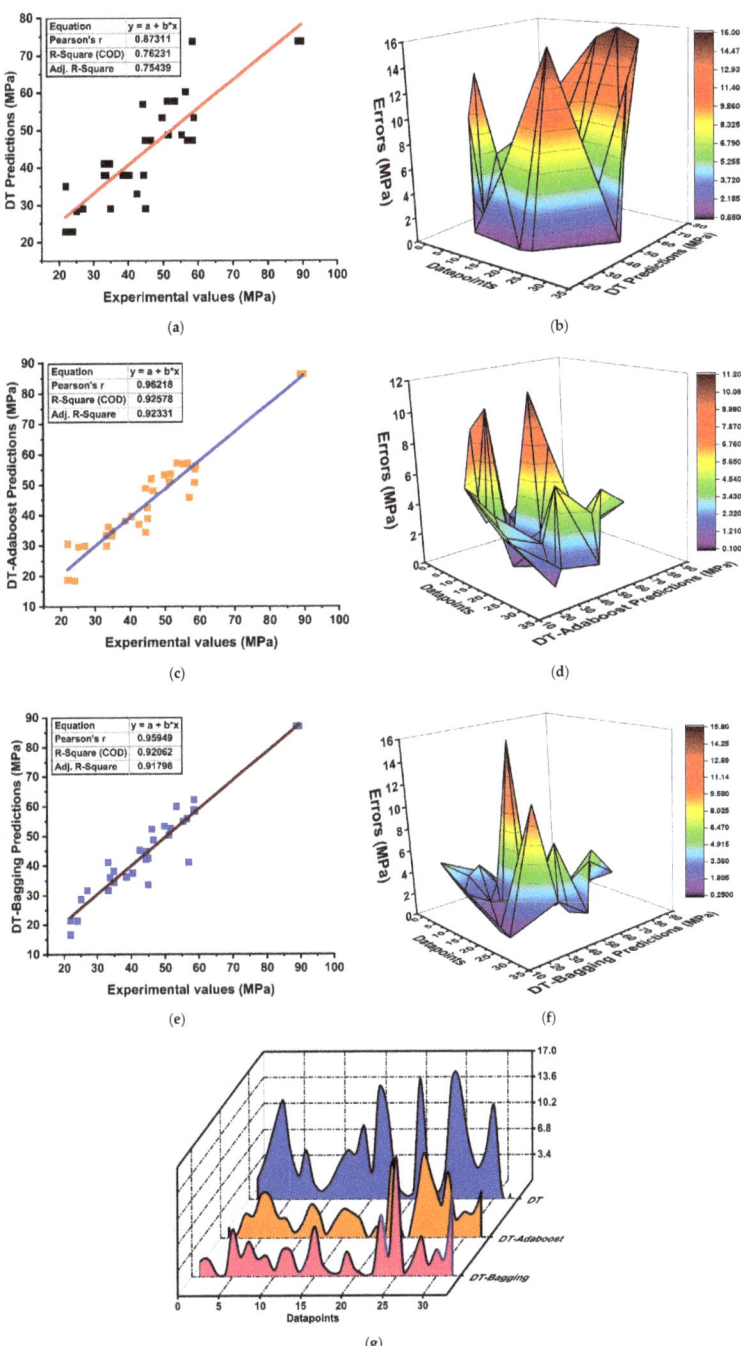

Figure 5. DT modeling; (**a**) non-ensemble model; (**b**) non-ensemble absolute-error distribution; (**c**) ensemble model of DT with boosting approach; (**d**) absolute errors of DT with boosting approach; (**e**) ensemble model of DT with bagging approach; (**f**) absolute errors of DT with bagging approach; (**g**) Comparison of models with errors using waterfall plot.

5.2. Support Vector Regression

ML with SVR was carried out to predict the mechanical properties of GPC, as shown in Figure 6. The predicted outcome with experimental data points as individual regression models depicts a strong relationship with $R^2 = 0.79$ due to its obstinate generalization capacity in making a robust performance, as shown in Figure 6a. Similarly, to the DT, SVR model accuracy can also be evaluated by its absolute-error distributions, as depicted in Figure 6b. It shows that the overall results of the predicted outcome lie close to the experimental values with minor data lying as outliers, but it does not devalue the accuracy of the model. In addition, in terms of statistical measures, SVM models show reduced average errors of about 5.69 MPa as compared to MLR (7.69 MPa). Likewise, the SVM model is ensembled and thus shows significant enhancements as depicted in Figure 6c,e with $R^2 = 0.90$ and $R^2 = 0.93$, respectively. Figure 6c,d represent the regression analysis of the boosting algorithm with its error distribution, showing that the boosting algorithm has an obstinate effect on forecasting the properties of concrete. Overall, the efficiency of the model can also be evaluated by its maximum (13.97 MPa), minimum (0.19 MPa), and average errors (4.14 MPa), and it is reported as a minimum compared to MLR. In addition, the bagging algorithm shows a similar trend by yielding a reasonable model with $R^2 = 0.93$ and its error distribution of 9.92 MPa (maximum), 0.08 MPa (minimum), and 3.76 MPa (average), as illustrated in Figure 6e,f. The overall comparison between SVR and bagging and boosting in terms of their absolute errors is shown in Figure 6g. The model with SVR demonstrates a significant and accurate prediction due to the strong learner in the model.

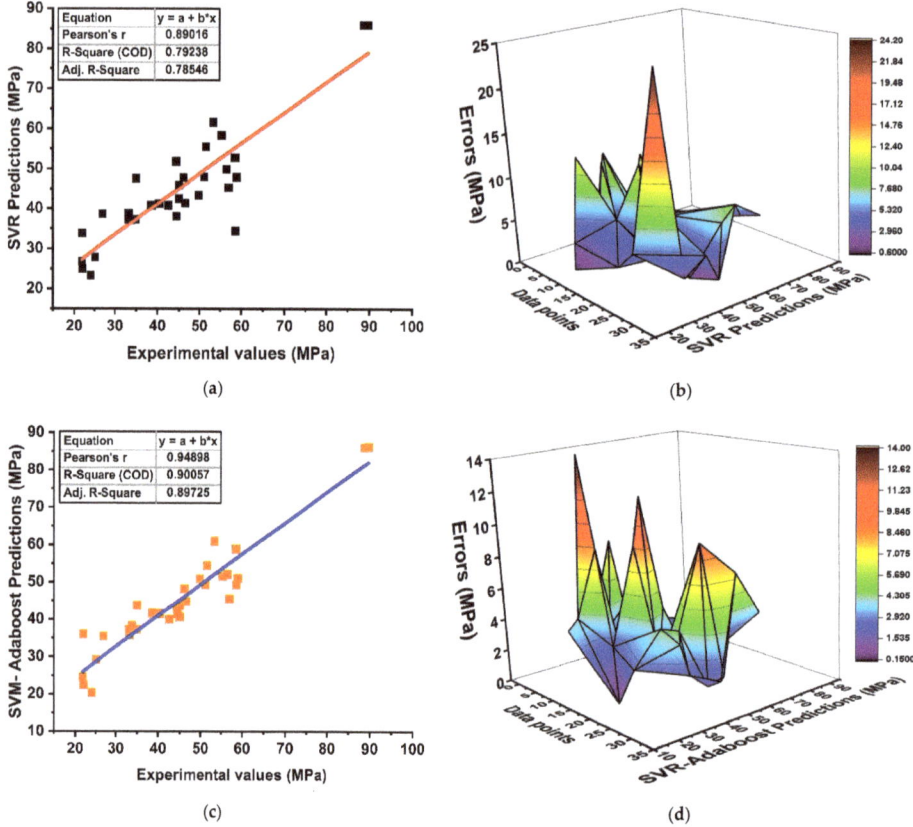

Figure 6. Cont.

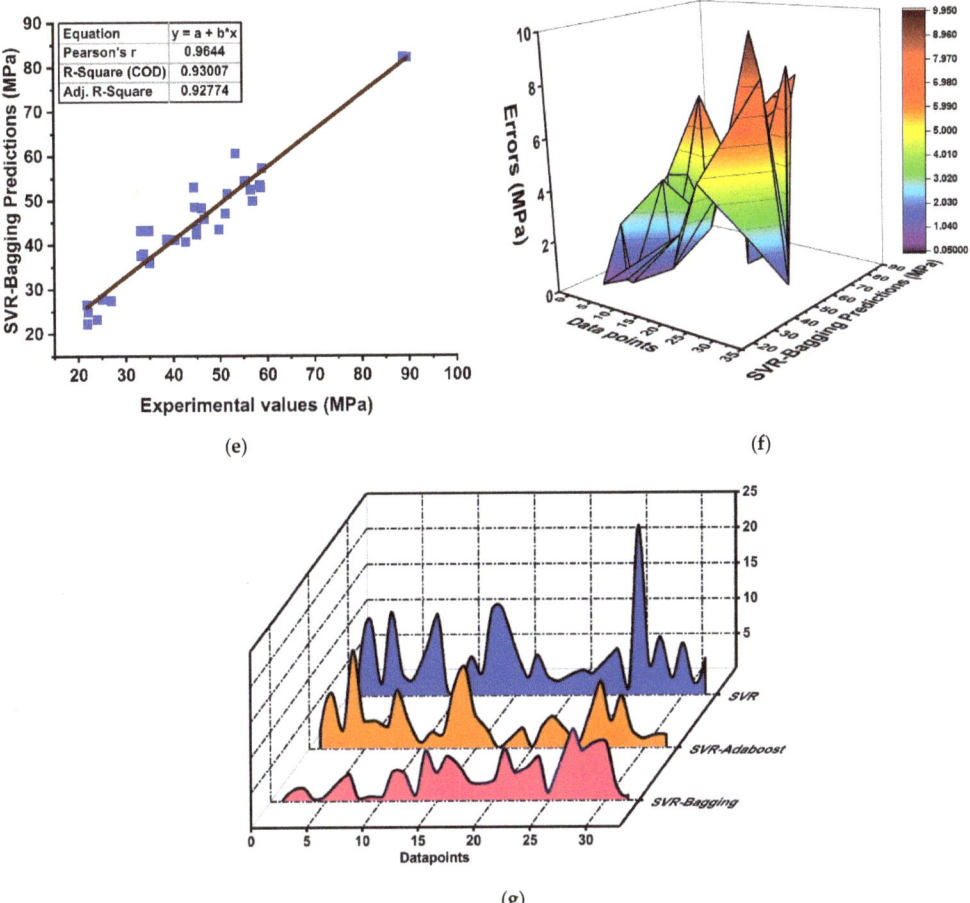

Figure 6. SVR modeling; (**a**) non-ensemble model; (**b**) non-ensemble absolute errors distribution; (**c**) ensemble model of SVR with boosting approach; (**d**) absolute errors of SVR with boosting approach; (**e**) ensemble model of SVR with bagging approach; (**f**) absolute errors of SVR with bagging approach; (**g**) Comparison of models with errors using waterfall plot.

5.3. Random Forest

The RF algorithm is a type of ensemble ML approach that incorporates the bagging method and random-feature-selection procedure to yield a predictive model. The predictive performance between the target and experimental results is depicted in Figure 7. The model illustrates a well-defined correlation with $R^2 = 0.938$ and is also assessed by its absolute error distributions as illustrated in Figure 7b. It can be seen that the RF-based model gives a lesser difference between prediction and experimental values with maximum, minimum, and average errors of about 10.54 MPa, 0.08 MPa and 3.217 MPa, respectively. Similarly, the forecasted results show that the influence of the strong learner in prediction is far better than individual approaches.

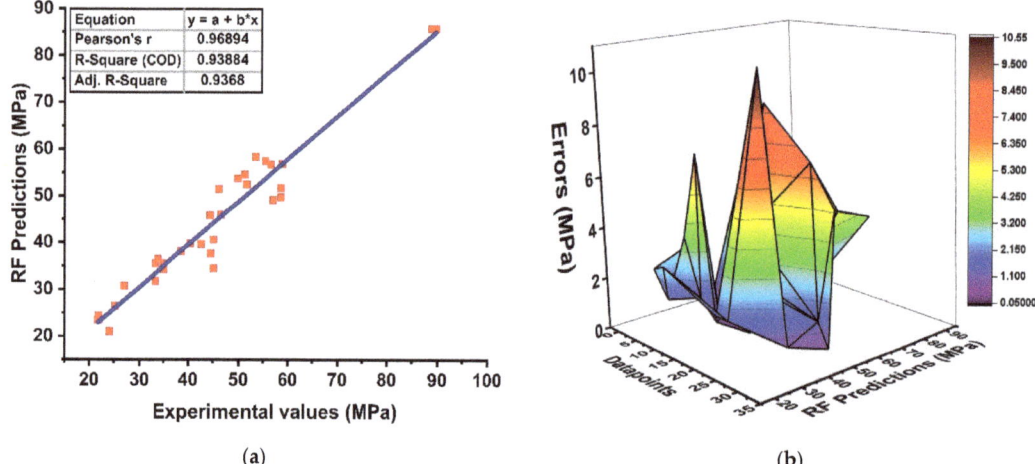

Figure 7. RF; (**a**) Relationship of experimental and predicted values; (**b**) Error-distribution result of model.

5.4. Cross-Validation Results

In order to assess a model, it must have the desired level of accuracy. To assure the accuracy of prediction models, it is necessary to perform this validation. The validation of this model was performed by using a ten-fold validation, as illustrated in Figure 8. This strategy is intended to limit the degree of bias involved in selecting the training data set at random during the training process throughout the training phase. It divides the data that are used to make the model into ten equal sections.

It uses nine out of ten subsets to design the robust learner and one set to authenticate the model. This approach yields an average error accuracy and is evaluated through statistical errors. The ten-fold cross-validation approach is said to demonstrate the generalization and dependability of the model performance, as demonstrated in Figure 8. The DT model with the ensemble approach via AdaBoost and bagging depicts good ten-fold R^2 values with an average values of $R^2 = 0.89$ and 0.879 for the AdaBoost and bagging approaches, as illustrated in Figure 8a. Similarly, the model shows a significant validation response by showing lesser RMSE and MAE errors with 8.99 MPa and 10.65 MPa for both ensemble models, respectively, as shown in Figure 8b,c. Moreover, the validation response via the SVR model in terms of R^2 shows an average error of 0.89 and 0.86 for the tenth k series for both models, as illustrated in Figure 8d. This depicts a strong accuracy of the models towards predictions. Likewise, the validation response in term of RMSE and MAE for the SVR model demonstrate the same response as for DT by showing lesser errors, as illustrated in Figure 8e,f. Additionally, the RF model depicts a comparable response to DT and SVR by adamantly representing a positive R^2 relation with predicted values and showing lesser errors.

5.5. Statistical Analysis of Models

The evaluation of the models is also performed by conducting statistical measures. Apart from R^2, the statistical check is significantly useful in the assessment of any model by measuring the numerical values, as depicted in Table 5. It can be seen that the individual model yields an MAE error of about 7.69 MPa, which is more than the ensemble models. DT with AdaBoost and bagging yields 53.3% and 58.12% more accurate models as compared to the individual. Similarly, RMSE and MSE show the similar response for the DT model. The SVR model shows that the ensemble model increases the efficiency of the models by

27.24%, 49.51%, and 28.99% for the AdaBoost model and by 33.92%, 60.8%, and 37.4% for the bagging model due to the incorporation of the weak learner in the making of a resilient model. Likewise, the RF model demonstrates a more efficient prediction model due to its lesser errors, as illustrated in Figure 7.

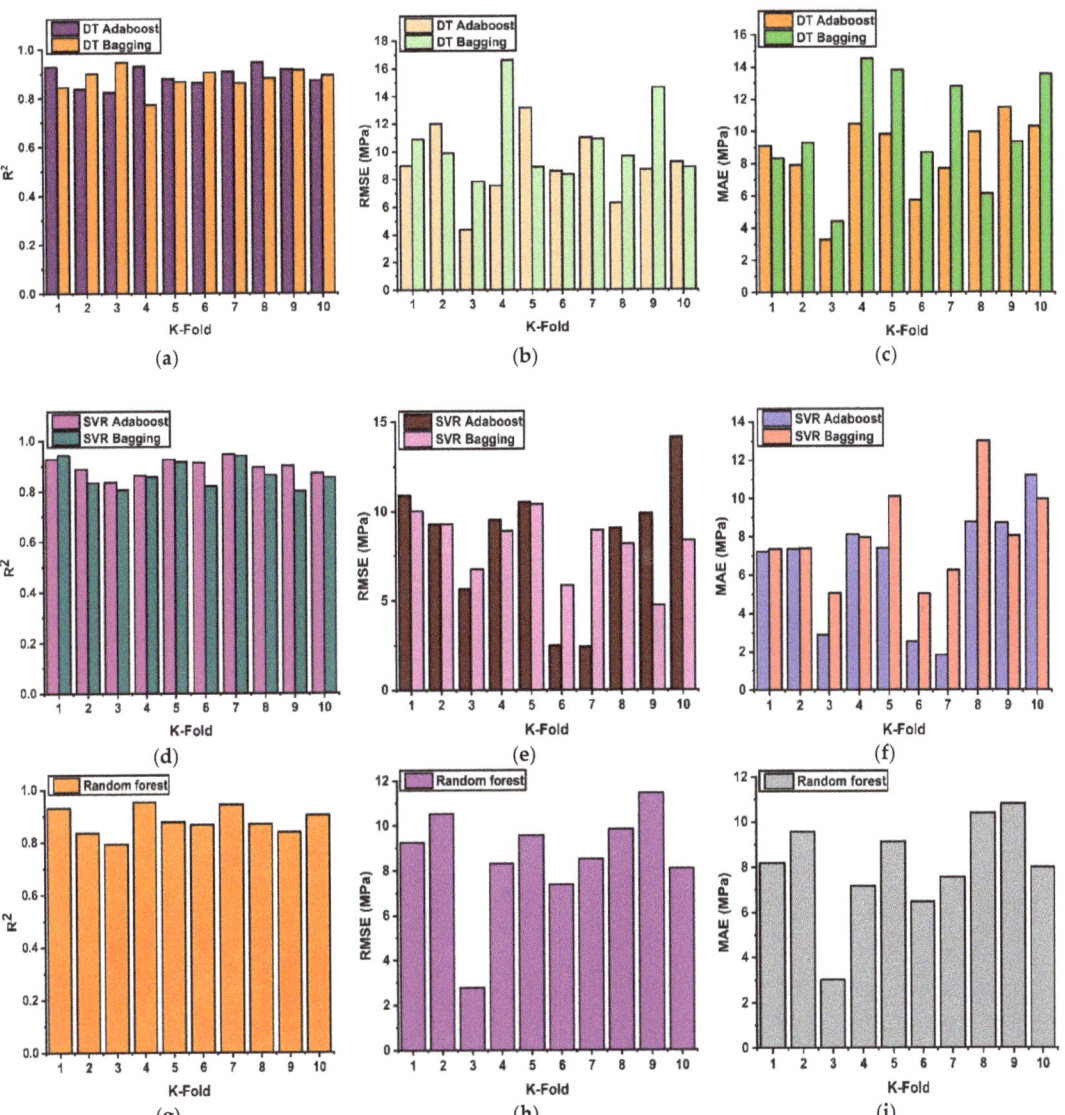

Figure 8. K-fold validations; (a) DT correlation with ensemble models; (b) DT RMSE errors with ensemble models; (c) DT MAE errors with ensemble models; (d) SVR correlation with ensemble models; (e) SVR RMSE errors with ensemble models; (f) SVR MAE errors with ensemble models; (g) RF correlation with ensemble models; (h) RF RMSE errors with ensemble models; (i) RF MAE errors with ensemble models.

Table 5. Statistical analysis.

Approaches Use	ML Methods	MAE	MSE	RMSE
Individual learner	DT	7.69	63.20	7.95
	SVR	5.69	55.20	7.43
Ensembling with AdaBoost	DT	3.59	20.70	4.55
	SVR	4.14	27.87	5.28
Ensembling with bagging	DT	3.22	21.52	4.64
	SVR	3.76	21.62	4.65
Ensemble model	RF	3.21	16.89	4.11

5.6. Permutation Features Analysis of Variables in Geopolymer Concrete

The permutation analysis depicts the influence of each variable on the target strength of GPC, and was conducted through the spyder notebook by using Python language in Anaconda software, as illustrated in Figure 9. The analysis results reveal that the GGBS, FA, and temperature (°C) have a significant effect on the strength of GPC due to the occurrence of major SiO_2, Al_2O_3, and CaO in the amorphous state [73–75]. Additionally, the presence of GGBS in concrete gives rise to binding phenomena in the presence of the alkaline medium. Moreover, when GGBS is combined with FA in an alkaline medium, it gives rise to additional calcium content that is responsible for the enhanced mechanical properties.

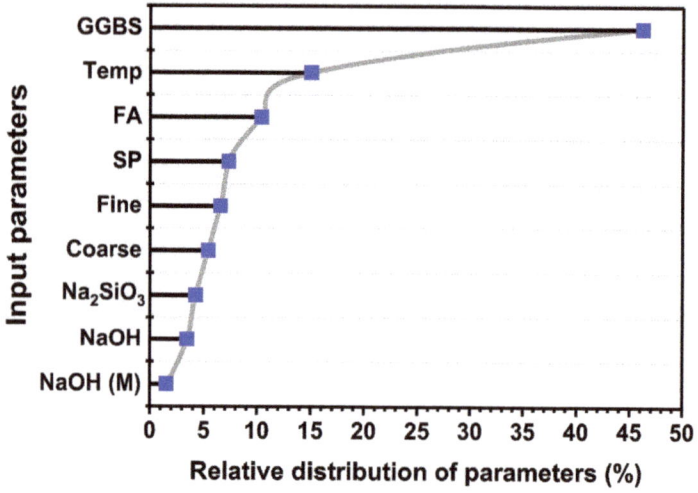

Figure 9. Permutation analysis of parameters in geopolymer concrete (GPC).

6. Conclusions

The aim of this research was to anticipate the strength of GPC using the individual and ensemble ML approaches. For prediction, two individual approaches, DT and SVR, and three ensemble techniques, bagging, AdaBoost, and RF regression were used, and the following conclusions are drawn from the analysis.

1. The DT as an individual approach yields a positive outcome with $R^2 = 0.76$. Nevertheless, the ensemble approaches with bagging and boosting depict precise results with $R^2 = 0.92$. These indications make it clear that the ensemble approach yields positive results due to its weak-learner incorporation.
2. SVR shows a similar response with ensemble approaches as compared to the individual approach. Moreover, the SVR model shows superior performance by depicting a good coefficient of determination with $R^2 = 0.90$ for boosting and $R^2 = 0.93$ for

bagging. Similarly, RFR yields better performance with $R^2 = 0.93$ for the testing set. This shows that the ensemble model yields robust performance as compared to non-ensemble approaches.

3. Cross-validation of the test set reveals lesser MAE, RMSE errors, and good average correlations of R^2 for the DT, SVR, and RF, indicating the accuracy of the model. Statistical-analysis results reveal lesser error for MAE, RMSE and MSE as compared to individual approaches.
4. The RF and SVR with bagging were superior to individual and ensemble approaches by showing $R^2 = 0.93$.
5. Permutation analysis of variables shows that FA, GGBS, and temperature have a major influence on the strength of GPC.

Supplementary Materials: The following supporting information can be downloaded at: https://www.mdpi.com/article/10.3390/ma15103478/s1, Table S1: Parameters selected based on the literature review and the published data used in the prediction of fly ash slag-based concrete [62–72].

Author Contributions: Conceptualization, M.N.A., K.K. and M.F.J.; Data curation, F.A.; Formal analysis, M.F.J. and F.A.; Funding acquisition, M.N.A., K.K., M.G.Q. and M.I.F.; Investigation, F.A.; Methodology, M.F.J. and F.A.; Project administration, M.N.A.; Resources, M.N.A., K.K. and M.G.Q.; Software, M.F.J. and F.A.; Supervision, M.N.A.; Validation, M.F.J.; Visualization, K.K., M.G.Q. and M.I.F.; Writing—original draft, M.N.A., K.K. and M.F.J.; Writing—review & editing, M.N.A. All authors have read and agreed to the published version of the manuscript.

Funding: This work was supported by the Deanship of Scientific Research, Vice Presidency for Graduate Studies and Scientific Research, King Faisal University, Saudi Arabia [Project No. AN000680]. The APC was funded by the same "Project No. AN000680".

Institutional Review Board Statement: Not applicable.

Informed Consent Statement: Not applicable.

Data Availability Statement: The data used in this research has been properly cited and reported in the main text.

Acknowledgments: The authors acknowledge the Deanship of Scientific Research, Vice Presidency for Graduate Studies and Scientific Research, King Faisal University, Saudi Arabia [Project No. AN000680]. The authors extend their appreciation for the financial support that has made this study possible.

Conflicts of Interest: The authors declare no conflict of interest.

References

1. Latawiec, R.; Woyciechowski, P.; Kowalski, K.J. Sustainable concrete performance—CO_2-emission. *Environments* **2018**, *5*, 27. [CrossRef]
2. Liu, G.; Yang, H.; Fu, Y.; Mao, C.; Xu, P.; Hong, J.; Li, R. Cyber-physical system-based real-time monitoring and visualization of greenhouse gas emissions of prefabricated construction. *J. Clean. Prod.* **2020**, *246*, 119059. [CrossRef]
3. Akbarzadeh Bengar, H.; Shahmansouri, A.A. A new anchorage system for CFRP strips in externally strengthened RC continuous beams. *J. Build. Eng.* **2020**, *30*, 101230. [CrossRef]
4. Benhelal, E.; Shamsaei, E.; Rashid, M.I. Novel modifications in a conventional clinker making process for sustainable cement production. *J. Clean. Prod.* **2019**, *221*, 389–397. [CrossRef]
5. Samimi, K.; Kamali-Bernard, S.; Akbar Maghsoudi, A.; Maghsoudi, M.; Siad, H. Influence of pumice and zeolite on compressive strength, transport properties and resistance to chloride penetration of high strength self-compacting concretes. *Constr. Build. Mater.* **2017**, *151*, 292–311. [CrossRef]
6. Taji, I.; Ghorbani, S.; de Brito, J.; Tam, V.W.Y.; Sharifi, S.; Davoodi, A.; Tavakkolizadeh, M. Application of statistical analysis to evaluate the corrosion resistance of steel rebars embedded in concrete with marble and granite waste dust. *J. Clean. Prod.* **2019**, *210*, 837–846. [CrossRef]
7. Tang, Z.; Li, W.; Ke, G.; Zhou, J.L.; Tam, V.W.Y. Sulfate attack resistance of sustainable concrete incorporating various industrial solid wastes. *J. Clean. Prod.* **2019**, *218*, 810–822. [CrossRef]
8. Ma, Z.; Liu, M.; Duan, Z.; Liang, C.; Wu, H. Effects of active waste powder obtained from C&D waste on the microproperties and water permeability of concrete. *J. Clean. Prod.* **2020**, *257*, 120518. [CrossRef]

9. Shah, S.N.; Mo, K.H.; Yap, S.P.; Yang, J.; Ling, T.C. Lightweight foamed concrete as a promising avenue for incorporating waste materials: A review. *Resour. Conserv. Recycl.* **2021**, *164*, 105103. [CrossRef]
10. Sun, C.; Chen, Q.; Xiao, J.; Liu, W. Utilization of waste concrete recycling materials in self-compacting concrete. *Resour. Conserv. Recycl.* **2020**, *161*, 104930. [CrossRef]
11. Tang, Q.; Ma, Z.; Wu, H.; Wang, W. The utilization of eco-friendly recycled powder from concrete and brick waste in new concrete: A critical review. *Cem. Concr. Compos.* **2020**, *114*, 103807. [CrossRef]
12. Farooq, F.; Jin, X.; Faisal Javed, M.; Akbar, A.; Izhar Shah, M.; Aslam, F.; Alyousef, R. Geopolymer concrete as sustainable material: A state of the art review. *Constr. Build. Mater.* **2021**, *306*, 124762. [CrossRef]
13. Qureshi, L.A.; Ali, B.; Ali, A. Combined effects of supplementary cementitious materials (silica fume, GGBS, fly ash and rice husk ash) and steel fiber on the hardened properties of recycled aggregate concrete. *Constr. Build. Mater.* **2020**, *263*, 120636. [CrossRef]
14. Vishnu, N.; Kolli, R.; Ravella, D.P. Studies on Self-Compacting geopolymer concrete containing flyash, GGBS, wollastonite and graphene oxide. *Mater. Today Proc.* **2020**, *43*, 2422–2427. [CrossRef]
15. Prusty, J.K.; Pradhan, B. Effect of GGBS and chloride on compressive strength and corrosion performance of steel in fly ash-GGBS based geopolymer concrete. *Mater. Today Proc.* **2020**, *32*, 850–855. [CrossRef]
16. Bajpai, R.; Choudhary, K.; Srivastava, A.; Sangwan, K.S.; Singh, M. Environmental impact assessment of fly ash and silica fume based geopolymer concrete. *J. Clean. Prod.* **2020**, *254*, 120147. [CrossRef]
17. Ma, C.K.; Awang, A.Z.; Omar, W. Structural and material performance of geopolymer concrete: A review. *Constr. Build. Mater.* **2018**, *186*, 90–102. [CrossRef]
18. Amran, Y.H.M.; Alyousef, R.; Alabduljabbar, H.; El-Zeadani, M. Clean production and properties of geopolymer concrete; A review. *J. Clean. Prod.* **2020**, *251*, 119679. [CrossRef]
19. Almutairi, A.L.; Tayeh, B.A.; Adesina, A.; Isleem, H.F.; Zeyad, A.M. Potential applications of geopolymer concrete in construction: A review. *Case Stud. Constr. Mater.* **2021**, *15*, e00733. [CrossRef]
20. Parathi, S.; Nagarajan, P.; Pallikkara, S.A. Ecofriendly geopolymer concrete: A comprehensive review. *Clean Technol. Environ. Policy* **2021**, *23*, 1701–1713. [CrossRef]
21. Kumar, R.; Verma, M.; Dev, N. Investigation on the Effect of Seawater Condition, Sulphate Attack, Acid Attack, Freeze–Thaw Condition, and Wetting–Drying on the Geopolymer Concrete. *Iran. J. Sci. Technol.-Trans. Civ. Eng.* **2021**, 1–31. [CrossRef]
22. Nnaemeka, O.F.; Singh, N.B. Durability properties of geopolymer concrete made from fly ash in presence of Kaolin. *Mater. Today Proc.* **2019**, *29*, 781–784. [CrossRef]
23. Guo, X.; Xiong, G. Resistance of fiber-reinforced fly ash-steel slag based geopolymer mortar to sulfate attack and drying-wetting cycles. *Constr. Build. Mater.* **2021**, *269*, 121326. [CrossRef]
24. Hassan, A.; Arif, M.; Shariq, M. Use of geopolymer concrete for a cleaner and sustainable environment—A review of mechanical properties and microstructure. *J. Clean. Prod.* **2019**, *223*, 704–728. [CrossRef]
25. Kotwal, A.R.; Kim, Y.J.; Hu, J.; Sriraman, V. Characterization and Early Age Physical Properties of Ambient Cured Geopolymer Mortar Based on Class C Fly Ash. *Int. J. Concr. Struct. Mater.* **2015**, *9*, 35–43. [CrossRef]
26. Pimraksa, K.; Chindaprasirt, P.; Rungchet, A.; Sagoe-Crentsil, K.; Sato, T. Lightweight geopolymer made of highly porous siliceous materials with various Na_2O/Al_2O_3 and SiO_2/Al_2O_3 ratios. *Mater. Sci. Eng. A* **2011**, *528*, 6616–6623. [CrossRef]
27. Hadi, M.N.S.; Zhang, H.; Parkinson, S. Optimum mix design of geopolymer pastes and concretes cured in ambient condition based on compressive strength, setting time and workability. *J. Build. Eng.* **2019**, *23*, 301–313. [CrossRef]
28. Ukritnukun, S.; Koshy, P.; Rawal, A.; Castel, A.; Sorrell, C.C. Predictive model of setting times and compressive strengths for low-alkali, ambient-cured, fly ash/slag-based geopolymers. *Minerals* **2020**, *10*, 920. [CrossRef]
29. Gholizadeh-Vayghan, A.; Nofallah, M.-H.; Khaloo, A. Technoeconomic Study of Alkali-Activated Slag Concrete with a Focus on Strength, CO_2 Emission, and Material Cost. *J. Mater. Civ. Eng.* **2021**, *33*. [CrossRef]
30. Songpiriyakij, S.; Kubprasit, T.; Jaturapitakkul, C.; Chindaprasirt, P. Compressive strength and degree of reaction of biomass- and fly ash-based geopolymer. *Constr. Build. Mater.* **2010**, *24*, 236–240. [CrossRef]
31. Puertas, F.; Martínez-Ramírez, S.; Alonso, S.; Vázquez, T. Alkali-activated fly ash/slag cements. Strength behavior and hydration products. *Cem. Concr. Res.* **2000**, *30*, 1625–1632. [CrossRef]
32. Olivia, M.; Nikraz, H. Properties of fly ash geopolymer concrete designed by Taguchi method. *Mater. Des.* **2012**, *36*, 191–198. [CrossRef]
33. Kurtoğlu, A.E.; Alzeebaree, R.; Aljumaili, O.; Niş, A.; Gülşan, M.E.; Humur, G.; Çevik, A. Mechanical and durability properties of fly ash and slag based geopolymer concrete. *Adv. Concr. Constr.* **2018**, *6*, 345–362. [CrossRef]
34. Podolsky, Z.; Liu, J.; Dinh, H.; Doh, J.H.; Guerrieri, M.; Fragomeni, S. State of the art on the application of waste materials in geopolymer concrete. *Case Stud. Constr. Mater.* **2021**, *15*, e00637. [CrossRef]
35. Kumar, M.L.; Revathi, V. Microstructural Properties of Alkali-Activated Metakaolin and Bottom Ash Geopolymer. *Arab. J. Sci. Eng.* **2020**, *45*, 4235–4246. [CrossRef]
36. Hassan, A.; Arif, M.; Shariq, M. Influence of microstructure of geopolymer concrete on its mechanical properties—A review. In *Lecture Notes in Civil Engineering*; Springer: Singapore, 2020; Volume 35, pp. 119–129.
37. Jiang, X.; Xiao, R.; Zhang, M.; Hu, W.; Bai, Y.; Huang, B. A laboratory investigation of steel to fly ash-based geopolymer paste bonding behavior after exposure to elevated temperatures. *Constr. Build. Mater.* **2020**, *254*, 119267. [CrossRef]
38. Salas, D.A.; Ramirez, A.D.; Ulloa, N.; Baykara, H.; Boero, A.J. Life cycle assessment of geopolymer concrete. *Constr. Build. Mater.* **2018**, *190*, 170–177. [CrossRef]

39. Zakka, W.P.; Abdul Shukor Lim, N.H.; Chau Khun, M. A scientometric review of geopolymer concrete. *J. Clean. Prod.* **2021**, *280*, 124353. [CrossRef]
40. Fang, G.; Ho, W.K.; Tu, W.; Zhang, M. Workability and mechanical properties of alkali-activated fly ash-slag concrete cured at ambient temperature. *Constr. Build. Mater.* **2018**, *172*, 476–487. [CrossRef]
41. Yazdi, M.A.; Liebscher, M.; Hempel, S.; Yang, J.; Mechtcherine, V. Correlation of microstructural and mechanical properties of geopolymers produced from fly ash and slag at room temperature. *Constr. Build. Mater.* **2018**, *191*, 330–341. [CrossRef]
42. Farooq, F.; Ahmed, W.; Akbar, A.; Aslam, F.; Alyousef, R. Predictive modeling for sustainable high-performance concrete from industrial wastes: A comparison and optimization of models using ensemble learners. *J. Clean. Prod.* **2021**, *292*, 126032. [CrossRef]
43. Soni, N.; Shukla, D.K. Analytical study on mechanical properties of concrete containing crushed recycled coarse aggregate as an alternative of natural sand. *Constr. Build. Mater.* **2021**, *266*, 120595. [CrossRef]
44. Assi, L.N.; Eddie Deaver, E.; Ziehl, P. Effect of source and particle size distribution on the mechanical and microstructural properties of fly Ash-Based geopolymer concrete. *Constr. Build. Mater.* **2018**, *167*, 372–380. [CrossRef]
45. Song, H.; Ahmad, A.; Farooq, F.; Ostrowski, K.A.; Maślak, M.; Czarnecki, S.; Aslam, F. Predicting the compressive strength of concrete with fly ash admixture using machine learning algorithms. *Constr. Build. Mater.* **2021**, *308*, 125021. [CrossRef]
46. Farooq, F.; Czarnecki, S.; Niewiadomski, P.; Aslam, F.; Alabduljabbar, H.; Ostrowski, K.A.; Śliwa-Wieczorek, K.; Nowobilski, T.; Malazdrewicz, S. A comparative study for the prediction of the compressive strength of self-compacting concrete modified with fly ash. *Materials* **2021**, *14*, 4934. [CrossRef]
47. Khan, M.A.; Memon, S.A.; Farooq, F.; Javed, M.F.; Aslam, F.; Alyousef, R. Compressive Strength of Fly-Ash-Based Geopolymer Concrete by Gene Expression Programming and Random Forest. *Adv. Civ. Eng.* **2021**, *2021*, 6618407. [CrossRef]
48. Ilyas, I.; Zafar, A.; Faisal Javed, M.; Farooq, F.; Aslam, F.; Musarat, M.A.; Vatin, N.I.; Fabbrocino, F. Forecasting Strength of CFRP Confined Concrete Using Multi Expression Programming. *Materials* **2021**, *14*, 7134. [CrossRef]
49. Javed, M.F.; Farooq, F.; Memon, S.A.; Akbar, A.; Khan, M.A.; Aslam, F.; Alyousef, R.; Alabduljabbar, H.; Rehman, S.K.U.; Ur Rehman, S.K.; et al. New prediction model for the ultimate axial capacity of concrete-filled steel tubes: An evolutionary approach. *Crystals* **2020**, *10*, 741. [CrossRef]
50. Farooq, F.; Amin, M.N.; Khan, K.; Sadiq, M.R.; Javed, M.F.; Aslam, F.; Alyousef, R. A comparative study of random forest and genetic engineering programming for the prediction of compressive strength of high strength concrete (HSC). *Appl. Sci.* **2020**, *10*, 7330. [CrossRef]
51. Nafees, A.; Javed, M.F.; Khan, S.; Nazir, K.; Farooq, F.; Aslam, F.; Musarat, M.A.; Vatin, N.I. Predictive Modeling of Mechanical Properties of Silica Fume-Based Green Concrete Using Artificial Intelligence Approaches: MLPNN, ANFIS, and GEP. *Materials* **2021**, *14*, 7531. [CrossRef]
52. Raza, F.; Alshameri, B.; Jamil, S.M. Assessment of triple bottom line of sustainability for geotechnical projects. *Environ. Dev. Sustain.* **2021**, *23*, 4521–4558. [CrossRef]
53. Khan, M.A.; Farooq, F.; Javed, M.F.; Zafar, A.; Ostrowski, K.A.; Aslam, F.; Malazdrewicz, S.; Maślak, M. Simulation of depth of wear of eco-friendly concrete using machine learning based computational approaches. *Materials* **2022**, *15*, 58. [CrossRef] [PubMed]
54. Ahmad, A.; Ostrowski, K.A.; Maślak, M.; Farooq, F.; Mehmood, I.; Nafees, A. Comparative Study of Supervised Machine Learning Algorithms for Predicting the Compressive Strength of Concrete at High Temperature. *Materials* **2021**, *14*, 4222. [CrossRef] [PubMed]
55. Aslam, F.; Farooq, F.; Amin, M.N.; Khan, K.; Waheed, A.; Akbar, A.; Javed, M.F.; Alyousef, R.; Alabduljabbar, H. Applications of Gene Expression Programming for Estimating Compressive Strength of High-Strength Concrete. *Adv. Civ. Eng.* **2020**, *2020*, 8850535. [CrossRef]
56. Song, H.; Ahmad, A.; Ostrowski, K.A.; Dudek, M. Analyzing the compressive strength of ceramic waste-based concrete using experiment and artificial neural network (Ann) approach. *Materials* **2021**, *14*, 4518. [CrossRef]
57. Ahmad, W.; Ahmad, A.; Ostrowski, K.A.; Aslam, F.; Joyklad, P.; Zajdel, P. Application of advanced machine learning approaches to predict the compressive strength of concrete containing supplementary cementitious materials. *Materials* **2021**, *14*, 5762. [CrossRef]
58. Xu, Y.; Ahmad, W.; Ahmad, A.; Ostrowski, K.A.; Dudek, M.; Aslam, F.; Joyklad, P. Computation of high-performance concrete compressive strength using standalone and ensembled machine learning techniques. *Materials* **2021**, *14*, 7034. [CrossRef]
59. Song, Y.; Zhao, J.; Ostrowski, K.A.; Javed, M.F.; Ahmad, A.; Khan, M.I.; Aslam, F.; Kinasz, R. Prediction of compressive strength of fly-ash-based concrete using ensemble and non-ensemble supervised machine-learning approaches. *Appl. Sci.* **2022**, *12*, 361. [CrossRef]
60. Ahmad, A.; Ahmad, W.; Aslam, F.; Joyklad, P. Compressive strength prediction of fly ash-based geopolymer concrete via advanced machine learning techniques. *Case Stud. Constr. Mater.* **2022**, *16*, e00840. [CrossRef]
61. Zou, Y.; Zheng, C.; Alzahrani, A.M.; Ahmad, W.; Ahmad, A.; Mohamed, A.M.; Khallaf, R.; Elattar, S. Evaluation of Artificial Intelligence Methods to Estimate the Compressive Strength of Geopolymers. *Gels* **2022**, *8*, 271. [CrossRef]
62. Yang, K.; Yang, C.; Magee, B.; Nanukuttan, S.; Ye, J. Establishment of a preconditioning regime for air permeability and sorptivity of alkali-activated slag concrete. *Cem. Concr. Compos.* **2016**, *73*, 19–28. [CrossRef]
63. Sumanth Kumar, B.; Sen, A.; Rama Seshu, D. Shear Strength of Fly Ash and GGBS Based Geopolymer Concrete. In *Lecture Notes in Civil Engineering*; Springer: Singapore, 2020; Volume 68, pp. 105–117.
64. Ullah, H.S.; Khushnood, R.A.; Farooq, F.; Ahmad, J.; Vatin, N.I.; Yehia, D.; Ewais, Z. Prediction of Compressive Strength of Sustainable Foam Concrete Using Individual and Ensemble Machine Learning Approaches. *Materials* **2022**, *15*, 3166. [CrossRef]
65. Jithendra, C.; Elavenil, S. Role of superplasticizer on GGBS based Geopolymer concrete under ambient curing. *Mater. Today Proc.* **2019**, *18*, 148–154. [CrossRef]
66. Verma, M.; Dev, N. Sodium hydroxide effect on the mechanical properties of flyash-slag based geopolymer concrete. *Struct. Concr.* **2021**, *22*, E368–E379. [CrossRef]

67. Deb, P.S.; Nath, P.; Sarker, P.K. The effects of ground granulated blast-furnace slag blending with fly ash and activator content on the workability and strength properties of geopolymer concrete cured at ambient temperature. *Mater. Des.* **2014**, *62*, 32–39. [CrossRef]
68. Ding, Y.; Shi, C.J.; Li, N. Fracture properties of slag/fly ash-based geopolymer concrete cured in ambient temperature. *Constr. Build. Mater.* **2018**, *190*, 787–795. [CrossRef]
69. Nath, P.; Sarker, P.K. Fracture properties of GGBFS-blended fly ash geopolymer concrete cured in ambient temperature. *Mater. Struct. Constr.* **2017**, *50*, 32. [CrossRef]
70. Karthik, A.; Sudalaimani, K.; Vijaya Kumar, C.T. Investigation on mechanical properties of fly ash-ground granulated blast furnace slag based self curing bio-geopolymer concrete. *Constr. Build. Mater.* **2017**, *149*, 338–349. [CrossRef]
71. Lee, N.K.; Lee, H.K. Setting and mechanical properties of alkali-activated fly ash/slag concrete manufactured at room temperature. *Constr. Build. Mater.* **2013**, *47*, 1201–1209. [CrossRef]
72. Mallikarjuna Rao, G.; Gunneswara Rao, T.D. A quantitative method of approach in designing the mix proportions of fly ash and GGBS-based geopolymer concrete. *Aust. J. Civ. Eng.* **2018**, *16*, 53–63. [CrossRef]
73. Rashad, A.M. Properties of alkali-activated fly ash concrete blended with slag. *Iran. J. Mater. Sci. Eng.* **2013**, *10*, 57–64.
74. Aydin, S.; Baradan, B. Effect of activator type and content on properties of alkali-activated slag mortars. *Compos. Part B Eng.* **2014**, *57*, 166–172. [CrossRef]
75. Imbabi, M.S.; Carrigan, C.; McKenna, S. Trends and developments in green cement and concrete technology. *Int. J. Sustain. Built Environ.* **2012**, *1*, 194–216. [CrossRef]

Article

Setting Behavior and Phase Evolution on Heat Treatment of Metakaolin-Based Geopolymers Containing Calcium Hydroxide

Byoungkwan Kim [1], Sujeong Lee [2,3,*], Chul-Min Chon [4] and Shinhu Cho [5]

1. Division of Advanced Nuclear Engineering, Pohang University of Science and Technology, Pohang 37673, Korea; kwan928@postech.ac.kr
2. Mineral Resources Research Division, Korea Institute of Geoscience and Mineral Resources, Daejeon 34132, Korea
3. Department of Resources Recycling Engineering, University of Science and Technology, Daejeon 34113, Korea
4. Geological Environment Division, Korea Institute of Geoscience and Mineral Resources, Daejeon 34132, Korea; femini@kigam.re.kr
5. Advanced Materials Research Team, Hyundai Motor Group, Uiwang 16082, Korea; s.cho@hyundai.com
* Correspondence: crystal2@kigam.re.kr; Tel.: +82-42-868-3125; Fax: +82-42-868-3418

Abstract: The setting behavior of geopolymers is affected by the type of source materials, alkali activators, mix formulations, and curing conditions. Calcium hydroxide is known to be an effective additive to shorten the setting period of geopolymers. However, there is still room for improvement in the understanding of the effect of calcium hydroxide on the setting and phase evolution of geopolymers. In this study, the setting behavior and phase evolution of geopolymer containing calcium hydroxide were investigated by XRD analysis. The setting time of the geopolymer was inconsistently shortened as the amount of calcium hydroxide increased. A low calcium hydroxide dose of up to 2% of the total mix weight could contribute to the enhancement of compressive strength of geopolymers besides a fast-setting effect. The C-S-H gel is rapidly precipitated at the early stage of reaction in geopolymers containing high calcium hydroxide with some of the calcium hydroxide remaining intact. The ex-situ high-temperature XRD analysis and Rietveld refinement results revealed that geopolymer and C-S-H gel transformed into Si-rich nepheline and wollastonite, respectively. The wollastonite was also observed in heat-treated geopolymers with a low calcium hydroxide dose. It is believed that C-S-H gel can be precipitated along with geopolymers regardless of how much calcium hydroxide is added.

Keywords: fast setting; calcium hydroxide; C-S-H; metakaolin; geopolymer

Citation: Kim, B.; Lee, S.; Chon, C.-M.; Cho, S. Setting Behavior and Phase Evolution on Heat Treatment of Metakaolin-Based Geopolymers Containing Calcium Hydroxide. *Materials* 2022, 15, 194. https://doi.org/10.3390/ma15010194

Academic Editors: F. Pacheco Torgal and Hubert Rahier

Received: 30 November 2021
Accepted: 22 December 2021
Published: 28 December 2021

Publisher's Note: MDPI stays neutral with regard to jurisdictional claims in published maps and institutional affiliations.

Copyright: © 2021 by the authors. Licensee MDPI, Basel, Switzerland. This article is an open access article distributed under the terms and conditions of the Creative Commons Attribution (CC BY) license (https://creativecommons.org/licenses/by/4.0/).

1. Introduction

Geopolymers are inorganic binder materials synthesized by combining amorphous aluminosilicate materials such as fly ash and metakaolin with alkaline solutions between ambient and low temperatures below 100 °C [1,2]. As compared to ordinary Portland cement (OPC) which is hardened via hydration reaction, the hardening of geopolymers is caused by polycondensations between aluminate and silicate species which are released from source materials and alkaline solution [1]. Randomly connected Si and Al tetrahedrons are linked by sharing oxygen anions form the network structure of geopolymers. The negative charge of structure created by Si substitution of Al is compensated by alkali metal ions such as Na and K [1]. Geopolymers exhibit outstanding performances which are hardly obtained in OPC such as high early-age strength, good chemical stability, thermal resistance, and immobilization of toxic elements [3–8].

Regardless of the source materials, it takes a longer period of time for geopolymers to set compared to OPC, which takes about 10 h for the final setting at ambient temperature [9]. The initial setting of fly ash-based geopolymer using low calcium fly ash (class F) takes more than a day at ambient temperature [10]. In the case of metakaolin-based geopolymer, the final setting time is about 20 h at room temperature, which is faster than fly ash-based geopolymer [11–13]. This is because the combination of several factors such as mix formulation, including a targeted Si/Al molar ratio in the mix, curing temperature, the reactivity of source material, and the pH of an alkaline activator affects the setting of geopolymers. If it is possible to reduce or control the setting time of geopolymers in the desired time, they could be used in a wider variety of fields where OPC and rapid-setting cement are used, such as in 3D printing and building maintenance. In order to control the setting of geopolymers, three setting control methods have been suggested: (1) the increasing of curing temperatures (heat curing), (2) the use of source materials with a high calcium content, and (3) the addition of calcium compounds [14–16]. Heat curing can significantly reduce setting time and accelerate geopolymerization [15], but it is difficult to apply to the production of site-poured concrete. The use of source material with high calcium content such as class C fly ash can lead to fast setting at room temperature. The initial and final setting times for fly ash-based geopolymers fabricated using class C fly ash and cured at room temperature are 30 min and 1 h, respectively [16]. However, it is difficult to quantitatively control the setting of geopolymers because setting time is only unaffected by the calcium content in the fresh mix. In addition, the bad workability of a mixture which is caused by fast setting needs to be improved for practical use [5].

The addition of calcium compounds such as calcium oxide, calcium hydroxide, and calcium rich by-product such as ground granulated blast furnace slag can also reduce the setting time of geopolymers [12,17–22]. The alkali activated materials using calcium rich by-products form two-dimensional structure [23,24]. It may affect the properties of geopolymer such as long-term properties and resistance to the chemical agents, but its effect on the geopolymer structure is unclear [23,24]. Considering the disadvantages of heat curing and the use of high calcium source materials, the addition of calcium compound is more beneficial since the appropriate amount of calcium compound is effective for fast setting as well as the enhancement of compressive strength [17,19–21]. The most effective calcium source is calcium hydroxide, though more understanding of the setting behavior and phase evolution of geopolymers containing calcium still needs to be achieved to fully understand its mechanisms [12,17–22].

The objective of this study is an investigation of the setting behavior and phase evolution of geopolymers containing calcium hydroxide. The appropriate amount of calcium hydroxide was added to the geopolymer which is fabricated using pure aluminosilicate material such as metakaolin. The stoichiometry of metakaolin-based geopolymers was designed with $NaSi_2AlO_6 \cdot 5H_2O$ and calcium hydroxide was added as 2%, 4%, 8%, and 16% by weight of the geopolymer mixture. The variations of setting behavior, compressive strength, and the phase evolution of geopolymers depending on the calcium hydroxide content were investigated. In addition, the Rietveld refinement based on X-ray diffraction was performed to indirectly understand the role of calcium in geopolymer structures and the high-temperature phase of Si-rich geopolymer after ex-situ heat treatment. The low-carnegieite and Si-rich nepheline were newly refined to understand the phase transition of geopolymer containing high Si content. The scientific findings of this study will contribute to the understanding of the setting behavior and phase evolution of geopolymers containing calcium hydroxide.

2. Materials and Methods

2.1. Characterization of Metakaolin

Geopolymers were synthesized from commercial metakaolin (MetaMax®, BASF, Ludwigshafen, Germany). The chemical composition of metakaolin was analyzed using X-ray fluorescence spectroscopy (Shimadzu Sequential XRF-1800, Shimadzu, Japan). Quantitative X-ray diffraction phase analysis was conducted employing the DIFFRAC.EVA V4.2 (Bruker-AXS, Karlsruhe, Germany), PDF-2 release 2016 (ICDD, Newtown Square, PA, USA), and TOPAS 5 software (Bruker-AXS). Calcium fluoride (CaF_2, 99.985%, Alfa aesar,

Ward Hill, MA, USA) was used as an internal standard to comprise 10.0000% of the sample weight. The mixture of metakaolin and calcium fluoride was ground in a micronizer mill (McCrone, Westmont, IL, USA) for 5 min. X-ray diffraction patterns were obtained using a D8 Advance diffractometer (Bruker-AXS, Karlsruhe, Germany) over a 2θ range from 5 to 65 with a step size 0.01° or 1 s/step. To evaluate the instrument-derived errors, an X-ray diffraction pattern of lanthanum hexaboride (LaB_6, SRM 660b, NIST, Gaithersburg, MD, USA) was also obtained employing the same diffraction conditions.

2.2. Synthesis of Metakaolin-Based Geopolymer

Metakaolin-based geopolymers were synthesized by mixing the metakaolin and alkaline activators to achieve the molar ratio of $Na_2O:Al_2O_3:SiO_2:H_2O$ = 1:1:4:10 in geopolymers (Table 1). Fumed silica (EH-5, Cabot corporation, Boston, MA, USA assay 99.9%), sodium hydroxide (NaOH, Daejung chemical, Siheung-si, Korea, assay 97.0%), and distilled water were mixed at 25 °C for ≥20 h before use and then added to metakaolin. Calcium hydroxide ($Ca(OH)_2$, Kyoto, Japan, Yakuri chemical, assay 96.0%) was added from 2% to 16% of the total mix weight just before the end of the geopolymer mixing process.

Table 1. Mix formulation of metakaolin-based geopolymers containing calcium hydroxide. Calcium hydroxide was added from 0% to 16% of the total mix weight.

Sample Name	Chemical Composition	Dosages of Calcium Hydroxide
Ca0		0%
Ca2		2%
Ca4	$Na_2O:Al_2O_3:SiO_2:H_2O$ = 1:1:4:10	4%
Ca8		8%
Ca16		16%

2.3. Measurement of Setting Time

Geopolymers for measuring the final setting time were synthesized by using a planetary mixer (ARE-310, Thinky, Tokyo, Japan). Metakaolin and alkaline activator were mixed for 4 min at 1400 RPM, and then defoamed for 3 min at 1700 RPM. The geopolymer mixture was poured into a 200 mL PP beaker to fill a height of 30 mm of the beaker subsequent to mixing. Setting time was measured in accordance with ASTM C191 (Standard Test Methods for Time of Setting of Hydraulic Cement by Vicat Needle) utilizing a Vicat needle apparatus.

2.4. Measurement of Compressive Strength

Geopolymers for measuring the 7-d compressive strength were prepared using a Kenwood mixer (Kenwood kitchen appliances, Hertfordshire, UK). Metakaolin and alkaline activator were mixed for 2 min at low speed, and then mixed for 5 min at high speed. The geopolymer mixtures were filled into cube molds (5 cm × 5 cm × 5 cm) according to ASTM C109 (Standard Test Method for Compressive Strength of Hydraulic Cement Mortars). One-third of the mold volume was filled and then tapped to remove air bubbles trapped during the mixing process. The molded geopolymers were cured at room temperature for 7 days after sealing to prevent the evaporation of moisture. The 7-d compressive strength of geopolymers was measured with MTS 815 rock mechanics test machine (MTS system corporation, Eden Prairie, MN, USA) at a loading rate of 5.5×10^{-3} mm/s.

2.5. X-ray Diffraction Analysis of Geopolymers

X-ray diffraction analysis for hardened geopolymers and heat-treated geopolymers was performed using a D8 Advance diffractometer. Diffraction patterns of hardened geopolymers were obtained over a range of 8–80° 2θ with a step size of 0.02°, and a scan speed of 0.2 s/step. For the ex-situ high-temperature X-ray diffraction analysis, geopolymer fragments were obtained from a fresh fracture of hardened geopolymer. The samples were crushed in agate mortar with the geopolymer powder then sieved to a size of

−150 + 106 µm. After placing about 5 g of sieved geopolymer powder in a nickel crucible, it was heated at 900 °C for 30 min at a heating rate of 5 °C/min using an electric furnace. The qualitative analysis of obtained X-ray diffraction patterns was conducted utilizing DIFFRAC.EVA V4.2 and PDF-2 Release 2016. Structure refinement employing the Rietveld of newly formed phases was performed using TOPAS 5 software.

3. Results and Discussion

3.1. Characteristics of Metakaolin

The main components of metakaolin were SiO_2 and Al_2O_3, although it contained roughly 1.5 wt% of TiO_2 as a major impurity which is identified as anatase in XRD analysis (Table 2 and Figure 1). Metakaolin was mainly composed of amorphous materials while containing crystalline impurities such as illite, quartz, and anatase (Figure 1). The center of amorphous diffraction hump of metakaolin was located at roughly 2θ 23° (Figure 1). The mix formulation of geopolymers was obtained by considering both bulk Si and Al contents in Table 2 as all reactive because the trace amount of crystalline SiO_2 in illite and quartz could be ignored.

Table 2. The bulk chemical composition of metakaolin in this study. XRD analysis proved that the TiO_2 content originated from anatase.

Oxide	SiO_2	Al_2O_3	Fe_2O_3	CaO	MgO	K_2O	Na_2O	TiO_2	MnO	P_2O_5	Others
wt%	52.64	43.84	0.26	0.03	0.09	0.16	0.35	1.49	0.01	0.08	0.93

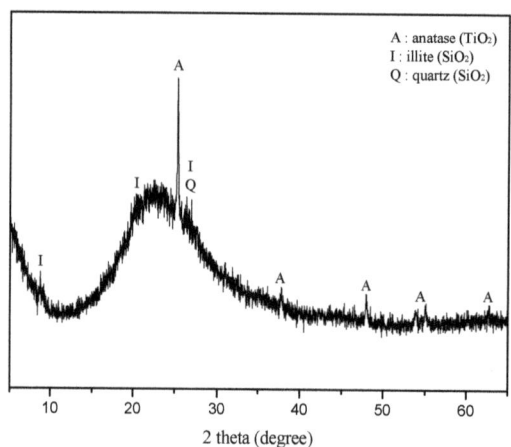

Figure 1. X-ray diffraction pattern of metakaolin in this study. Metakaolin was composed of amorphous and several crystalline phases such as anatase, illite, and quartz as impurities. The center of the amorphous hump was located at 2θ 22.9°.

3.2. Setting Behavior and Compressive Strength of Metakaolin-Based Geopolymers Containing Calcium Hydroxide

The final setting time of Ca0 was not able to be continuously measured due to the essential limitations of the Vicat needle test, which is performed manually. Although the exact final setting time of pure geopolymer cannot be determined here, it can be concluded that the hardening of the geopolymer having the targeted chemical composition was accomplished within 19 h post mixing as proved in previous studies (Figure 2). The setting time substantially decreased by adding calcium hydroxide. There was, however, no significant difference in the initial setting time between 2.3 h for Ca0 and 2.1 h for Ca2. Then, Ca2 showed a faster setting duration than Ca0 after 2.6 h (Figure 2). When the content of calcium was doubled in Ca4, the initial setting time was dramatically reduced to 1.3 h (Figure 2). It took just 1.7 h to reach the final setting in Ca4 (Figure 2). Geopolymers containing high amounts of calcium hydroxide (Ca8 and Ca16) were too rapidly hardened even during mixing to measure the setting time.

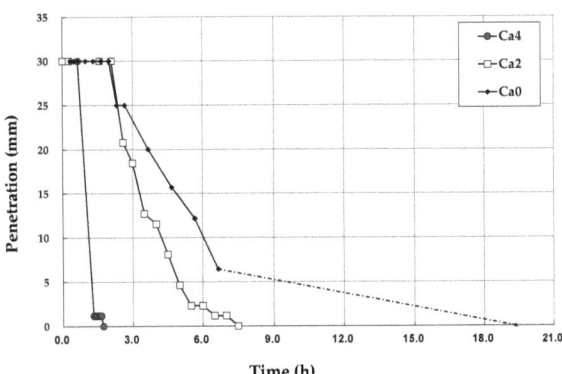

Figure 2. Setting behavior of metakaolin-based geopolymers. The addition of 4% calcium hydroxide promoted a fast set by interlocking coagulate products from the outset.

The addition of calcium hydroxide contributed to the superior mechanical property of calcium-containing geopolymers. The compressive strength of Ca2 was 76 MPa, which was higher than 68 MPa of the Ca0 (Table 3). In other words, the addition of 2% calcium hydroxide appears less effective in decreasing the initial setting time while beneficial for a fast final set. By contrast, 4% calcium hydroxide inclusion promoted a fast set by interlocking coagulate products from the outset. The polycondensation of geopolymer gel occurs slowly [11–13] and the setting time of pure geopolymers is mainly controlled by the alumina content, because aluminate is more readily soluble in metakaolin [25]. Condensation between silicate entities occurs slower than aluminate-silicate condensation so the setting time increases with the increasing SiO_2/Al_2O_3 ratio of the initial mixture [25]. According to the dynamic rheological measurement study during geopolymerization [26], the size of alkali cations affects the dissolution and condensation rates of aluminate and silicate entities. The role of calcium in fast set geopolymers should be deliberately investigated due to the fact that sodium is fully added to compensate the negative charge in the formulation (Table 1) and coexists with extra cation and calcium in the mixture.

Table 3. The 7-d compressive strength of metakaolin-based geopolymer. The addition of 2% calcium hydroxide improved the mechanical property of Ca2.

Sample Name	7-d Compressive Strength (MPa)
Ca0	68 (±4.58)
Ca2	76 (±7.21)

The Ca4 produced by the Kenwood mixer hardened too quickly to be molded for strength measurement. It set faster than the one synthesized using the plenary mixer which is known to be much more effective in geopolymer mixing. A high-shear mixer such as the plenary mixer contributes to obtain high strength geopolymers to ensure full reactivity [27]. Aluminate and silicate species probably dissolved more slowly in the mixture than one produced by the plenary mixer for the Vicat test. As a result, polycondensation of entities may be retarded but nevertheless contain high calcium content which possibly played a crucial role in decreasing the setting time in Ca4. The calcium hydroxide promotes the dissolution of metakaolin in geopolymer activated by alkali silicate solution [12]. The 5- and

6-coordinated Al present in the metakaolin convert to 4-coordinated Al during the geopolymerization [28]. In geopolymers containing calcium hydroxide, 5- and 6-coordinated Al is rapidly converted into 4-coordinated Al in the beginning of the geopolymerization. The percentage of 4-coordinated Al is higher than that of the geopolymer without calcium by at least 10% [11]. Thus, calcium hydroxide contributes to the fast dissolution of metakaolin and the promotion of geopolymer gel formation resulting in the fast setting of the geopolymer [11]. However, what is finally formed apart from geopolymers in extra alkali cations-containing geopolymers should be thoroughly investigated.

The number of network modifiers in the geopolymer structure may increase due to the addition of calcium hydroxide if calcium plays a role of network modifier. It may result in an increase in the number of non-bridging oxygens in the geopolymer network locally and may lead to depolymerization of geopolymers. Calcium can play different structural roles such as network modifier or charge compensator in a $CaO-Na_2O-Al_2O_3-SiO_2$ glass system, but it has an affinity for network modifiers [29,30]. It is known that network modifiers reduce the number of strong bonds in the glass and lower the melting point. The deteriorated consistency of calcium-containing Ca2 and Ca4 reflects that calcium did not contribute to the depolymerization. On the contrary, C-S-H gel can be precipitated easily if there is only a small amount of calcium in the silicate solution because the solubility product of C-S-H gel which can be formed by the interaction of calcium and silicate has a low value of 5.5×10^{-49} at room temperature [31,32]. Therefore, the increase in compressive strength can be attributed to the fast dissolution of metakaolin, the rapid formation of the geopolymer, and the precipitation of C-S-H gel concurrently. Although the microstructure of C-S-H gel was not confirmed in this study, its characteristic needle-like or reticular structure of C-S-H phases possibly played a part in the improvement of compressive strength of Ca2 compared with Ca0.

3.3. X-ray Diffraction Analysis of Metakaolin-Based Geopolymer Containing Calcium Hydroxide

The XRD patterns of Ca0, Ca2, and Ca4 showed the broadening amorphous hump which indicates the geopolymer structure (Figure 3) [33]. On the other hand, a new cuspidal diffraction peak was recorded at about 2θ 29° in Ca8 and Ca16 which is higher 2θ compared with the amorphous hump of Ca0 (Figure 3). This cuspidal reflection at about 2θ 29° represents the C-S-H gel [34]. The crystalline calcium hydroxide, portlandite, was recognized as well in Ca16. (Figure 3). It is due to the common ion effect of calcium hydroxide and high pH environment in the alkaline activator. These conditions lead to a decrease in solubility of calcium hydroxide and result in the reduction of C-S-H gel precipitation [35–37]. In addition, the solubility of calcium hydroxide decreases with increasing temperature [38]. Geopolymerization is an exothermic reaction and releases the highest heat at the early stage of the reaction in which metakaolin is dissolved in the alkaline activator [39]. The exothermic heat released during dissolution of metakaolin can increase as calcium hydroxide content increases, as calcium hydroxide contributes to the fast dissolution of metakaolin [12,37]. As a result, the solubility of calcium hydroxide probably decreased in the geopolymer containing higher calcium hydroxide, with Ca16 here remaining intact.

The XRD analysis results of Ca16 showed C-S-H gel is rapidly precipitated at the early stage of reaction and some of the calcium hydroxide remains inert in hardened geopolymers (Figure 4). The Ca16 at the first day of curing was composed of geopolymer presenting an amorphous hump at 2θ 23°, with C-S-H gel presenting another hump at 2θ 23°, unreactive metakaolin, and portlandite (Figure 4). At three and seven days of curing, the shape of the amorphous humps of geopolymer and C-S-H gel became more symmetrical, and the intensity of portlandite peaks decreased (Figure 4). This showed that the dissolution of metakaolin and calcium hydroxide continued with the increase in the curing period (Figure 4). The continued dissolution of calcium hydroxide may supply calcium to the C-S-H gel with a low Ca/Si ratio formed at 1 day of curing, so that the C-S-H gel at a longer curing period probably has a higher Ca/Si ratio and the atomic ordering will be increased. Meanwhile, the XRD patterns of three and seven days of curing showed little difference and the intensities of undissolved calcium hydroxide were also similar (Figure 4). Therefore, it can be concluded that C-S-H gel is rapidly precipitated at the early stage of reaction and some of the calcium hydroxide remains inert due to the common ion effect, high pH environment, and exothermic heat of geopolymerization when added as an excessive amount to the geopolymer mixture.

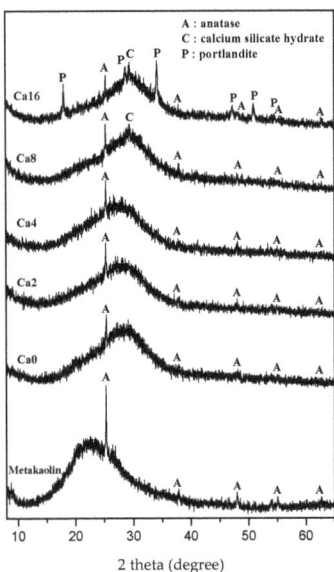

Figure 3. XRD patterns of pure geopolymer and geopolymers containing calcium hydroxide. The newly formed C-S-H gel and unreacted calcium hydroxide were observed in Ca8 and Ca16.

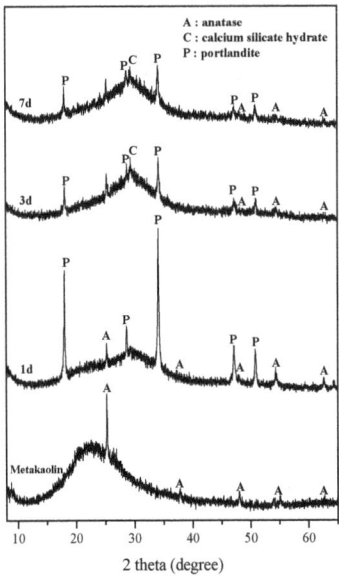

Figure 4. XRD patterns of Ca16 cured for one, three, and seven days. C-S-H gel is rapidly precipitated at the early stage of reaction and some of the calcium hydroxide remains intact.

3.4. Ex-Situ High-Temperature X-ray Diffraction Analysis of Metakaolin-Based Geopolymers Containing Calcium Hydroxide

The crystalline phases of heat-treated metakaolin-based geopolymer varies with its chemical formulation and the cation of alkaline activator [40]. The geopolymers using potassium hydroxide as an alkaline activator transform into kalsilite ($KAlSiO_4$) and leucite ($KAlSi_2O_6$) depending on the Si/Al ratio at temperatures above 1000 °C [40]. In the case of sodium hydroxide being used as an alkaline activator, geopolymers transform into nepheline ($NaAlSiO_4$), jadeite ($NaAlSi_2O_6$), and albite ($NaAlSi_3O_8$) [41–43] depending on the Si content at temperatures above 800 °C. The high-temperature crystalline phase can be assumed to be jadeite because the Ca0 is Na-based and its Si/Al molar ratio was two. Jadeite is involved in a high-pressure phase and exhibits strong diffraction peaks at 2θ 30° and 31°. This is inconsistent with the strong diffraction peak at 2θ 23° of the newly formed phase. Rietveld refinement of XRD pattern presented that Si-rich nepheline and low-carnegieite were formed in heat-treated Ca0 geopolymer at 900 °C (Figure 5).

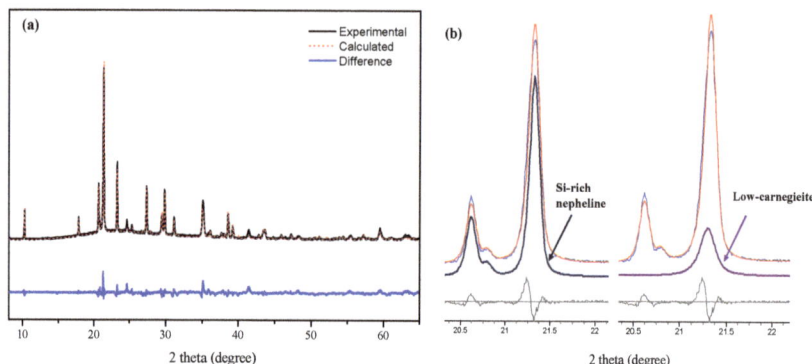

Figure 5. Rietveld refinement result of pure geopolymer heat treated at 900 °C (**a**); the crystal information of Si-rich nepheline and low-carnegieite was refined. The overlapped diffraction pattern of Si-rich nepheline and low-carnegieite was confirmed at 2θ 21.2° (**b**).

As mentioned earlier, the Si/Al molar ratio of nepheline is 1.0, yet several studies reported that nepheline was formed as a high-temperature crystalline phase even in geopolymers with a Si/Al molar ratio of 1.0 or higher [44,45]. Metakaolin-based geopolymers demonstrating a Si/Al molar ratio between 1.15 and 2.15 were crystallized to nepheline (PDF #00-035-0424) after heating at 900 °C, and the diffraction intensities of nepheline decreased as the Si/Al molar ratio of the geopolymer increased [44]. Another study reported on the formation of nepheline when the metakaolin-based geopolymers with a Si/Al molar ratio of 1.8 were heated at 1000 °C [44]. Nevertheless, the Si/Al molar ratio of geopolymers in these studies (Si/Al molar ratio = 1.15≥) [44,45] was higher than that of nepheline (Si/Al molar ratio = 1). The intensity ratio of the obtained peaks of this phase were also inconsistent with that of nepheline reported in [44,45]. These studies showed that nepheline could be formed via heat-treatment of geopolymers with a Si/Al ratio of more than 1. In conclusion, the high-temperature crystalline phase of geopolymers in this study is estimated as a Si-rich nepheline from the previous research results as well as the refinement analysis in this study.

The Ca0 showed the strongest XRD pattern at 2θ 21° while geopolymers containing calcium hydroxide represented it at 2θ 23° (Figure 6a). These differences may be due to the overlap of Si-rich nepheline and low-carnegieite diffraction patterns. Nepheline and low-carnegieite can be formed concurrently when the metakaolin-based geopolymer is heated at 900 °C, and the strongest diffraction peaks are overlapped at 2θ 21° [42]. Nepheline and carnegieite are polymorphisms with the same chemical composition, but with different crystal structures. The Si/Al molar ratio of geopolymers that were synthesized in this study was higher than low-carnegieite. In addition, it is known that low-carnegieite containing high Si content does not exist. According to a previous study, when metakaolin and Na_2CO_3 are milled and then heated at 600–800 °C, carnegieite can be formed [46]. Hence, the formation of low-carnegieite is presumably caused by the reaction between Na which is released from the alkaline activator, and metakaolin which is not involved in (or unreacted) geopolymerization at high-temperature. As a result of Rietveld refinement of Si-rich nepheline and low-carnegieite, it was confirmed that two crystalline diffraction patterns were overlapped at 2θ 21° (Figure 5b).

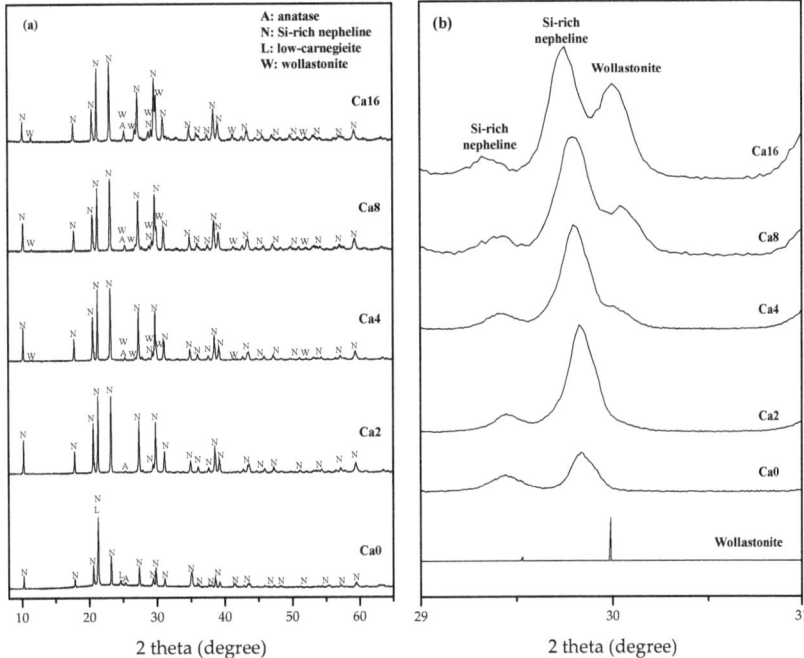

Figure 6. XRD analysis results of pure geopolymer and geopolymers containing calcium hydroxides post heat treatment (a) and 2θ range from 29° to 31° (b); Si-rich nepheline and low-carnegieite were observed in Ca0 geopolymers as high temperature crystalline phases. The low reflection of wollastonite was confirmed in Ca4.

The diffraction pattern of wollastonite ($CaSiO_3$, PDF #00-027-0088) was consistent with a newly formed phase in Ca4, Ca8, and Ca16 geopolymers (Figure 6). It is known as the high-temperature crystalline phase that is formed when C-S-H(I) gel is heated at 800 °C and its main diffraction pattern of wollastonite is observed at 2θ 30° [47]. The low reflection of wollastonite was confirmed in Ca4 geopolymer which did not show the C-S-H reflection at room temperature XRD analysis (Figure 6b). Considering these aspects, C-S-H gel was formed in all geopolymers containing calcium hydroxide, though the amount of precipitated C-S-H gel was small and the asymmetric amorphous hump of geopolymer concealed its reflection.

4. Conclusions

The effect of calcium hydroxide on the setting behavior and compressive strength of metakaolin-based geopolymers was investigated. The ratio of metakaolin, fumed silica, NaOH, and distilled water in the geopolymers was kept in the formulations ($Na_2O:Al_2O_3:SiO_2:H_2O$ = 1:1:4:10) in order to confirm the effect of calcium. The setting of geopolymers was accelerated in proportion to the dosages of calcium hydroxide in up to 4% of the total mix weight. The addition of 2% calcium hydroxide to the mixture was beneficial for the improvement of compressive strength of the geopolymers. The fast setting and enhancement of compressive strength is presumably caused by the fast dissolution of metakaolin and precipitation of C-S-H gel. In the XRD analysis, the cuspidal diffraction peak of C-S-H gel was observed in geopolymers containing 8% and 16% calcium hydroxide. The C-S-H gel is rapidly precipitated at the early stage of reaction and some of the calcium hydroxide remains intact due to the reduction of solubility and common ion effect. The ex-situ high-temperature XRD analysis and Rietveld refinement results showed that geopolymer with a Si/Al ratio of 2.0 and C-S-H gel transformed into Si-rich nepheline and wollastonite, respectively.

Author Contributions: B.K. contributed to data analysis, visualization and original draft preparation; S.L. contributed to conception, design, data analysis, writing—review and editing and funding acquisition; C.-M.C. contributed to data analysis and original draft preparation; S.C. contributed to conception and funding acquisition. All authors have read and agreed to the published version of the manuscript.

Funding: This work is supported by a research grant from the Hyundai Motor Group; project title: Development of geopolymer composites for 3D printing.

Institutional Review Board Statement: Not applicable.

Informed Consent Statement: Not applicable.

Data Availability Statement: Not applicable.

Conflicts of Interest: The authors declare no conflict of interest.

References

1. Davidovits, J. Geopolymers: Inorganic polymeric new materials. *J. Therm. Anal. Calorim.* **1991**, *37*, 1633–1656. [CrossRef]
2. Hardjito, D.; Rangan, B.V. Development and properties of low-calcium fly ash-based geopolymer concrete. In *Research Report GC 1*; Faculty of Engineering, Curtin University of Technology: Perth, Australia, 2005. Available online: http://hdl.handle.net/20.500.11937/5594 (accessed on 28 November 2021).
3. Rostami, H.; Brendley, W. Alkali ash material: A novel fly ash-based cement. *Environ. Sci. Technol.* **2003**, *37*, 3454–3457. [CrossRef] [PubMed]
4. Bakharev, T. Durability of geopolymer materials in sodium and magnesium sulfate solutions. *Cem. Concr. Res.* **2005**, *35*, 1233–1246. [CrossRef]
5. Provis, J.L.; van Deventer, J.S.J. *Geopolymers: Structure, Processing, Properties and Industrial Applications*; Woodhead: Oxford, UK, 2009.
6. Palomo, A.; Fernández-Jiménez, A. Alkaline activation, procedure for transforming fly ash into new materials. Part I: Applications. In Proceedings of the World of Coal Ash (WOCA) Conference, Denver, CO, USA, 9–12 May 2011; pp. 1–14.
7. Vickers, L.; Van Riessen, A.; Rickard, W.D. *Fire-Resistant Geopolymers: Role of Fibres and Fillers to Enhance Thermal Properties*; Springer: Singapore, 2015.
8. Lee, S.; van Riessen, A.; Chon, C.M.; Kang, N.H.; Jou, H.T.; Kim, Y.J. Impact of activator type on the immobilisation of lead in fly ash-based geopolymer. *J. Hazard. Mater.* **2016**, *305*, 59–66. [CrossRef]
9. Aggoun, S.; Cheikh-Zouaoui, M.; Chikh, N.; Duval, R. Effect of some admixtures on the setting time and strength evolution of cement pastes at early ages. *Constr. Build. Mater.* **2008**, *22*, 106–110. [CrossRef]
10. Nath, P.; Sarker, P.K.; Rangan, V.B. Early Age Properties of Low-calcium Fly Ash Geopolymer Concrete Suitable for Ambient Curing. *Procedia Eng.* **2015**, *125*, 601–607. [CrossRef]
11. Chen, X.; Sutrisno, A.; Zhu, L.Y.; Struble, L.J. Setting and nanostructural evolution of metakaolin geopolymer. *J. Am. Ceram. Soc.* **2017**, *100*, 2285–2295. [CrossRef]
12. Chen, X.; Sutrisno, A.; Struble, L.J. Effects of calcium on setting mechanism of metakaolin-based geopolymer. *J. Am. Ceram. Soc.* **2018**, *101*, 957–968. [CrossRef]
13. Diffo, B.K.; Elimbi, A.; Cyr, M.; Manga, J.D.; Kouamo, H.T. Effect of the rate of calcination of kaolin on the properties of metakaolin-based geopolymers. *J. Asian Ceram. Soc.* **2015**, *3*, 130–138. [CrossRef]
14. Hardjito, D.; Cheak, C.C.; Ing, C.H.L. Strength and setting times of low calcium fly ash-based geopolymer mortar. *Mod. Appl. Sci.* **2008**, *2*, 3–11. [CrossRef]
15. Mo, B.H.; Zhu, H.; Cui, X.M.; He, Y.; Gong, S.Y. Effect of curing temperature on geopolymerization of metakaolin-based geopolymers. *Appl. Clay Sci.* **2014**, *99*, 144–148. [CrossRef]

16. Mohamed, R.; Abd Razak, R.; Abdullah, M.M.A.B.; Shuib, R.K.; Mortar, N.A.M.; Zailani, W.W.A. Investigation of heat released during geopolymerization with fly ash based geopolymer. *IOP Conf. Ser. Mater. Sci. Eng.* **2019**, *551*, 012093. [CrossRef]
17. Temuujin, J.; Van Riessen, A.; Williams, R. Influence of calcium compounds on the mechanical properties of fly ash geopolymer pastes. *J. Hazard. Mater.* **2009**, *167*, 82–88. [CrossRef] [PubMed]
18. Granizo, M.L.; Alonso, S.; Blanco-Varela, M.T.; Palomo, A. Alkaline activation of metakaolin: Effect of calcium hydroxide in the products of reaction. *J. Am. Ceram. Soc.* **2002**, *85*, 225–231. [CrossRef]
19. Kim, E. Understanding Effects of Silicon/Aluminum Ratio and Calcium Hydroxide on Chemical Composition, Nanostructure and Compressive Strength for Metakaolin Geopolymers. Master's Thesis, University of Illinois at Urbana-Champaign, Champaign, IL, USA, 2012.
20. Buchwald, A.; Hilbig, H.; Kaps, C. Alkali-activated metakaolin-slag blends—performance and structure in dependence of their composition. *J. Mater. Sci.* **2007**, *42*, 3024–3032. [CrossRef]
21. Sankar, K.; Stynoski, P.; Al-Chaar, G.K.; Kriven, W.M. Sodium silicate activated slag-fly ash binders: Part I–Processing, microstructure, and mechanical properties. *J. Am. Ceram. Soc.* **2018**, *101*, 2228–2244. [CrossRef]
22. Sankar, K.; Sutrisno, A.; Kriven, W.M. Slag-fly ash and slag-metakaolin binders: Part II—Properties of precursors and NMR study of poorly ordered phases. *J. Am. Ceram. Soc.* **2019**, *102*, 3204–3227. [CrossRef]
23. Davidovits, J. *Geopolymer Chemistry and Applications*, 4th ed.; Geopolymer Institute: Saint-Quentin, France, 2015.
24. Korniejenko, K.; Figiela, B.; Miernik, K.; Ziejewska, C.; Marczyk, J.; Hebda, M.; Cheng, A.; Lin, W.T. Mechanical and Fracture Properties of Long Fiber Reinforced Geopolymer Composites. *Materials* **2021**, *14*, 5183. [CrossRef] [PubMed]
25. De Silva, P.; Sagoe-Crenstil, K.; Sirivivatnanon, V. Kinetics of geopolymerization: Role of Al_2O_3 and SiO_2. *Cem. Concr. Res.* **2007**, *37*, 512–518. [CrossRef]
26. Steins, P.; Poulesquen, A.; Diat, O.; Frizon, F. Structural evolution during geopolymerization from an early age to consolidated material. *Langmuir* **2012**, *28*, 8502–8510. [CrossRef] [PubMed]
27. Kriven, W.M.; Bell, J.L.; Gordon, M. Microstructure and microchemistry of fully-reacted geopolymers and geopolymer matrix composites. *Ceram. Trans.* **2003**, *153*, 227–250.
28. Duxson, P.; Lukey, G.C.; Separovic, F.; van Deventer, J.S.J. Effect of Alkali Cations on Aluminum Incorporation in Geopolymeric Gels. *Ind. Eng. Chem. Res.* **2005**, *44*, 832–839. [CrossRef]
29. Cormier, L.; Neuville, D.R. Ca and Na environments in Na_2O–CaO–Al_2O_3–SiO_2 glasses: Influence of cation mixing and cation-network interactions. *Chem. Geol.* **2004**, *213*, 103–113. [CrossRef]
30. Lee, S.K.; Sung, S. The effect of network-modifying cations on the structure and disorder in peralkaline Ca–Na aluminosilicate glasses: O-17 3QMAS NMR study. *Chem. Geol.* **2008**, *256*, 326–333. [CrossRef]
31. Suzuki, K.; Nishikawa, T.; Ito, S. Formation and carbonation of C-S-H in water. *Cem. Concr. Res.* **1985**, *15*, 213–224. [CrossRef]
32. Gartner, E.M.; Jennings, H.M. Thermodynamics of calcium silicate hydrates and their solutions. *J. Am. Ceram. Soc.* **1987**, *70*, 743–749. [CrossRef]
33. Williams, R.P.; Hart, R.D.; Van Riessen, A. Quantification of the extent of reaction of metakaolin-based geopolymers using X-ray diffraction, scanning electron microscopy, and energy-dispersive spectroscopy. *J. Am. Ceram. Soc.* **2011**, *94*, 2663–2670. [CrossRef]
34. Si, R.; Guo, S.; Dai, Q. Influence of calcium content on the atomic structure and phase formation of alkali-activated cement binder. *J. Am. Ceram. Soc.* **2019**, *102*, 1479–1494. [CrossRef]
35. Skoog, D.A.; West, D.M.; Holler, F.J.; Crouch, S.R. *Fundamentals of Analytical Chemistry*; Cengage Learning: Belmont, CA, USA, 2014.
36. De Silva, P.; Glasser, F. Pozzolanic activation of metakaolin. *Adv. Cem. Res.* **1992**, *4*, 67–178. [CrossRef]
37. Alventosa, K.M.; White, C.E. The effects of calcium hydroxide and activator chemistry on alkali-activated metakaolin pastes. *Cem. Concr. Res.* **2021**, *145*, 106453. [CrossRef]
38. Johnston, J.; Grove, C. The solubility of calcium hydroxide in aqueous salt solutions. *J. Am. Chem. Soc.* **1931**, *53*, 3976–3991. [CrossRef]
39. Zhang, Z.; Wang, H.; Provis, J.L.; Bullen, F.; Reid, A.; Zhu, Y. Quantitative kinetic and structural analysis of geopolymers. Part 1. The activation of metakaolin with sodium hydroxide. *Thermochim. Acta* **2012**, *539*, 23–33. [CrossRef]
40. Bell, J.L.; Driemeyer, P.E.; Kriven, W.M. Formation of Ceramics from Metakaolin-Based Geopolymers. Part II: K-Based Geopolymer. *J. Am. Ceram. Soc.* **2009**, *92*, 607–615. [CrossRef]
41. Buchwald, A.; Vicent, M.; Kriegel, R.; Kaps, C.; Monzó, M.; Barba, A. Geopolymeric binders with different fine fillers—Phase transformations at high temperatures. *Appl. Clay Sci.* **2009**, *46*, 190–195. [CrossRef]
42. Kuenzel, C.; Grover, L.M.; Vandeperre, L.; Boccaccini, A.R.; Cheeseman, C.R. Production of nepheline/quartz ceramics from geopolymer mortars. *J. Eur. Ceram. Soc.* **2013**, *33*, 251–258. [CrossRef]
43. Rovnanik, P.; Šafránková, K. Thermal Behaviour of Metakaolin/Fly Ash Geopolymers with Chamotte Aggregate. *Materials* **2016**, *9*, 535. [CrossRef] [PubMed]
44. Duxson, P.; Lukey, G.C.; van Deventer, J.S.J. The thermal evolution of metakaolin geopolymers: Part 2—Phase stability and structural development. *J. Non-Cryst. Solids* **2007**, *353*, 2186–2200. [CrossRef]
45. Lambertin, D.; Boher, C.; Dannoux-Papin, A.; Galliez, K.; Rooses, A.; Frizon, F. Influence of gamma ray irradiation on metakaolin based sodium geopolymer. *J. Nucl. Mater.* **2013**, *443*, 311–315. [CrossRef]

46. Kubo, Y.; Yamaguchi, G.; Kasahara, K. Kinetic and electron optical studies of the reaction processes between kaolinite and sodium carbonate. *Am. Mineral.* **1968**, *53*, 917–928.
47. Rodriguez, E.T.; Garbev, K.; Merz, D.; Black, L.; Richardson, I.G. Thermal stability of CSH phases and applicability of Richardson and Groves' and Richardson C-(A)-SH (I) models to synthetic CSH. *Cem. Concr. Res.* **2017**, *93*, 45–56. [CrossRef]

Article

Mechanical and Microstructural Characterization of Quarry Rock Dust Incorporated Steel Fiber Reinforced Geopolymer Concrete and Residual Properties after Exposure to Elevated Temperatures

Muhammad Ibraheem [1], Faheem Butt [1,*], Rana Muhammad Waqas [1], Khadim Hussain [1], Rana Faisal Tufail [2], Naveed Ahmad [1], Ksenia Usanova [3] and Muhammad Ali Musarat [4,*]

[1] Department of Civil Engineering, University of Engineering and Technology, Taxila 47050, Pakistan; Ibrahim.123456@yahoo.com (M.I.); rana.waqas@uettaxila.edu.pk (R.M.W.); hussainkhadim173@gmail.com (K.H.); naveed.ahmad@uettaxila.edu.pk (N.A.)
[2] Department of Civil Engineering, Wah Campus, COMSATS University Islamabad, Wah Cantt 47040, Pakistan; faisal.tufail@ciitwah.edu.pk
[3] Department of Construction of Unique Buildings and Constructions, Peter the Great St. Petersburg Polytechnic University, 195291 St. Petersburg, Russia; plml@mail.ru
[4] Department of Civil and Environmental Engineering, Universiti Teknologi PETRONAS, Bandar Seri Iskandar 32610, Malaysia
* Correspondence: faheem.butt@uettaxila.edu.pk (F.B.); muhammad_19000316@utp.edu.my (M.A.M.)

Citation: Ibraheem, M.; Butt, F.; Waqas, R.M.; Hussain, K.; Tufail, R.F.; Ahmad, N.; Usanova, K.; Musarat, M.A. Mechanical and Microstructural Characterization of Quarry Rock Dust Incorporated Steel Fiber Reinforced Geopolymer Concrete and Residual Properties after Exposure to Elevated Temperatures. *Materials* 2021, 14, 6890. https://doi.org/10.3390/ma14226890

Academic Editor: Claudio Ferone

Received: 3 October 2021
Accepted: 4 November 2021
Published: 15 November 2021

Publisher's Note: MDPI stays neutral with regard to jurisdictional claims in published maps and institutional affiliations.

Copyright: © 2021 by the authors. Licensee MDPI, Basel, Switzerland. This article is an open access article distributed under the terms and conditions of the Creative Commons Attribution (CC BY) license (https://creativecommons.org/licenses/by/4.0/).

Abstract: The purpose of this research is to study the effects of quarry rock dust (QRD) and steel fibers (SF) inclusion on the fresh, mechanical, and microstructural properties of fly ash (FA) and ground granulated blast furnace slag (SG)-based geopolymer concrete (GPC) exposed to elevated temperatures. Such types of ternary mixes were prepared by blending waste materials from different industries, including QRD, SG, and FA, with alkaline activator solutions. The multiphysical models show that the inclusion of steel fibers and binders can enhance the mechanical properties of GPC. In this study, a total of 18 different mix proportions were designed with different proportions of QRD (0%, 5%, 10%, 15%, and 20%) and steel fibers (0.75% and 1.5%). The slag was replaced by different proportions of QRD in fly ash, and SG-based GPC mixes to study the effect of QRD incorporation. The mechanical properties of specimens, i.e., compressive strength, splitting tensile strength, and flexural strength, were determined by testing cubes, cylinders, and prisms, respectively, at different ages (7, 28, and 56 days). The specimens were also heated up to 800 °C to evaluate the resistance of specimens to elevated temperature in terms of residual compressive strength and weight loss. The test results showed that the mechanical strength of GPC mixes (without steel fibers) increased by 6–11%, with an increase in QRD content up to 15% at the age of 28 days. In contrast, more than 15% of QRD contents resulted in decreasing the mechanical strength properties. Incorporating steel fibers in a fraction of 0.75% by volume increased the compressive, tensile, and flexural strength of GPC mixes by 15%, 23%, and 34%, respectively. However, further addition of steel fibers at 1.5% by volume lowered the mechanical strength properties. The optimal mixture of QRD incorporated FA-SG-based GPC (QFS-GPC) was observed with 15% QRD and 0.75% steel fibers contents considering the performance in workability and mechanical properties. The results also showed that under elevated temperatures up to 800 °C, the weight loss of QFS-GPC specimens persistently increased with a consistent decrease in the residual compressive strength for increasing QRD content and temperature. Furthermore, the microstructure characterization of QRD blended GPC mixes were also carried out by performing scanning electron microscopy (SEM), X-ray diffraction (XRD), and energy dispersive spectroscopy (EDS).

Keywords: microstructural characterization; quarry rock dust; geopolymer concrete; steel fibers; workability; mechanical strength; elevated temperature; residual compressive strength

1. Introduction

There has been a significant increase in construction activities around the globe to fulfill the growing infrastructural needs. Ordinary Portland cement concrete (OPC) is the most important material generally used in all construction activities. The ordinary Portland cement is manufactured by the consumption of fuel and conversion of raw materials, during which an enormous amount of CO_2 is released into the atmosphere. According to a study, one ton of carbon dioxide is released into the environment during the production of one ton of cement [1]. Further, an enormous amount of waste is produced from different industries, such as slag from steel or iron industries, ceramic wastes from ceramic industries, red mud from alumina industries, and fly ash from thermal power plants. It is challenging for researchers and environmentalists to find an alternative to traditional OPC concrete and manage or dispose of these industrial wastes. Therefore, it is necessary to find the best solutions to effectively utilize these industrial byproducts/wastes to minimize land and air pollution. One of the solutions is alkali-activated cement or geopolymer concrete (GPC) that is produced by alkali activation of different byproducts/waste materials and minerals, i.e., fly ash, slag, rice husk ash (RHA), waste ceramic materials, etc. [2,3]. A review of alternative binders reveals that gypsum, geopolymer, and starch can be good alternatives to lime and magnesium-based binders for building materials made of bio-composites [4]

GPC is one of the better and feasible solutions to decrease or completely avoid using traditional ordinary Portland cement concrete. Further, it promotes industrial wastes/byproducts to produce environmentally friendly binder material [2,3]. Due to the early compressive strength, better chemical resistance, and low permeability, GPC has presented itself as a good alternative to the traditional binders [5,6]. It can be manufactured from basic geological materials (such as metakaolin) or industrial pozzolanic materials (slag, fly ash, RHA, and ceramic wastes), which consist of a large amount of alumina (Al_2O_3) and silica (SiO_2) [7–10]. Since SiO_2 and Al_2O_3 are the main oxides in the GPC production, industrial waste materials such as fly ash, slag, copper, and zinc SG can be used as an aluminosilicate source in GPC synthesis. The FA has been widely used to produce geopolymer binders due to its wide availability, durability, and high pozzolanic properties [11–15]. Several studies highlight low calcium FA-based GPC production under elevated temperature curing for short periods [16–18]. However, the results are not promising under ambient curing conditions due to the slower polymerization process. The polymerization process leads to the formation of calcium aluminate silicate hydrate and sodium aluminate silicate hydrate compounds [19]. It was found that the optimum curing temperature was 60 °C for a curing duration of 19 to 24 h, depending on the type of binder contents for the activation of the polymerization process in GPC. Different structural members indicate better mechanical properties of GPC with inclusion of steel fibers and binders [20]. It was also concluded in the previous studies that heat curing limits the usage of GPC to precast structural members only. Therefore, it is imperative to study the feasibility of using ambient cured cast-in-situ GPC. Researchers have endorsed the use of alccofine in GPC to achieve encouraging results at ambient curing conditions [19]. Some researchers have also tried to enhance the reactivity of FA in the basic environment (i.e., at ambient temperature) by increasing the fineness of FA [21] and by the addition of calcium-containing materials [22,23] such as SG [24], alccofine [19], etc. It has been reported that SG blended FA-based GPC specimens showed good mechanical properties, and resistance to elevated temperature [25,26] and sodium sulphate attack. However, it showed substantial deterioration in magnesium sulphate attack [27] and exhibited increased shrinkage [28]. Several studies have been carried out to investigate the effect of SG on the fresh, mechanical, and durability properties of GPC. However, the development of GPC mixes by blending quarry rock dust (QRD) wastes as a binder has rarely been explored. The QRD is a waste material of rock quarries that is produced during the coarse aggregates manufacturing process [29]. A portion of this unwanted waste is often used on site as a filling material for the quarry pit [30]. Recently, QRD has been used in GPC and OPC specimens as a partial replacement of fine aggregates, i.e., sand [29,31]. QRD has also been used as a partial replacement of cement in conventional concrete and showed improved strength results with 20% replacement of QRD with the OPC. The effect of QRD as a binder on properties of FA and SG-based GPC at ambient as well as at elevated temperatures has also been recently investigated [32] and used in columns with steel fibers. The columns were tested under concentric and eccentric loading [33]. The compressive strength of control OPC specimens (without QRD) was 50.23 MPa; whereas the compressive strength of OPC mixtures with 10%, 20%, and 30% replacement of QRD were increased by 4.8%, 8.2%, and 5.92%, respectively. Similarly, the flexural strength of control specimen was 5.12 MPa and it was increased by 107%, 110%, and 106% for 10%, 20%, and 30% replacement of QRD, respectively, at 28 days [30]. It therefore would be interesting to explore the suitability of QRD as a binder in GPC mixes which is the objective of the present study.

GPC is a promising material for the construction industry with environmental benefits and equally good engineering properties. However, one of the drawbacks of GPC in large-scale structural applications is the low ductility [34,35]. There are different types of fibers, i.e., steel, nylon, polypropylene, and polyethylene, that can improve the ductility, flexural, and tensile properties of GPC blends. However, steel fibers (SF) have been the best due to their better ductile and thermal properties at elevated temperatures [36,37]. Genesa et al. [35] explored the basic characteristics of SF-based GPC. It was observed that compressive as well as splitting tensile strength was increased by a margin of 8.5% and 61.6%, respectively, with the fiber volume fraction of 1%. It has also been reported that the splitting tensile and flexural strength of GPC mixes with 0.5% steel fibers by volume were increased by 19–38% and 13–44%, respectively, than the plain samples [38]. Another study reported that compressive and flexural strengths were increased by 3.4% and 31.5%, respectively, by adding 1.2% by volume SF [39].

Recently, research proved that GPC has potential to be used as thermal barriers [40,41]. The structure may be exposed to open fire as well as closed fire, so it must possess thermal stability and fire resistance. The literature shows that fiber-reinforced geopolymers can be excellent materials for thermal and fire-resistant applications [40,42].

The actual behavior of concrete exposed to elevated temperatures depends on many factors such as the properties of materials, heating rate, maximum temperature, exposure period, cooling method, and loading level at the time of cooling [43,44]. It has been found that SF has shown a higher retaining capacity of its original mechanical properties during fire due to its higher melting temperature [45]. It has also been reported that GPC specimens have good resistance against fire due to the presence of nanopores in abundance that allows bonded water to migrate and evaporate without destroying the aluminosilicate network [5,46,47]. Investigations proved that GPC loses its strength after being exposed to a temperature of about 400 °C depending on the raw material. The drop in compressive strength (in percentage) of fly-ash-based GPC is 35%, 44%, 50%, and 75% when exposed to elevated temperatures, i.e., 400 °C, 600 °C, 800 °C and 1000 °C, respectively [48]. Similar to fire resistance, GPC specimens also have frost resistance [49].

Currently, many countries are facing land and air pollution problems. A huge amount of industrial wastes/byproducts are produced globally. The disposal of these wastes in dump yards is linked with high costs and a negative impact on the environment. There is a need to work on creating better and feasible solutions that can productively use industrial wastes. The QRD wastes, also known as limestone dust, dolomite or silica powders, are produced during the quarrying of the large parent mass rock to produce aggregates. These wastes are nonbiodegradable and cause environmental pollution creating health hazards. Therefore, it is best if such wastes can be recycled and used in construction activities in order to help preserve natural resources and the environment.

Several studies are available on fiber-reinforced FA and SG-based GPC at ambient and heat curing conditions [18,24,50]. However, the publications on QRD as a geopolymer binder in fly ash and slag-based GPC are rather scarce. Therefore, the present study has been undertaken to examine the properties of fiber-reinforced fly ash and slag based GPC mixtures with QRD incorporation at room and elevated temperatures. The effects of QRD incorporation on fresh, mechanical, and microstructural properties have been investigated. The optimum percentage of QRD addition has been worked out considering the performance in workability and mechanical properties. Moreover, weight loss and residual compressive strength were also investigated after heating the specimens at elevated temperatures, i.e., 400 °C and 800 °C.

2. Materials and Methods
2.1. Experimental Program

An experimental program was designed to achieve the objective of finding an optimum mix of ambient cured, ternary blended GPC comprising fly ash, slag and QRD, reinforced with steel fibers. The low calcium FA (Super fine ash, Matrixx, Karachi, Pakistan), SG (ground granulated blast furnace slag Grade-80, Dewan Cement (PVT) Limited, Karachi, Pakistan), and QRD were utilized as a binder to produce GPC mixtures. The QRD wastes were obtained from the Margallah hills quarries near Taxila, Pakistan. The QRD is collected at the bottom of aggregate crushers during the formation of the coarse aggregates. It was grinded to achieve the required size equivalent to the OPC particles, which can be sieved through a 45 µm sieve. The OPC type II cement conforming to ASTM C-150 [51] was used for control specimens of conventional concrete, the properties of which are provided in Table 1. The fly ash (FA) was collected from Karachi, Pakistan through combustion process of coal (steam coal) in thermal power plants. Table 2 shows the chemical composition of FA, SG, and QRD, determined from X-ray fluorescence (XRF) analysis (Fecto cement factory, Taxila, Pakistan). The alkaline activator used in this study consists of sodium silicate and sodium hydroxide. The molarity of sodium hydroxide was kept as 12M. It was prepared one day before the application by mixing 98% pure flakes with tap water. The sodium silicate solution has a modulus ratio (MR) of SiO_2 to Na_2O between 1.90 and 2.01. The chemical composition of sodium silicate is shown in Table 3. The local natural river sand was used as fine aggregates. Crushed stone aggregates available locally in the size of 10 mm and 20 mm was used as a coarse aggregate. The fineness modulus of coarse aggregate conformed to ASTM-C136-06 and specific gravity satisfied ASTM-C127-07. The coarse aggregates (CA) were obtained from the Margallah hills quarry near Taxila, Pakistan. Table 4 shows the properties of coarse and fine aggregates. The commercially available hooked end hard-drawn wire SF (MasterFiber® S 65, BASF, Karachi, Pakistan), conforming to ASTM A820 [52], type 1, were used to improve tensile as well as flexural strength of the GPC. The SF provided best results under impact load than the remaining fibers. The specifications of hooked end steel fibers are presented in Table 5. The alkaline solution used in GPC mixes has a sticky characteristic. Therefore, their use makes the GPC mixes more viscous than the ordinary concrete. A naphthalene-based superplasticizer (SP) (Chemrite-SP 200, Imporient Chemicals (PVT) LTD, Lahore, Pakistan) confirming to ASTM C494 was used to increase the workability of GPC mixes [53]. The different materials used in this study are shown in Figure 1.

Table 6 shows a total of 18 mixtures were designed with different proportions of QRD (0%, 5%, 10%, 15%, and 20% by weight of binder) and steel fibers (0.75% and 1.5% by volume). The SG was replaced by different proportions of QRD in FA, and SG-based GPC mixes to study the effect of QRD incorporation. As shown in Table 7, three mix types comprising OPC concrete group serving as the control mix group; another group GPC-A of three GPC mixes without QRD; while the remaining four GPC groups viz. GPC-B, GPC-C, GPC-D, and GPC-E, with 5%, 10%, 15%, and 20% QRD, partially replacing SG (by weight of binder) and keeping all the other ingredients the same in the groups. Further, each group comprises three mix types with 0%, 0.75%, and 1.5% (by volume of composites) SF, thus making a total of 18 mix types in the six groups.

Figure 1. The materials used in the present study to produce QRD incorporated FA and SG-based GPC (QFS-GPC) reinforced with steel fibers; (**a**) fly ash, (**b**) ground granulated blast furnace slag, (**c**) steel fibers, (**d**) quarry rock dust at site, (**e**) aggregates, (**f**) alkaline solution, and (**g**) superplasticizer.

2.2. Mixing and Casting Procedure

The mixtures were prepared based on a unit volume of one cubic meter. The quantity of binder content was kept fixed at 400 kg/m^3 in all the mixes. A total of eighteen mixtures were designed: three OPC-based concrete as control specimens and fifteen GPC specimens with fly ash, slag, and QRD as the source binding materials. The amount of FA was kept constant at 200 kg/m^3 in all the GPC mixes, whereas SG was replaced with QRD at 0%, 5%, 10%, 15%, and 20% by weight of the binder. A total of 540 specimens were cast, consisting of 270 cubes (150 × 150 × 150 mm) for compressive strength, weight loss, and residual compressive strength tests; 162 cylinders (150 mm dia, 300 mm height) for splitting tensile strength tests; and 108 prisms (100 × 100 × 500 mm) for flexural strength tests. The alkaline activator solution was kept at 200 kg/m^3 with a sodium silicate to sodium hydroxide ratio of 1.5 and alkaline solution to binder ratio (A/B) of 0.5 in all GPC mixes. The steel fibers with both ends hooked were used in the mixes with varying contents of 0.75% and 1.5% by volume of the concrete. The specifications of mix design proportions and mix designations are shown in Tables 6 and 7.

All the mixes were prepared in a mechanical mixer of 0.15 m^3 capacity, as shown in Figure 2a. Before mixing ingredients, aggregates were prepared to the saturated surface dry (SSD) condition. The sodium hydroxide solution was blended a day before [19] the application and mixed with SS solution about 30 min before its use [24] to improve the reactivity of the solution. Firstly, the coarse and fine aggregates and binders (fly ash, slag, and QRD) were mixed in dry condition thoroughly in the mixer for 2 min. The SF was then added to the dry mixture, and mixing was continued for another 2 min, ensuring adequate and homogenous dispersion of fibers in the mix. After that, the premixed alkaline solution was incorporated gradually into the mixer. Then, mixing was continued for another 2–3 min to achieve a uniform homogeneous mixture. Finally, superplasticizer and remaining water were added in the mix to achieve the required workability in the range of 50–89 mm and mixing was continued for another 2–3 min. The freshly prepared steel fibers reinforced GPC mix is shown in Figure 2b. The flow chart elaborating the mixing sequence of GPC mixes is shown in Figure 3.

The newly mixed concrete was instantly cast into different molds, i.e., cylinders, cubes, and prisms. All specimens were placed in a room at ambient temperature. After 24 h, all the samples were demolded and kept in the ambient curing conditions for 7, 28, and 56 days. Three specimens were used for testing each mix, and an average result was reported. The designated specimens (150 mm cubes) were exposed to elevated temperatures (400 °C and 800 °C) after 56 days of curing to investigate the weight loss and residual compressive strength. Before placing the samples in the kiln, they were dehydrated in an oven for 24 h at 105 ± 5 °C to remove any free water, thus preventing the samples from a possible explosion in the kiln during the heating procedure due to very high pore water pressure resulting from the superheated water.

Figure 2. (a) The mixing of ingredients in a mechanical mixer. (b) The freshly prepared SF-reinforced GPC mixture.

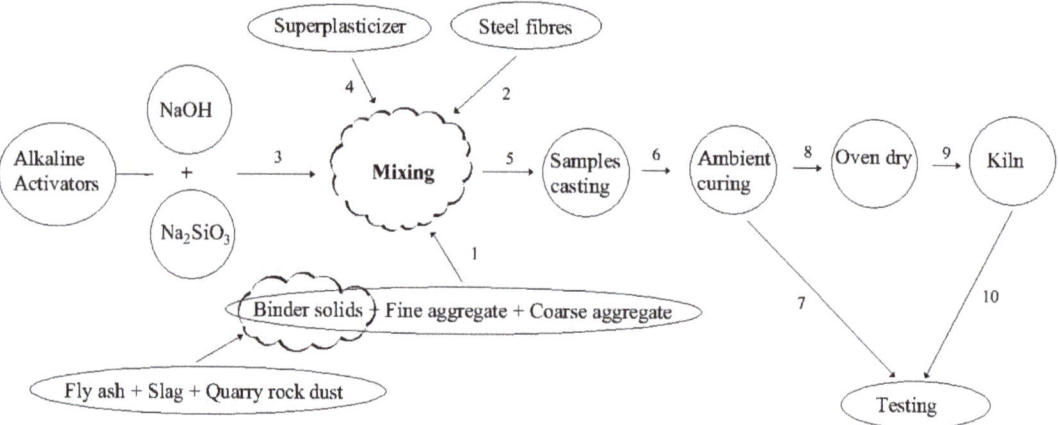

Figure 3. The mix production to the testing sequence of GPC specimens. 1—mixing of Raw materials; 2—adding Steel fibers; 3—Adding alkaline solution; 4—Adding superplasticizer; 5—Casting of samples; 6—Curing of samples at room temperature; 7—Compressive, tensile and flexural testing of samples; 8—Oven dry of samples; 9—Heating of samples at elevated temperature; 10—Mechanical tests.

2.3. Experimental Setup

A series of tests were carried out to determine the fresh properties (workability), mechanical properties (compressive, split tensile, and flexural strengths), residual properties after exposure to elevated temperature (weight loss and residual compressive strength), and microstructural properties (from X-ray diffraction analysis, scanning electron microscope images, and energy dispersive spectroscopy). Workability is defined as the ease of placement and compaction of freshly mixed concrete. The slump cone test is commonly used to check the workability of freshly made concrete conforming to ASTM C143M-15a [54] The mechanical properties of specimens, i.e., compressive, splitting tensile, and flexural strengths were determined by testing cubes, cylinders, and prisms, respectively, at different ages (7, 28, and 56 days). The specimens were heated to 800 °C to evaluate the resistance of specimens to elevated temperature in terms of residual compressive strength and weight loss. The microstructural properties from scanning electron microscopy (SEM), X-ray diffraction (XRD), and energy dispersive spectroscopy (EDS) were investigated.

A universal testing machine (UTM) of 3000 KN capacity was used for compressive and splitting tensile strength tests of the specimens at various ages, i.e., 7, 28, and 56 days, according to ASTM C39/C39M-03 [55] and C496/C496M−11, respectively. The flexural strength test was performed on prismatic samples at the ages of 28 and 56 days under third point loading according to ASTM C1609/C1609M-19a [56]. The workability of mixes was measured by using a slump cone test conforming to ASTM C143M-15a [54]. After 56 days of curing, three samples (150 mm cubes) from each mix group were subjected to elevated temperatures (400 °C and 800 °C) in an automatic controlled electric furnace of 1000 °C capacity following ASTM E 119 [57]. The samples were then cooled at room temperature, and residual properties, i.e., compressive strength and weight loss, were determined. The microstructural characterization of the specimens was evaluated by X-ray diffraction (XRD) analysis, energy dispersive spectroscopy (EDS), and scanning electron microscopy (SEM). The XRD test was performed to provide fundamental information on the geopolymer crystal structure. A crystal structure is one of the important aspects of materials since many properties depend on it. The X-ray diffraction (XRD) test was performed using JEOL JDX-3532 (JEOL, Tokyo, Japan), with a step size of 0.025° and 2θ range from 10–75° to analyze the geopolymer structure to find whether it is crystalline or amorphous. The powder method was used to evaluate the degree of crystalline structure in the polymer. The sample was scanned with a voltage of 40 kV and a current of 30 mA using a copper X-ray tube (Cu-Kα radiation). It has been reported that diffraction occurs only if electromagnetic radiation interacts with periodic structures. The non-crystalline portion in GPC mixes simply scatters the X-ray beam to give a continuous background, whereas the crystalline portion causes diffraction lines (peaks) that are not continuous. Miller indices of the peaks describe the information about the planes of diffraction [19,58].

The EDS test was carried out to study the chemical composition of the GPC mixtures. The energy dispersive spectroscopy (EDS) test was performed for the area using JSM-5910 (Oxford Instruments, Abingdon, UK) to find the chemical composition of geopolymers. For the purpose of testing, a sample of GPC is taken out from the middle of the cube specimen cured for 28 days and then ground to a powder with the help of mortar and pestle. The powdered sample is then oven-dried to remove moisture. The EDS data is acquired using INCA software at five different spectrums (locations) on the SEM image. SEM was undertaken to study the fracture surface of the GPC specimens. The SEM analysis was carried out using JSM-6490 (JEOL, Tokyo, Japan) to study the microstructural behavior of GPC specimens when subjected to an elevated temperature of 800 °C. Samples were collected from the failed specimens in the compressive strength test from the adjacent parts of failure surfaces.

Table 1. The general characteristics of type II OPC used in the study.

Oxides	Results (%)	Physical Characteristics	Results
CaO	64.2	Specific surface	322 m^2/kg
SiO$_2$	22.0	Consistency	30%
Al$_2$O$_3$	5.50	Initial setting time	1 h 42 min
Fe$_2$O$_3$	3.50	Final setting time	3 h 55 min
SO$_3$	2.90	Specific gravity	3.5
MgO	2.50	Soundness	No soundness
K$_2$O	1.00	Color	Grey
Na$_2$O	0.20	-	-
LOI	0.64	-	-

Table 2. The chemical composition of QRD, SG, and FA used in the present study.

Oxides	QRD	SG	FA
SiO_2	9.35%	34.38%	57–65%
Al_2O_3	1.64%	12.98%	28–32%
Fe_2O_3	1.03%	1.29%	1–4% max
CaO	47.13%	37.33%	1–2%
MgO	1.25%	5.59%	0.50%
K_2O	0.20%	0.82%	-
Na_2O	−0.11%	0.29%	1.5 max%
SO_3	0.08%	0.23% max	4%
$SiO_2:Al_2O_3$	5.70	2.64	2.03
LOI	38.65%	3.4%	2.9%

Table 3. The chemical composition of Na_2SiO_3.

Composition	Percentage
Na_2O	8.93%
SiO_2	29.8%
Water	61.78%
Density (kg/m^3)	1400

Table 4. The properties of coarse and fine aggregates.

Coarse Aggregate		Fine Aggregate (Sand)	
Moisture content	1.0%	Fineness modulus	2.72
Specific gravity	2.66	Specific gravity	2.74
Water absorption	0.8%	Water absorption	1.25%

Table 5. The specifications of hooked end steel fibers.

Entity	Specification
Length	35 mm
Diameter	0.55
Aspect ratio	64
Tensile strength	1345 MPa

Table 6. The mix designations of OPC and GPC mixes.

Mix ID	Mix Composition
OPC-F0	100% cement
OPC-F0.75	100% cement + 0.75% steel fibers control mixes
OPC-F1.5	100% cement + 1.5% steel fibers traditional concrete
GPC-AF0	50%FA + 50% SG
GPC-AF0.75	50%FA + 50% SG + 0.75% steel fibers
GPC-AF1.5	50%FA + 50% SG + 1.5% steel fibers
GPC-BF0	50% FA + 45% SG + 5% QRD
GPC-BF0.75	50% FA + 45% SG + 5% QRD + 0.75% steel fibers
GPC-BF1.5	50% FA + 45% SG + 5% QRD + 1.5% steel fibers
GPC-CF0	50% FA + 40% SG + 10% QRD
GPC-CF0.75	50% FA + 40% SG + 10% QRD + 0.75% steel fibers
GPC-CF1.5	50% FA + 40% SG + 10% QRD + 1.5% steel fibers
GPC-DF0	50% FA + 35% SG + 15% QRD
GPC-DF0.75	50% FA + 35% SG + 15% QRD + 0.75% steel fibers
GPC-DF1.5	50% FA + 35% SG + 15% QRD + 1.5% steel fibers
GPC-EF0	50% FA + 30% SG + 20% QRD
GPC-EF0.75	50% FA + 30% SG + 20% QRD + 0.75% steel fibers
GPC-EF1.5	50% FA + 45% SG + 5% QRD + 1.5% steel fibers

Table 7. The detail of mix proportions of OPC and GPC mixtures.

Group ID	Mix No.	Mix ID	B	C	FA	Binders SG	QRD	SF	AL/B Ratio	W/C Ratio	Molarity of SH	SS/SH Ratio	SH	SS	S	CA 20 mm	CA 10 mm	SPs	Water
OPC	1	OPC-0F	400	400	-	-	-	-	-	0.35	-	-	-	-	680	751	340	10	140
	2	OPC-0.75F	400	400	-	-	-	58.5	-	0.35	-	-	-	-	680	752	340	10	140
	3	OPC-1.5F	400	400	-	-	-	117	-	0.35	-	-	-	-	680	753	340	10	140
GPC-A	4	GPC-A0F	400	-	200	200	0	-	0.5	-	12	1.5	80	120	680	751	340	11	35
	5	GPC-A0.75F	400	-	200	200	0	58.5	0.5	-	12	1.5	80	120	680	752	340	18	35
	6	GPC-A1.5F	400	-	200	200	0	117	0.5	-	12	1.5	80	120	680	753	340	20	35
GPC-B	7	GPC-B0F	400	-	200	180	20	-	0.5	-	12	1.5	80	120	680	754	340	12	35
	8	GPC-B0.75F	400	-	200	180	20	58.5	0.5	-	12	1.5	80	120	680	755	340	17	35
	9	GPC-B1.5F	400	-	200	180	20	117	0.5	-	12	1.5	80	120	680	756	340	21	35
GPC-C	10	GPC-C0F	400	-	200	160	40	-	0.5	-	12	1.5	80	120	680	757	340	14	35
	11	GPC-C0.75F	400	-	200	160	40	58.5	0.5	-	12	1.5	80	120	680	758	340	20	35
	12	GPC-C1.5F	400	-	200	160	40	117	0.5	-	12	1.5	80	120	680	759	340	22	35
GPC-D	13	GPC-D0F	400	-	200	140	60	-	0.5	-	12	1.5	80	120	680	760	340	14.5	35
	14	GPC-D0.75F	400	-	200	140	60	58.5	0.5	-	12	1.5	80	120	680	761	340	21	35
	15	GPC-D1.5F	400	-	200	140	60	117	0.5	-	12	1.5	80	120	680	762	340	23	35
GPC-E	16	GPC-E0F	400	-	200	120	80	-	0.5	-	12	1.5	80	120	680	763	340	14.5	35
	17	GPC-E0.75F	400	-	200	120	80	58.5	0.5	-	12	1.5	80	120	680	764	340	22	35
	18	GPC-E1.5F	400	-	200	120	80	117	0.5	-	12	1.5	80	120	680	765	340	24	35

Note: W (water); B (binder); C (cement); OPC (ordinary portland cement); SF (steel fibers); AL (alkaline solution); QRD (quarry rock dust); SG (ground granulated blast furnace); FA (fly ash); SH (sodium hydroxide); SS (sodium silicate); SP (superplasticizers); S (sand); CA (coarse aggregates).

3. Results and Discussion

3.1. Workability

The slump values of OPC- and QRD-blended FA-SG-based GPC (QFS-GPC) mixtures are shown in Figure 4. It can be observed from Figure 4 that all GPC mixtures without steel fibers (GPC-A0F, GPC-B0F, GPC-C0F, GPC-D0F, and GPC-E0F) have lower slump values than OPC-based mixes due to the combined effect of slag and QRD particles along with higher viscosity of alkaline solutions. It can also be observed that QRD content has a negative effect on the workability of GPC mixes. The workability of QFS-GPC specimens persistently decreased with the increase in QRD content from 0% to 20%. The slump values of QFS-GPC mixes, i.e., GPC-B0F, GPC-C0F, GPC-D0F, and GPC-E0F are 25%, 29%, 31%, and 51% lower, respectively, than their counterpart without QRD, i.e., GPC-A0F. Previous studies have also reported this decreasing trend of the slump with the increased QRD [30]. This decreasing trend of slump can be attributed to the angular shape particles of QRD [59] that restrain the flowability of the mixture, contrary to the spherical-shaped particles of FA [34] that make concrete more flowable. The workability of GPC mixes was observed to be lesser than the corresponding OPC-based mixes. The slump values of GPC mixes viz. GPC-A0F, GPC-B0F, GPC-C0F, GPC-D0F, and GPC-E0F were 11%, 33%, 37%, 39%, and 57% lower than the OPC-based control mix. It has been investigated that workability of FA-based GPC decreases by increasing SG content [24].

It can be noticed from Figure 4 that the workability of steel-fibers-reinforced concrete mixtures is lower than their counterparts, i.e., plain specimens (without fibers). The slump values of OPC-0.75F and OPC-1.5F are 16% and 36% lower, respectively, than OPC-0F (without steel fibers). Similarly, the slump values of GPC mixes, GPC-A0.75F and GPC-A1.5F, are 13% and 41% lower than their counterpart plain samples of GPC-A0F. The trend of decreasing workability in GPC mixes increases with the increase in QRD content. The slump values of group E mixes, i.e., GPC-E0.75F and GPC-E1.5F, are 51% and 58% lower than GPC-E0F. The addition of SF with 1.5% by volume decreased the workability up to 60% then 0.75% by volume of SF. This can be due to uneven scattering of fibers that may have hindered the movement of mixture particles. Moreover, the fibers absorb more binder (cement, FA, SG, or QRD) mortar due to the large surface area, which increases the viscosity of mixes resulting in low slump values. Therefore, the optimum SF content from the above finding is 0.75% considering workability.

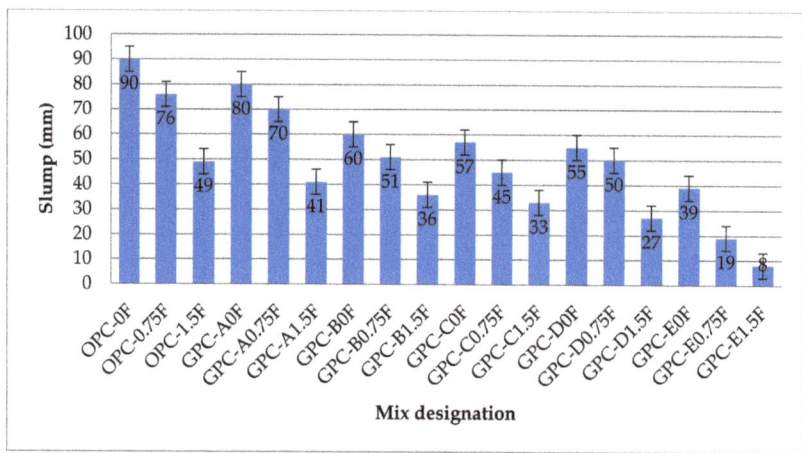

Figure 4. The slump values of GPC and OPC mixes.

The rheology of a GPC mix is generally not similar to that of an OPC mix. Hence, the slump values of GPC do not resemble the same level of workability in OPC mixtures [24].

Based on compaction, slump values of GPC are classified as: highly workable (90 mm and above), medium workable (50–89 mm), and low workable (less than 50 mm) [56]. According to this criterion, GPC mixtures with 0% and 0.75% SF contents are medium workable except GPC-E0F; and those with 1.5% SF content are classified as low workable.

3.2. Compressive Strength

Compressive strength is an important property of concrete that is connected to other mechanical properties as well. According to ACI 318 M-11 [59], the 28 day compressive strength needs to be at least 28 MPa for basic engineering applications, while it should be 35 MPa for corrosion protection of deform steel bars in concrete. In this study, all samples were tested at the age of 7, 28, and 56 days according to ASTM C39/C39M [55], as shown in Figure 5. The mean values of the compressive strength test results of OPC and GPC mixes obtained from three identical samples are shown in Figure 6. Generally, FA- and SG-based GPC mixes exhibit higher compressive strength values than the OPC-based mixes [16]. It can be observed from Figure 6 that the compressive strength of GPC mixes GPC-B0F, GPC-C0F, and GPC-D0F, increased by increasing the QRD replacement level up to 15% (i.e., for 5%, 10%, and 15% QRD replacement).

Figure 5. (a) The compressive testing of cubes; and (b) the failure samples.

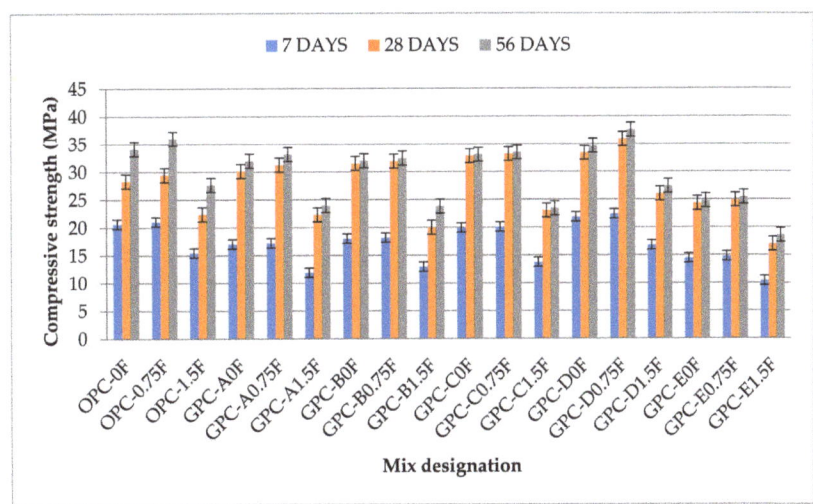

Figure 6. The compressive strength values of GPC and OPC mixes.

The 28 day compressive strength of QFS-GPC mixes viz. GPC-B0F (5% QRD), GPC-C0F (10% QRD), and GPC-D0F (15% QRD) are 5%, 9%, and 11% higher than the GPC-

AOF (without QRD). This increase in strength due to QRD addition can be attributed to the increased quantity of lime produced in the geopolymerization process due to SG replacement with QRD. From the XRF analysis of SG and QRD, CaO content in SG is 37.33%, whereas QRD composes 47.13% of CaO content as shown in Table 2. The replacement of SG with QRD ultimately resulted in increasing the CaO contents in the geopolymer matrix. It has been reported that calcium-containing materials such as SG, alccofine, and QRD accelerate the rate of polymerization at ambient temperature (room temperature) and reduce the pore sizes [60]. The inclusion of the calcium-containing materials has increased the compressive strength of the QRD blended FA-SG-based GPC (QFS-GPC) mixes. Hence, GPC mixes with QRD replacement level up to 15% would produce compacted geopolymer matrix compound, which will increase the compressive strength of the specimens at early ages. However, when the amount of QRD is increased further from 15% to 20% as in mix GPC-E0F, it decreases the workability of the mix drastically as shown in Figure 4, making it difficult to handle during placement. Hence, extra water or superplasticizer was added to the mix GPC-E0F to increase the workability that ultimately resulted in decreasing the compressive strength by an amount of 19% compared with GPC-A0F (without QRD content). This phenomenon has also been reported by Hake et al., 2018 [61].

The 28 day compressive strength of QFS-GPC mixes viz. GPC-B0F, GPC-C0F, and GPC-D0F are 11%, 16%, and 18% higher than the control OPC mix (OPC-0F). However, the compressive strength of GPC-E0F is 14% lower than the OPC-0F. After 28 days, the compressive strength of GPC-D0F (33.4 MPa) is almost 18% higher than the control mix OPC-0F. Therefore, mix GPC-D0F can be considered as an optimum mixture without any fiber reinforcement considering the compressive strength.

It was observed that the effect of SF addition on the compressive strength is very low compared with the flexural and tensile strengths. The compressive strengths of GPC and OPC mixes were increased in the range of 2–8% by adding 0.75% steel fibers (by volume) than their counterparts (without steel fibers). When the fraction of SF was further increased from 0.75% to 1.5% in all GPC and OPC mixtures, the strength was further decreased by 20–30% of the counterparts without fibers. This decrease in strength can be due to the uneven dispersion of fibers causing insufficient compaction and non-uniformity of the mix.

3.3. Splitting Tensile Strength

It is an important mechanical characteristic of concrete that is used in designing some reinforced concrete structural members. The splitting strength testing setup of cylindrical samples is shown in Figure 7a and determined at the ages of 7, 28, and 56 days according to ASTM C496 [62]. The results of splitting tensile strength values of OPC and GPC specimens are shown in Figure 8.

The splitting tensile strength of OPC-0F (1.91 MPa) and GPC-D0F (1.94 MPa) at the age of 7 days were maximum among OPC and GPC specimens, respectively. The splitting tensile strength of QFS-GPC specimens without fibers viz. GPC-B0F, GPC-C0F, and GPC-D0F are 3%, 5%, and 6% higher, respectively, than the GPC control mix GPC-A0F (without QRD). However, GPC-E0F shows a decrease in splitting tensile strength than GPC-A0F. This decrease in strength can be attributed to the decreased workability of the mix due to increased QRD content, thus making it more difficult to handle during placement. The additional water or admixture (superplasticizer) was added to the mix (GPC-E0F) to increase the workability, ultimately decreasing the splitting tensile strength.

The values of splitting tensile strength of all GPC mixes without fibers, i.e., GPC-A0F, GPC-B0F, GPC-C0F, GPC-D0F, and GPC-E0F, are 15%, 12%, 10%, 9%, and 21%, respectively lower than the control OPC mix OPC-0F; which shows that GPC mixes are weak in tensile strength than the OPC mixes. An increase in splitting tensile strength was also observed with the increase in curing age. The maximum splitting tensile strength in non-fiber mixes was observed for the mix GPC-D0F with values of 2.27 MPa and 2.36 MPa, respectively, at 28 and 56 days. These higher strength values can be due to more compactness in the presence of optimum calcium content than the other non-fiber GPC mixes.

(a)   (b)

Figure 7. (a) The splitting tensile load application on a cylindrical specimen; and (b) the failure of the specimen after the test.

It can be observed from Figure 8 that the addition of SF by 0.75 fraction of volume resulted in increasing the splitting tensile strength of GPC mixes viz. GPC-A0.75F, GPC-B0.75F, GPC-C0.75F, GPC-D0.75F, and GPC-E0.75F by 13%, 13%, 16%, 31%, and 12%, respectively, than their counterparts without fibers. This increase can be attributed to the relatively strong bonding and matrix between the aggregate and SF at 0.75% fraction by volume. The mix GPC-D0.75F achieved the maximum splitting tensile strength with 15% QRD and 0.75% SF. The results are also in agreement with the previous studies [60]. However, the increase in SF content from 0.75% to 1.5% decreased the splitting tensile strength. The possible cause can be uneven dispersion of fibers in the mixes for more than 0.75% SF, which caused low workability of the mixes. Extra water and superplasticizer were used during the mixing procedure resulting in a decrease in the strength.

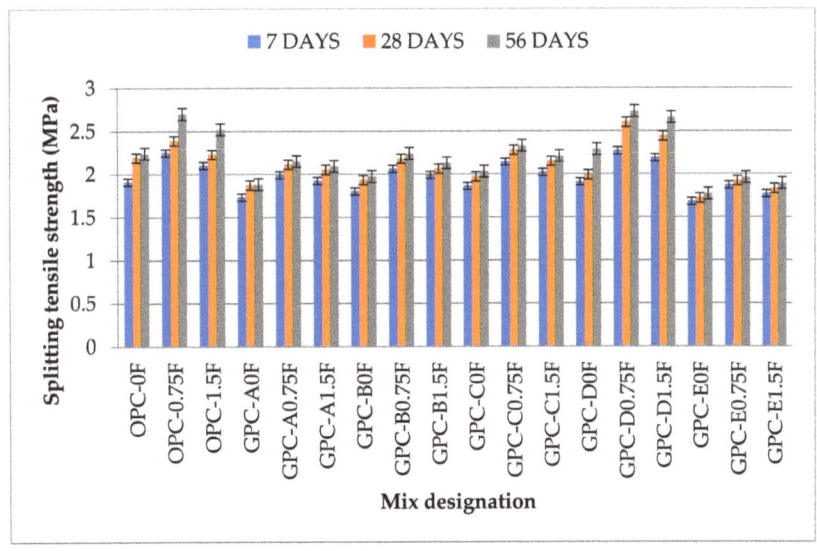

Figure 8. The splitting tensile strength values of GPC and OPC mixes.

3.4. Flexural Strength

The flexural strength test is carried out to find the indirect tensile strength of concrete, also known as the modulus of rupture (MOR). It is an important property that affects concrete's shear strength, bending characteristics, and brittleness ratio in structural concrete design. In this study, the flexural strength was determined by using prismatic specimens (100 × 100 × 500 mm) at the age of 28 and 56 days according to ASTM C1609 [56], the standard test method used for the flexural performance of fiber-reinforced concrete. The flexural strength testing of prismatic samples under center point loading is shown in Figure 9. It can be noticed from Figure 10 that flexural strength of QFS-GPC mixes viz GPC-B0F, GPC-C0F, and GPC-D0F, are 3%, 7%, and 10% higher, respectively, than GPC-A0F (without QRD). This higher strength indicates that the flexural strength of QRD-blended specimens increased with the increase in QRD content up to 15%. The maximum flexural strength of non-fiber specimens at the age of 28 days was obtained by the mix GPC-D0F (3.65 MPa) with 15% QRD content. However, the flexural strength of GPC-E0F (with 20% QRD content) decreased by 18% from GPC-A0F (without QRD). The test results also showed that flexural strength of GPC mixes without fibers, i.e., GPC-A0F, GPC-B0F, GPC-C0F, and GPC-E0F were 8%, 5%, 2%, and 25% lower than the OPC control mix, i.e., OPC-0F. However, the flexural strength of GPC-D0F was 2% higher than the mix OPC-0F. Hence, GPC-D0F can be considered as an optimum mix considering the flexural strength. The test results also showed that the flexural strength of all GPC mixes increased with age. However, this rate of strength gain is slower than the OPC mix samples.

Figure 9. (a) Flexural testing setup for Prismatic specimens of length 500 mm; and (b) the failure of a specimen after the test.

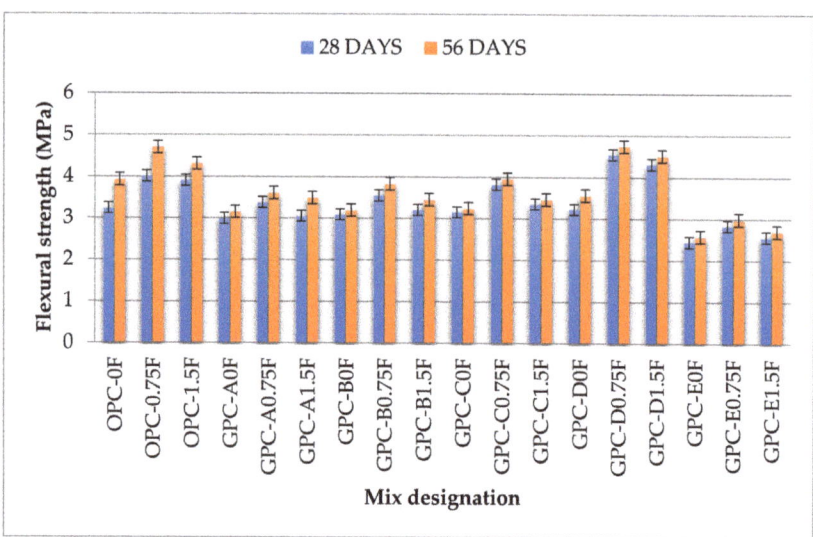

Figure 10. The flexural strength values of GPC and OPC mixes.

The addition of SF improved the flexural strength and improved the post-cracking behavior (also called crack bridging effect) for all GPC and OPC mixes. The flexural strength of GPC mixes viz. GPC-A0.75F, GPC-B0.75F, GPC-C0.75F, GPC-D0.75F, and GPC-

E0.75F are 13%, 15%, 20%, 38%, and 16% higher than their counterparts without fibers. It was also observed that the increase in fiber content from 0.75% to 1.5% by volume decreased the flexural strength of GPC mixes. This decrease in strength can be due to the uneven dispersion of steel fibers causing insufficient compaction and non-uniformity of the mix. Experimental research proved that the addition of 0.5% SF in oil palm shell (OPS)-based GPC improves flexural strength by approximately 13–44% [38].

3.5. Weight Loss and Residual Compressive Strength

The specimens from each mix group were subjected to elevated temperatures (400 °C and 800 °C) at the age of 56 days to measure the weight loss and residual compressive strength. The oven-dried samples were placed in an electric furnace of 1000 °C heating capacity. The specimens were exposed to an elevated temperature at 8 °C/min heating rate until the target temperature was reached. The specimens were kept for 1 h at the required temperature, i.e., 400 °C and 800 °C. After heating, the specimens were cooled at room temperature. The weight before and after the exposure was measured to determine the weight loss of specimens. The results for weight loss of specimen mixes are presented in Figure 11. The replacement of SG with QRD resulted in an increase in the weight loss of QRD-blended mixes. The weight loss observed in GPC mixes viz. GPC-A0F, GPC-B0F, GPC-C0F, GPC-D0F, and GPC-E0F at 800 °C were 2.34%, 3.13%, 4.23%, 5.5%, and 6.35%, respectively. The weight loss values were increased with the increase in QRD content. The increase in weight loss may be due to the higher loss on ignition (LOI) of QRD (38.65%) compared with the FA (2.9%) and SG (3.4%). The weight loss values of QRD-blended GPC mixes are higher at 800 °C than 400 °C since the release of gases in the form of carbon dioxide (CO_2), due to thermal decomposition of concrete, occurred at 800 °C. The addition of 0.75% steel fibers in all GPC mixes, i.e., GPC-A0.75F, GPC-B0.75F, GPC-C0.75F, GPC-D0.75F, and GPC-E0.75F, resulted in a reduction in weight loss by an amount of 2.32%, 2.67%, 4.11%, 4.63%, and 6.04%, respectively, compared with their counterparts without fibers. This reduction in weight loss by fibers can be due to the high heat-absorbing capacity of SF. The lowest weight loss was observed in the OPC mix OPC-0.75F and the GPC mix GPC-A0.75F. The reduced weight loss in these specimens can be due to lesser calcium content and SF presence that resisted the decomposition process.

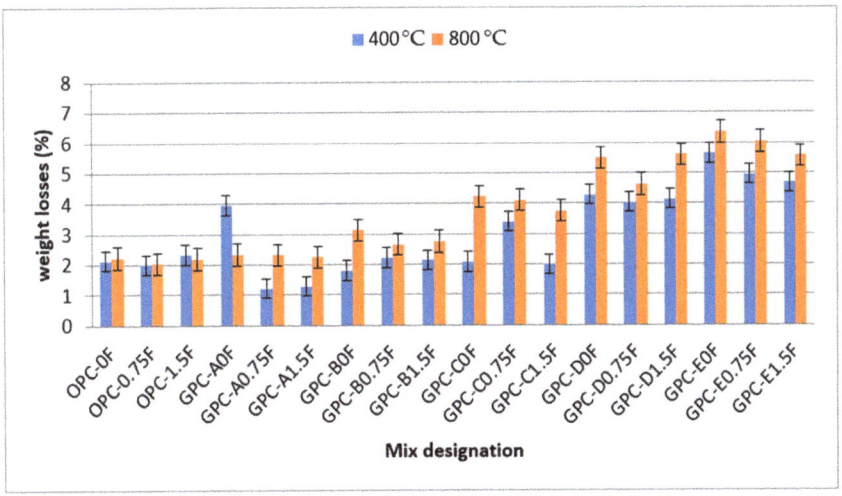

Figure 11. The weight loss values of GPC and OPC mixes after heating the specimens at elevated temperature.

The residual compressive strength of OPC and GPC mixes was investigated to be inversely proportional to the temperature and QRD content. The compressive strength of all plain GPC mixes viz. GPC-A0F, GPC-B0F, GPC-C0F, GPC-D0F, and GPC-E0F at 400 °C were decreased by 9%, 13%, 23%, 31%, and 33%, respectively, from the compressive strength at room temperature as shown in Figure 6. The strength of the OPC control mix OPC-0F at 400 °C was decreased by 60%, indicating that GPC mixes are more fire-resistant than the traditional OPC mixes at 400 °C. It can also be seen from Figure 12 that the residual strength of QRD-blended mixes decreased with the increase in QRD content.

The compressive strength of all plain GPC mixes, i.e., GPC-A0F, GPC-B0F, GPC-C0F, GPC-D0F, and GPC-E0F at 800 °C decreased by 57%, 61%, 67%, 71%, and 74%, respectively. The drop in compressive strength increased with the increase in temperature from 400 °C to 800 °C. This strength drop is due to the presence of a large amount of $Ca(OH)_2$ and $CaCO_3$ in QRD content that is dehydrated and decomposed, respectively, and converted into CaO at temperatures in the range of 600–700 °C [63]. As a result of dehydration and decomposition, H_2O and CO_2 are released, causing volume shrinkage and a significant decrease in compressive strength. It can also be one of the reasons that the matrix starts fusing and melting into a near homogeneous phase at 800 °C, which could include the formation of new products [64], resulting in volume reduction. It was observed during fire that spalling of GPC samples having high content of QRD occurred after being exposed to a temperature of about 400 °C.

It was noticed that adding 0.75% SF in OPC- and QRD-blended GPC mixes reduced the loss in compressive strength at high temperatures. The compressive strength of all fiber-reinforced GPC mixes, i.e., GPC-A0.75F, GPC-B0.75F, GPC-C0.75F, GPC-D0.75F, and GPC-E0.75F at 800 °C were decreased by 42%, 43%, 40%, 51%, and 63%, respectively. However, the increase in the volume fraction of SF from 0.75% to 1.5% did not show any improvement in the residual compressive strength at elevated temperatures. This negligible effect on compressive strength could be due to the poor dispersal of SF (1.5% by volume) in highly viscous GPC mixtures.

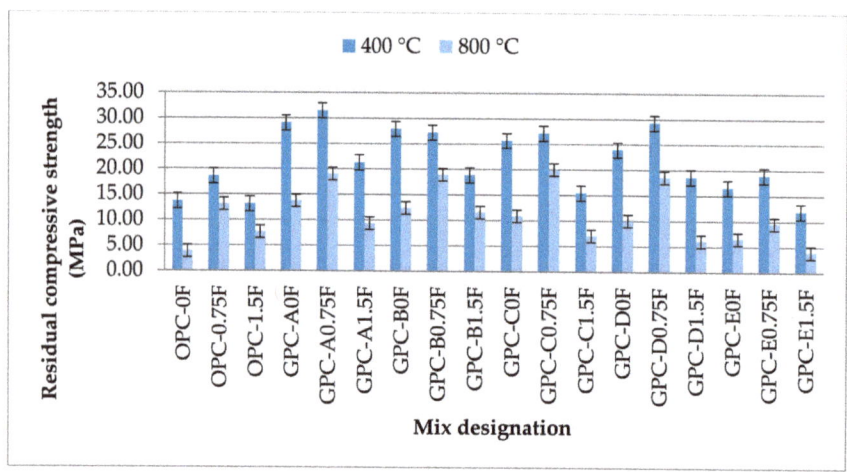

Figure 12. The residual compressive strength values of GPC and OPC mixes after heating the specimens at elevated temperatures.

3.6. X-ray Diffraction (XRD)

Figure 13 shows the XRD pattern (2θ = 10–75°) of OPC and GPC mixes (without steel fibers) observed after 28 days of ambient curing. The most significant zone where the reactions occur in the mix is in the range of 2θ = 20–30°. For the specimen GPC-D0F, sharp diffraction peaks are more in this range than all the other samples, including the control mix; which shows that GPC-D0F is highly crystalline. Similarly, in the temperature range 2θ = 40–50°, OPC and GPC-D0F show peaks representing crystalline phases while the other samples, viz. GPC-A0F, GPC-B0F, GPC-C0F, and GPC-E0F are in amorphous phases. Limited periodicity of atoms in the range of 2θ = 60–70° was also present in the samples GPC-C0F, GPC-D0F, and GPC-E0F due to the presence of QRD. The unreacted fly ash and QRD contains crystalline phases such as quartz (SiO_2), mullite ($Al_6Si_2O_{13}$), and maghemite and hematite (Fe_2O_3). Some studies on geopolymer materials indicate that a small amount of quartz may have a positive effect on the mechanical properties, and other minerals may have a detrimental effect on the geopolymer [65]. Hence, it can be observed from XRD diffractogram that an increase in the QRD content up to 15% in all GPC mixes resulted in an increase in the compressive strength due to the formation of crystalline phases. This increase is because mechanical properties (compressive, tensile, and flexural) of concrete mixtures increase in the presence of a high amount of calcium-rich species at ambient curing temperature [19,24].

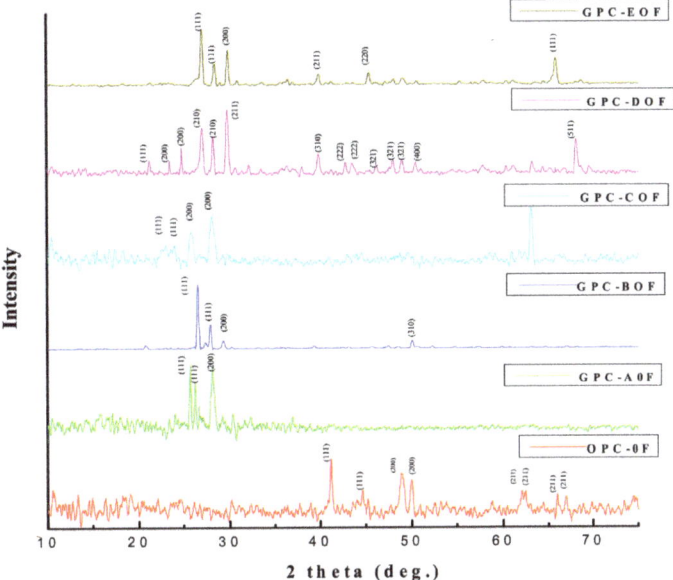

Figure 13. The XRD pattern of OPC and GPC specimens after 28 days.

3.7. Scanning Electron Microscopy (SEM) and Energy Dispersive Spectroscopy (EDS)

Figures 14–19 show the results of SEM micrographs of OPC- and QRD-incorporated GPC specimens.

The microstructure of GPC-D0F prepared with 15% QRD and activated by alkaline solution is denser and less porous than the remaining mixes, i.e., GPC-A0F, GPC-B0F, GPC-C0F, GPC-E0F, and OPC-0F; which shows that GPC-D0F is more compacted. Due to this reason, the strength of GPC-D0F is higher among all mixes. Further, there are no cracks and unreacted particles of FA and SG in the structure of GPC-D0F due to the presence of sufficient CaO in QRD. The SEM image of GPC-E0F shows that by increasing QRD from 15% to 20%, the structure of the hardened mix became porous, causing cracks, which eventually decreased the mechanical strength of the GPC specimens with more than 15% QRD. There are a lot of unreacted particles of FA and SG in the specimens of GPC-A0F, GPC-B0F, and GPC-C0F; which could be the cause of a decrease in the compressive strength. The spherical-shaped FA and angular-shaped SG and QRD particles are in the fuse condition after exposure to elevated temperatures, which decreased the mechanical properties.

Figures 20–25 show the results of EDS. The presence of elements such as Ca, Si, Al, C, and Fe indicates calcium aluminosilicate hydrate (CASH) in almost all specimens. Therefore, the formation of CASH in GPC-D0F makes the microstructure more compacted and dense, ultimately improving its mechanical properties. The presence of high calcium content in QRD and SG increases the geopolymerization process at ambient temperature which enhances the compressive strength of GPC specimens.

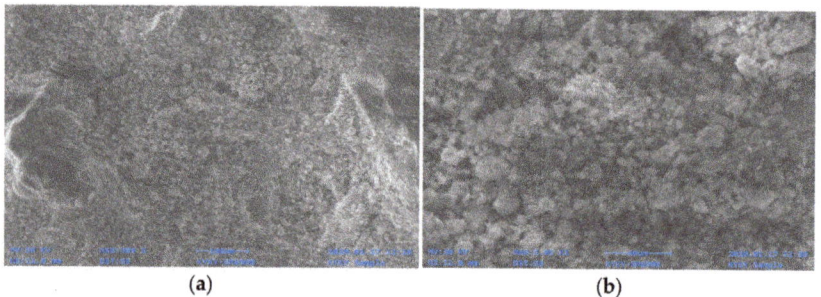

Figure 14. The SEM image of GPC-A0F mix specimen; (**a**) magnification 309×; and (**b**) magnification 1.05×.

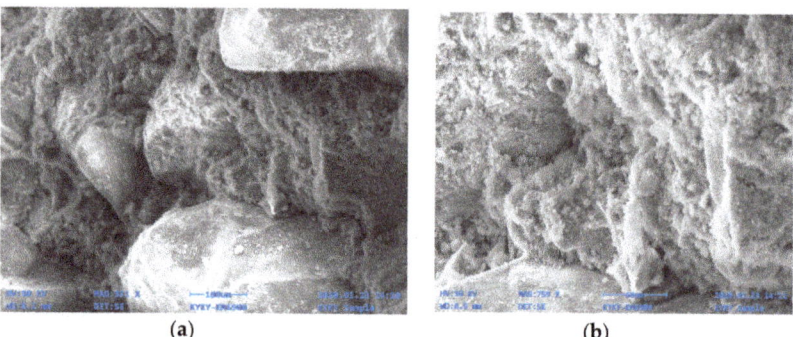

Figure 15. The SEM image of GPC-B0F mix specimen; (**a**) magnification 351×; and (**b**) magnification 759×.

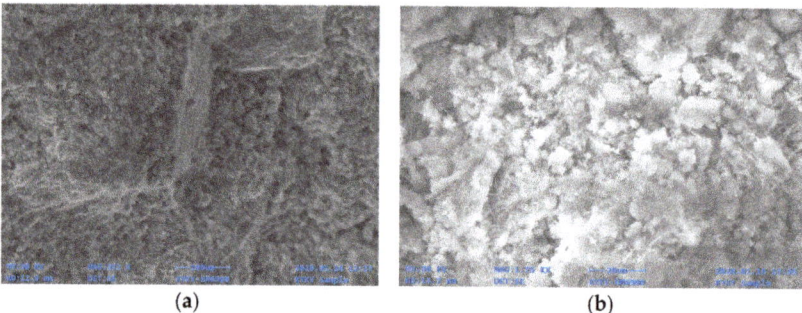

Figure 16. SEM image of GPC-C0F mix specimen; (**a**) magnification 352×; and (**b**) magnification 1.75×.

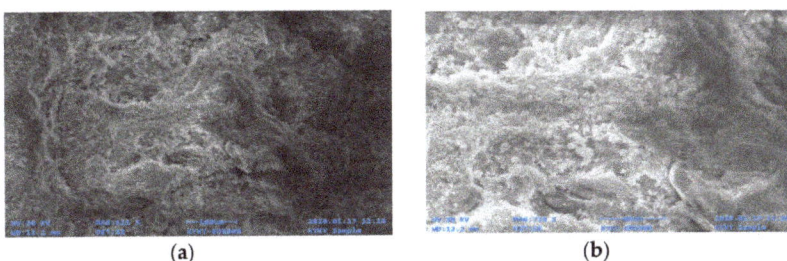

Figure 17. SEM image of GPC-D0F mix specimen; (**a**) magnification 321×; (**b**) magnification 738×.

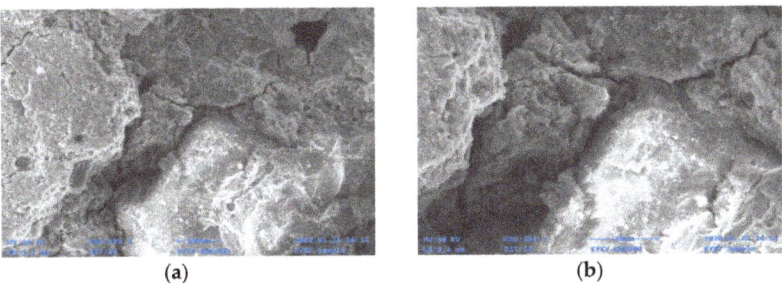

Figure 18. SEM image of GPC-E0F mix specimen; (**a**) magnification 293×; and (**b**) magnification 583×.

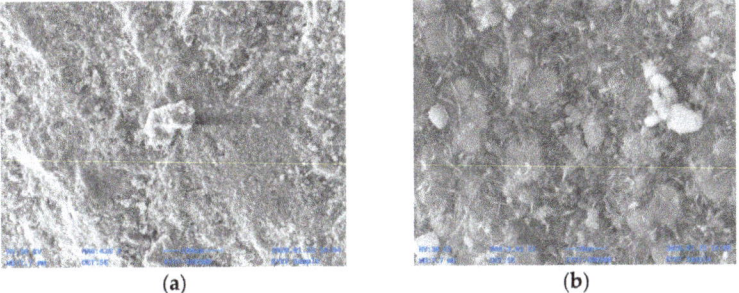

Figure 19. SEM image of OPC-0F mix specimen; (**a**) magnification 428×; and (**b**) magnification 3.01×.

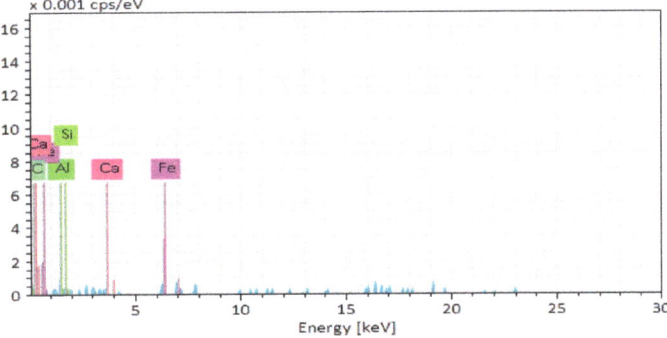

Figure 20. EDS graph of GPC-A0F mix specimen.

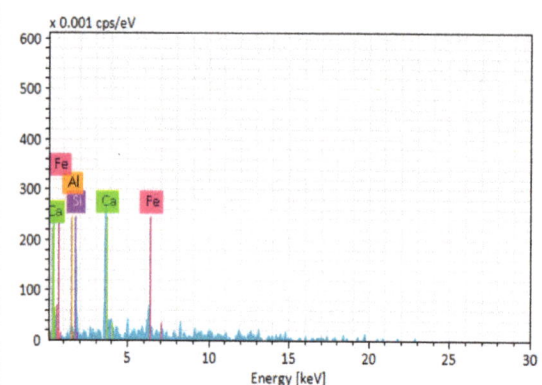

Figure 21. EDS graph of GPC-B0F mix specimen.

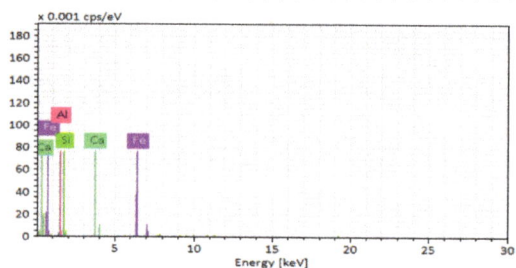

Figure 22. EDS graph of GPC-C0F mix specimen.

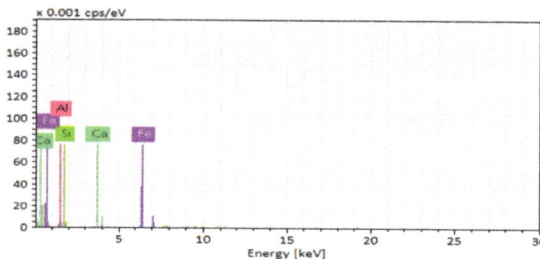

Figure 23. EDS graph of GPC-D0F mix specimen.

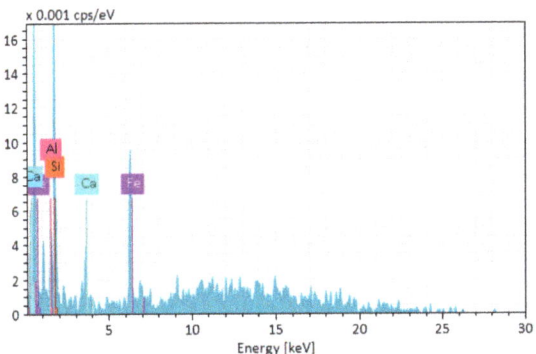

Figure 24. EDS graph of GPC-E0F mix specimen.

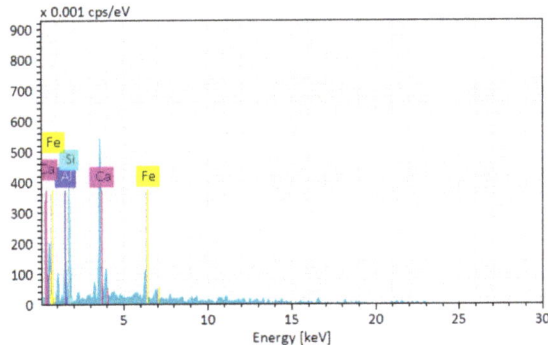

Figure 25. EDS graph of OPC-0F mix specimen.

4. Conclusions

This paper presented the results of an experimental study conducted to evaluate the influence of QRD and SF inclusion on fresh, mechanical, and residual properties of FA- and SG-based geopolymer concrete at ambient and elevated temperatures. The following key conclusions have been drawn from this study:

1. The workability of GPC mixes decreased by increasing the QRD content and by incorporating SF. QRD has a negative effect on the workability of GPC mixes.
2. The increase in QRD content from 0% to 15% resulted in an increase in the compressive strength of all GPC mixes at 28 and 56 days. The maximum increase in compressive strength was noticed for the 15% replacement level of QRD.
3. The addition of 0.75% SF increased the compressive strength of both OPC and GPC mixes but decreased their workability.
4. The tensile and flexural strength of GPC mixes increased at early ages (maximum up to 28 days) with the increase in QRD content and incorporation of SF. However, a reduction was observed in tensile and flexural strength with the increase of the volume fraction of SF from 0.75% to 1.5%.
5. After heating the GPC specimens at elevated temperatures, the weight loss consistently increased, and residual compressive strength respectively decreased with the increase in QRD content. However, the inclusion of SF reduced the loss in compressive strength of GPC mixes after exposure to elevated temperature.
6. The results of XRD analysis showed that the crystallinity of the geopolymer structure increased by increasing QRD content up to 15% of the total binder content.
7. The SEM analysis exhibited that increasing the QRD content up to 15% improved the mechanical properties due to a dense, less porous, and more compacted microstructure.

8. The EDS analysis showed that high content of calcium compounds improved mechanical properties of GPC specimens

It is worth noting that during the mixing operations, it is difficult to handle GPC at the field due to exothermic reaction and the harmful effect of alkaline solutions on the human body and cloth during use. Therefore, efforts are needed to produce ambient cured GPC using solid activators instead of alkaline solutions, given its wider acceptance in the field.

Author Contributions: Conceptualization, M.I. and F.B.; methodology, M.I., F.B. and R.M.W.; software, M.I. and K.H.; validation, F.B., M.I. and N.A.; formal analysis, M.I. and R.F.T.; investigation, M.I. and K.U.; data curation, M.I.; writing—original draft preparation, M.I.; writing—review and editing, F.B., R.F.T. and M.A.M.; supervision, F.B. All authors have read and agreed to the published version of the manuscript.

Funding: The research is partially funded by the Ministry of Science and Higher Education of the Russian Federation as part of the World-class Research Center program: Advanced Digital Technologies (contract No. 075-15-2020-934 dated 17 October 2020).

Institutional Review Board Statement: Not applicable.

Informed Consent Statement: Not applicable.

Data Availability Statement: Not applicable.

Conflicts of Interest: The authors declare that they have no competing interests.

Abbreviations

QRD	quarry rock dust
SG	ground granulated blast furnace slag
FA	fly ash
SP	superplasticizer
SF	steel fiber
W/C	water-cement ratio

References

1. Tamás, F.; Balázs, G. *Properties of Concrete*, 4th ed.; Neville, A.M., Ed.; Longman Scientific & Technical Ltd.: London, UK, 1995; ISBN 0-582-23070-5.
2. Turner, L.K.; Collins, F.G. Carbon dioxide equivalent (CO2-e) emissions: A comparison between geopolymer and OPC cement concrete. *Constr. Build. Mater.* **2013**, *43*, 125–130. [CrossRef]
3. McLellan, B.C.; Williams, R.P.; Lay, J.; van Riessen, A.; Corder, G.D. Costs and carbon emissions for geopolymer pastes in comparison to ordinary portland cement. *J. Clean. Prod.* **2011**, *19*, 1080–1090. [CrossRef]
4. Bumanis, G.; Vitola, L.; Pundiene, I.; Sinka, M.; Bajare, D. Gypsum, Geopolymers, and Starch—Alternative Binders for Bio-Based Building Materials: A Review and Life-Cycle Assessment. *Sustainability* **2020**, *12*, 5666. [CrossRef]
5. Duxson, P.; Fernández-Jiménez, A.; Provis, J.L.; Lukey, G.C.; Palomo, A.; van Deventer, J.S.J. Geopolymer technology: The current state of the art. *J. Mater. Sci.* **2006**, *42*, 2917–2933. [CrossRef]
6. Komnitsas, K.A. Potential of geopolymer technology towards green buildings and sustainable cities. *Procedia Eng.* **2011**, *21*, 1023–1032. [CrossRef]
7. Nematollahi, B.; Ranade, R.; Sanjayan, J.; Ramakrishnan, S. Thermal and mechanical properties of sustainable lightweight strain hardening geopolymer composites. *Arch. Civ. Mech. Eng.* **2017**, *17*, 55–64. [CrossRef]
8. Phummiphan, I.; Horpibulsuk, S.; Rachan, R.; Arulrajah, A.; Shen, S.L.; Chindaprasirt, P. High calcium fly ash geopolymer stabilized lateritic soil and granulated blast furnace slag blends as a pavement base material. *J. Hazard. Mater.* **2018**, *341*, 257–267. [CrossRef]
9. Boca Santa, R.A.A.; Soares, C.; Riella, H.G. Geopolymers with a high percentage of bottom ash for solidification/immobilization of different toxic metals. *J. Hazard. Mater.* **2016**, *318*, 145–153. [CrossRef]
10. Nematollahi, B.; Sanjayan, J.; Shaikh, F.U.A. Synthesis of heat and ambient cured one-part geopolymer mixes with different grades of sodium silicate. *Ceram. Int.* **2015**, *41*, 5696–5704. [CrossRef]
11. Wardhono, A.; Gunasekara, C.; Law, D.W.; Setunge, S. Comparison of long term performance between alkali activated slag and fly ash geopolymer concretes. *Constr. Build. Mater.* **2017**, *143*, 272–279. [CrossRef]
12. Singh, B.; Rahman, M.R.; Paswan, R.; Bhattacharyya, S.K. Effect of activator concentration on the strength, ITZ and drying shrinkage of fly ash/slag geopolymer concrete. *Constr. Build. Mater.* **2016**, *118*, 171–179. [CrossRef]

13. Abdulkareem, O.A.; Mustafa Al Bakri, A.M.; Kamarudin, H.; Khairul Nizar, I.; Saif, A.e.A. Effects of elevated temperatures on the thermal behavior and mechanical performance of fly ash geopolymer paste, mortar and lightweight concrete. *Constr. Build. Mater.* **2014**, *50*, 377–387. [CrossRef]
14. Soutsos, M.; Boyle, A.P.; Vinai, R.; Hadjierakleous, A.; Barnett, S.J. Factors influencing the compressive strength of fly ash based geopolymers. *Constr. Build. Mater.* **2016**, *110*, 355–368. [CrossRef]
15. Nguyen, K.T.; Ahn, N.; Le, T.A.; Lee, K. Theoretical and experimental study on mechanical properties and flexural strength of fly ash-geopolymer concrete. *Constr. Build. Mater.* **2016**, *106*, 65–77. [CrossRef]
16. Junaid, M.T.; Khennane, A.; Kayali, O.; Sadaoui, A.; Picard, D.; Fafard, M. Aspects of the deformational behaviour of alkali activated fly ash concrete at elevated temperatures. *Cem. Concr. Res.* **2014**, *60*, 24–29. [CrossRef]
17. Junaid, M.T.; Khennane, A.; Kayali, O. Performance of fly ash based geopolymer concrete made using non-pelletized fly ash aggregates after exposure to high temperatures. *Mater. Struct.* **2014**, *48*, 3357–3365. [CrossRef]
18. Wallah, S.; Rangan, B.V. *Low-Calcium Fly Ash-Based Geopolymer Concrete: Long-Term Properties*; Curtin University of Technology: Bentley, WA, Australia, 2006.
19. Parveen; Singhal, D.; Junaid, M.T.; Jindal, B.B.; Mehta, A. Mechanical and microstructural properties of fly ash based geopolymer concrete incorporating alccofine at ambient curing. *Constr. Build. Mater.* **2018**, *180*, 298–307. [CrossRef]
20. Mermerdaş, K.; Algın, Z.; Oleiwi, S.M.; Nassani, D.E. Optimization of lightweight GGBFS and FA geopolymer mortars by response surface method. *Constr. Build. Mater.* **2017**, *139*, 159–171. [CrossRef]
21. Somna, K.; Jaturapitakkul, C.; Kajitvichyanukul, P.; Chindaprasirt, P. NaOH-activated ground fly ash geopolymer cured at ambient temperature. *Fuel* **2011**, *90*, 2118–2124. [CrossRef]
22. Rashad, A.M. A comprehensive overview about the influence of different admixtures and additives on the properties of alkali-activated fly ash. *Mater. Des.* **2014**, *53*, 1005–1025. [CrossRef]
23. Temuujin, J.; van Riessen, A.; Williams, R. Influence of calcium compounds on the mechanical properties of fly ash geopolymer pastes. *J. Hazard. Mater.* **2009**, *167*, 82–88. [CrossRef] [PubMed]
24. Nath, P.; Sarker, P.K. Effect of GGBFS on setting, workability and early strength properties of fly ash geopolymer concrete cured in ambient condition. *Constr. Build. Mater.* **2014**, *66*, 163–171. [CrossRef]
25. Altun, Y.; Doğan, M.; Bayramlı, E. The effect of red phosphorus on the fire properties of intumescent pine wood flour–LDPE composites. *Fire Mater.* **2016**, *40*, 697–703. [CrossRef]
26. Waqas, R.M.; Butt, F.; Zhu, X.; Jiang, T.; Tufail, R.F. A Comprehensive Study on the Factors Affecting the Workability and Mechanical Properties of Ambient Cured Fly Ash and Slag Based Geopolymer Concrete. *Appl. Sci.* **2021**, *11*, 8722. [CrossRef]
27. Ismail, I.; Bernal, S.A.; Provis, J.L.; Hamdan, S.; van Deventer, J.S.J. Microstructural changes in alkali activated fly ash/slag geopolymers with sulfate exposure. *Mater. Struct.* **2012**, *46*, 361–373. [CrossRef]
28. Chi, M.; Huang, R. Binding mechanism and properties of alkali-activated fly ash/slag mortars. *Constr. Build. Mater.* **2013**, *40*, 291–298. [CrossRef]
29. Venu Madhav, T.; Ramana Reddy, I.V.; Ghorpade, V.G.; Jyothirmai, S. Compressivestrength study of geopolymer mortar using quarry rock dust. *Mater. Lett.* **2018**, *231*, 105–108. [CrossRef]
30. Venkata Sairam Kumar, N.; Sai Ram, K.S. Experimental study on properties of concrete containing crushed rock dust as a partial replacement of cement. *Mater. Today: Proc.* **2018**, *5*, 7240–7246. [CrossRef]
31. Meisuh, B.K.; Kankam, C.K.; Buabin, T.K. Effect of quarry rock dust on the flexural strength of concrete. *Case Stud. Constr. Mater.* **2018**, *8*, 16–22. [CrossRef]
32. Hussain, K.; Butt, F.; Alwetaishi, M.; Waqas, R.M.; Aslam, F.; Ibraheem, M.; Xulong, Z.; Ahmad, N.; Tufail, R.F.; Musarat, M.A.; et al. Effect of Quarry Rock Dust as a Binder on the Properties of Fly Ash and Slag-Based Geopolymer Concrete Exposed to Ambient and Elevated Temperatures. *Appl. Sci.* **2021**, *11*, 9192. [CrossRef]
33. Waqas, R.M.; Butt, F. Behavior of Quarry Rock Dust, Fly Ash and Slag Based Geopolymer Concrete Columns Reinforced with Steel Fibers under Eccentric Loading. *Appl. Sci.* **2021**, *11*, 6740. [CrossRef]
34. Sayyad, A.S.; Patankar, S.V. Effect of Steel Fibres and Low Calcium Fly Ash on Mechanical and Elastic Properties of Geopolymer Concrete Composites. *Indian J. Mater. Sci.* **2013**, *2013*, 1–8. [CrossRef]
35. Ganesan, N.; Indira, P. Engineering properties of steel fibre reinforced geopolymer concrete. *Adv. Concr. Constr.* **2013**, *1*, 305. [CrossRef]
36. Yunsheng, Z.; Wei, S.; Zongjin, L.; Xiangming, Z.; Eddie; Chungkong, C. Impact properties of geopolymer based extrudates incorporated with fly ash and PVA short fiber. *Constr. Build. Mater.* **2008**, *22*, 370–383. [CrossRef]
37. Lin, C.; Kayali, O.; Morozov, E.V.; Sharp, D.J. Influence of fibre type on flexural behaviour of self-compacting fibre reinforced cementitious composites. *Cem. Concr. Compos.* **2014**, *51*, 27–37. [CrossRef]
38. Islam, A.; Alengaram, U.J.; Jumaat, M.Z.; Ghazali, N.B.; Yusoff, S.; Bashar, I.I. Influence of steel fibers on the mechanical properties and impact resistance of lightweight geopolymer concrete. *Constr. Build. Mater.* **2017**, *152*, 964–977. [CrossRef]
39. Al-mashhadani, M.M.; Canpolat, O.; Aygörmez, Y.; Uysal, M.; Erdem, S. Mechanical and microstructural characterization of fiber reinforced fly ash based geopolymer composites. *Constr. Build. Mater.* **2018**, *167*, 505–513. [CrossRef]
40. Łach, M.; Korniejenko, K.; Mikuła, J. Thermal insulation and thermally resistant materials made of geopolymer foams. *Procedia Eng.* **2016**, *151*, 410–416. [CrossRef]

41. Shill, S.K.; Al-Deen, S.; Ashraf, M.; Hutchison, W. Resistance of fly ash based geopolymer mortar to both chemicals and high thermal cycles simultaneously. *Constr. Build. Mater.* **2020**, *239*, 117886. [CrossRef]
42. Silva, G.; Salirrosas, J.; Ruiz, G.; Kim, S.; Nakamatsu, J.; Aguilar, R. Evaluation of fire, high-temperature and water erosion resistance of fiber-reinforced lightweight pozzolana-based geopolymer mortars. In Proceedings of the IOP Conference Series: Materials Science and Engineering, Bangkok, Thailand, 17–19 May 2019; p. 012016.
43. Crook, D.; Murray, M. Regain of strength after firing of concrete. *Mag. Concr. Res.* **1970**, *22*, 149–154. [CrossRef]
44. Mendes, A.F.B. *Fire Resistance of Concrete Made with Portland and Slag Blended Cements*; Monash University: Clayton, VIC, Australia, 2010.
45. Chen, B.; Liu, J. Residual strength of hybrid-fiber-reinforced high-strength concrete after exposure to high temperatures. *Cem. Concr. Res.* **2004**, *34*, 1065–1069. [CrossRef]
46. Poon, C.S.; Shui, Z.; Lam, L. Compressive behavior of fiber reinforced high-performance concrete subjected to elevated temperatures. *Cem. Concr. Res.* **2004**, *34*, 2215–2222. [CrossRef]
47. Suhaendi, S.L.; Horiguchi, T. Effect of short fibers on residual permeability and mechanical properties of hybrid fibre reinforced high strength concrete after heat exposition. *Cem. Concr. Res.* **2006**, *36*, 1672–1678. [CrossRef]
48. Turkey, F.A.; Beddu, S.B.; Ahmed, A.N.; Al-Hubboubi, S. A review—Behaviour of geopolymer concrete to high temperature. *Mater. Today: Proc.* **2021**. [CrossRef]
49. Bilek, V.; Sucharda, O.; Bujdos, D. Frost Resistance of Alkali-Activated Concrete—An Important Pillar of Their Sustainability. *Sustainability* **2021**, *13*, 473. [CrossRef]
50. Shaikh, F.U.A.; Hosan, A. Mechanical properties of steel fibre reinforced geopolymer concretes at elevated temperatures. *Constr. Build. Mater.* **2016**, *114*, 15–28. [CrossRef]
51. American Society for Testing and Materials. *Standard Specification for Portland Cement*; ASTM C 150; ASTM: West Conshohocken, PA, USA, 2009.
52. American Society for Testing and Materials. *Standard Specification for Steel Fibers for Fiber Reinforced Concrete*; ASTM: West Conshohocken, PA, USA, 2011.
53. American Society for Testing and Materials. *Standard Specification for Chemical Admixtures for Concrete*; ASTM: West Conshohocken, PA, USA, 2013.
54. American Society for Testing and Materials. *Standard Test Method for Slump of Hydraulic-Cement Concrete*; C143/C143M-12; ASTM: West Conshohocken, PA, USA, 2012.
55. American Society for Testing and Materials. *Standard Test Method for Compressive Strength of Cylindrical Concrete Specimens*; ASTM C39/C39M-14; ASTM: West Conshohocken, PA, USA, 2014.
56. American Society for Testing and Materials. *Standard Test Method for Flexural Performance of Fiber-Reinforced Concrete (Using Beam with Third-Point Loading)*; ASTM-C-1609; ASTM: West Conshohocken, PA, USA, 2012.
57. Ahmed, G.N.; Hurst, J.P. Modeling the thermal behavior of concrete slabs subjected to the ASTM E119 standard fire condition. *J. Fire Prot. Eng.* **1995**, *7*, 125–132. [CrossRef]
58. Reddy, M.S.; Dinakar, P.; Rao, B.H. Mix design development of fly ash and ground granulated blast furnace slag based geopolymer concrete. *J. Build. Eng.* **2018**, *20*, 712–722. [CrossRef]
59. American Concrete Institute. *Building Code Requirements for Structural Concrete (ACI 318-11) and Commentary*; ACI: Farmington Hills, MI, SUA, 2011.
60. Dutta, D.; Ghosh, S. Effect of lime stone dust on geopolymerisation and geopolymeric structure. *Int. J. Emerg. Technol. Adv. Eng.* **2012**, *2*, 757–763.
61. Hake, S.L.; Damgir, R.M.; Patankar, S.V. Temperature Effect on Lime Powder-Added Geopolymer Concrete. *Adv. Civ. Eng. 2018*, *2018*, 1–5. [CrossRef]
62. Oluokun, F.A.; Burdette, E.G.; Deatherage, J.H. Splitting tensile strength and compressive strength relationships at early ages. *Mater. J.* **1991**, *88*, 115–121.
63. Omer, S.A.; Demirboga, R.; Khushefati, W.H. Relationship between compressive strength and UPV of GGBFS based geopolymer mortars exposed to elevated temperatures. *Constr. Build. Mater.* **2015**, *94*, 189–195. [CrossRef]
64. Škvára, F.; Jílek, T.; Kopecký, L. Geopolymer materials based on fly ash. *Ceramics-Silik* **2005**, *49*, 195–204.
65. Tchakoute, H.; Rüscher, C.; Djobo, J.Y.; Kenne, B.; Njopwouo, D. Influence of gibbsite and quartz in kaolin on the properties of metakaolin-based geopolymer cements. *Appl. Clay Sci.* **2015**, *107*, 188–194. [CrossRef]

Article

Development and Characteristics of Aerated Alkali-Activated Slag Cement Mixed with Zinc Powder

Taewan Kim [1], Choonghyun Kang [2,*] and Kiyoung Seo [3]

[1] Department of Civil Engineering, Pusan National University, Busan 46241, Korea; ring2014@naver.com
[2] Department of Ocean Civil Engineering, Gyeongsang National University, Tongyeong 53064, Korea
[3] HK Engineering and Consultants, Busan 46220, Korea; aricari@hanmail.net
* Correspondence: chkang@gnu.ac.kr; Tel.: +82-55-772-9124

Abstract: Experiments on the development and properties of aerated concrete based on alkali-activated slag cement (AASC) and using Zn powder (ZP) as a gas agent were carried out. The experiments were designed for water-binding material (w/b) ratios of 0.35 and 0.45, curing temperatures of 23 ± 2 °C and 40 ± 2 °C, and ZP of 0.25%, 0.50%, 0.75%, and 1.0%. ZP generates hydrogen (H_2) gas in AASC to form pores. At a w/b of 0.35, the curing temperature had little effect on the pore size by ZP. However, a w/b of 0.45 showed a clear correlation that the pore diameter increased as the curing temperature increased. The low w/b of 0.35 showed a small change in the pore size according to the curing temperature due to the faster setting time than 0.45 and the increased viscosity of the paste. Therefore, at a termination time exceeding at least 60 min and a w/b of 0.45 or more, it was possible to increase the size and expansion force of the pores formed by the ZP through the change of the curing temperature. ZP showed applicability to the manufacture of AASC-based aerated concrete, and the characteristics of foaming according to the curing temperature, w/b ratio, and ZP concentration were confirmed.

Keywords: zinc powder; gas agent; alkali-activated slag cement; hydrogen gas; aerated concrete

1. Introduction

Alkali-activated slag cement (AASC) has attracted significant attention as an eco-friendly material compared to ordinary Portland cement (OPC) [1–6]. Many studies have shown that AASC has high strength and durability [7–10]. AASC is widely applicable in various members of construction, and aerated concrete is one of them. Recently, even aerated cement/concrete to which AASC is applied has been researched and developed. Aerated concrete is developed and manufactured to improve the characteristics of cement/concrete, such as weight reduction, water permeability, and thermal insulation. The materials used for foaming are a foaming agent [11–14], metallic powder (Al, Zn) [15–17], or hydrogen peroxide (H_2O_2) [18–21]. Aluminum powder (Al powder) is the most commonly used material for foaming, as mentioned in several previous studies [16,17,22]. Al powder shows low usage and a high foaming effect. The foaming approach applies to a simple method for generating gas to form pores inside a specimen. However, the size, distribution, and quantity of the air foams are affected by various conditions. Several factors, such as binder material, admixture material, curing temperature, mixing ratio, and foaming agent, are considered [23–25]. Although aerated concrete can be manufactured quickly and easily, the compressive strength and durability are properties that need improvement. Recently, several studies on aerated concrete using various foaming agents for alkali-activated cement or geopolymer have been conducted [22,24,26–28]. For expanded cement/concrete using zinc powder (ZP), studies on OPC-based [23] and fly-ash-based geopolymer [29] or magnesium phosphate [15] have been reported; however, studies on AASC are still insufficient.

Therefore, in this study, an AASC-based experiment was designed to analyze the mechanical properties and bubble formation characteristics of aerated concrete. In this study, considering the high early-age strength characteristic reported as one of the characteristics of AASC, we intend to improve the mechanical characteristics that decrease after bubble formation. Herein, ZP was used as the foaming agent, different from the existing Al powder. It has been reported that the hydrogen gas (H_2) generated by the reaction of ZP is about half in quantity compared with aluminum powder and causes a relatively slow foaming reaction [29]. However, ZnO is potentially applicable in various fields, and several studies are currently being conducted on the multifunctionality of ZnO in cement/concrete. The zinc oxide applied to cement is a material reported to be effective in antibacterial concrete [30,31], radiation shielding [32], water purification [33], anticorrosion of steel [34], and photocatalysis [35,36].

The Equation (1) shows that ZP powder reacts with water to produce zinc oxide (ZnO) and H_2. The generated H_2 expands inside the cement, forming pores.

$$Zn + H_2O \rightarrow ZnO + H_2 \quad (1)$$

The ZnO is insoluble in water but soluble in alkaline or acidic environments [37,38]. By examining the various effects of cement using ZnO reported currently, it is expected that the effect of zinc will be improved even more if the application field is expanded and the optimum zinc oxide concentration and mixing conditions are found. Therefore, it is thought that aerated concrete using Zn powder can be broadly applied to multifunctional concrete that basically includes the effect of zinc oxide. For this, it is judged that experiments and studies on the mixing conditions and characteristics of aerated concrete mixed with ZP should be conducted. Furthermore, at the same time, we would like to present the purpose for the development and application of a new metal powder that can replace aluminum powder, which is the existing gas agent used in the production of aerated concrete. AASC has high alkalinity, fast setting, and high strength. Thus, hydration reactants are formed through the dissolution of ZnO in the high alkali environment of AASC. Several studies have been conducted on OPC mixed with ZnO. However, few studies have investigated the application of ZnO to AASC. Moreover, studies on AASC using ZP are rare. A study was recently conducted using 0.3–0.8% ZP in magnesium phosphate cement (MPC) [15]. Additionally, a study exists on porous cement manufacturing using ZP for rapid MPC. Similarly, ZP can be applied to AASC based on the MPC result with rapid setting high strength performance. Therefore, this study has two purposes. The first is the development of aerated cement using AASC, which is attracting attention as an eco-friendly, low-carbon cement. In addition, AASC has faster setting and higher early-age strength than OPC-based cement. This is expected to improve the mechanical performance degradation problem of OPC-based aerated cement to some extent. The second is to examine the influence and characteristics of ZP on aerated AASC. This is because studies on aerated cement using ZP in AASC are very rare. From the results of this experiment, we will examine the development and basic characteristics of aerated AASC using ZP, and plan additional experiments on the size, distribution, and durability of bubbles in subsequent studies. The experiments and analysis on the mechanical properties according to the concentration, formulation, and curing conditions of ZP are first performed. Furthermore, based on the results of the mechanical properties, follow-up studies on antibacterial, photocatalysis, and anticorrosion of steel are planned. For this investigation, the compressive strength, X-ray diffractometer (XRD) spectra, scanning electron microscope (SEM) images, water absorption rate, and ultrasonic pulse velocity (UPV) were measured and analyzed.

2. Materials and Methods

2.1. Materials

Table 1 shows the results of X-ray fluorescence analysis on the chemical composition of the slag used in this study. For alkali activator, bead type sodium hydroxide (NaOH, purity $\geq$ 98%) and liquid type sodium silicate (Na_2SiO_3, Ms = 2.1) were used at a 10%

concentration of binder weight (10% NaOH + 10% Na$_2$SiO$_3$). Before mixing, the activator was added to water, stirred well, and allowed to stand at room temperature in the laboratory for 6 h and then used. The ZP has a gray color, the specific gravity of 7.14, a pH of 6.95–7.37, and an average particle size of 4.0 μm with a purity ≥99.0%.

Table 1. Chemical components and physical properties used in slag.

	Chemical Components (%)							Density (g/cm^3)	Fineness (cm^2/kg)	LOI (%)
	SiO$_2$	Al$_2$O	Fe$_2$O$_3$	MgO	CaO	K$_2$O	SO$_3$			
Slag	34.57	10.88	0.61	4.19	44.56	0.37	3.94	2.89	4200	0.96

2.2. Experiments

The experimental design considered the effect of three variables: the water-binder ratio (w/b), curing temperature, and ZP content. Here, the binder consists only of 100% slag. Furthermore, the w/b ratio was selected to exclude the effect of superplasticizer. If it is less than w/b = 0.35, mixing and molding of the specimen are difficult, and if w/b = 0.45, material separation occurs due to excessive fluidity. Therefore, the final w/b ratio of 0.35 and 0.45 was selected in the range where the superplasticizer was not used. The authors selected three curing temperatures: 23 ± 2 °C, 40 ± 2 °C, and 60 ± 2 °C through preliminary experiments. However, at a temperature of 60 ± 2 °C, the specimen was not properly formed due to the rapid expansion of ZP. As a result, the curing temperature of the final experimental plan was selected as 23 ± 2 °C and 40 ± 2 °C. Finally, the ZP content was a total of five concentrations, including 0.0%, 0.25%, 0.50%, 0.75%, and 1.00% of the weight of the binder and a mixture without ZP. A total of 20 mixtures were produced. Table 2 summarizes the ratio for the detailed mixture.

Table 2. Mix properties.

w/b	Curing Temperature (°C)	ZP Contents (%)
0.35	23 ± 2 °C	0.00
		0.25
		0.50
		0.75
		1.00
0.45		0.00
		0.25
		0.50
		0.75
		1.00
0.35	40 ± 2 °C	0.00
		0.25
		0.50
		0.75
		1.00
0.45		0.00
		0.25
		0.50
		0.75
		1.00

Mixing was conducted following the instrument and method of ASTM C305 [39]. The mixed samples were poured into a 50 × 50 × 50 mm^3 cube metal mold, compacted, and stored in a chamber at 23 ± 2 °C or 40 ± 2 °C and relative humidity (RH) of 90 ± 5% for 24 h. The mold was then removed and stored in the chamber at 23 ± 2 °C and RH of 90 ± 5% until 28 d. The compressive strength was measured after 1, 3, 7, and 28 d, and the average of the measured values of the three samples was used. Microstructural analysis was performed using an XRD with a step size of 0.017° (2θ) from 5° to 60° and a scanning electron microscope/backscattered electron (SEM/BSE) at 15 kV in high vacuum mode. For the physical properties, water absorption rate and UPV were measured.

Table 3 shows the setting time and flow values of the mixture without ZP at 23 ± 2 °C. The setting time was measured following the Gillmore needle test made in ASTM C266 [40]. The flow value of the paste was measured using the flow table apparatus of ASTM C230 [41]. For the mixture with a w/b of 0.35, the initial and final setting times were 20 and 25 min, respectively, and the total setting time is 45 min, setting in less than one hour. However, for the mixture with a w/b of 0.45, the initial and final setting times were 55 and 65 min, respectively, and the total setting time was 120 min, with a more rapid setting. The total setting time of the mixture with a w/b of 0.45 increased about two times compared with that with a w/b of 0.35. This increment is because the slag hydration reaction was delayed due to the dilution of the concentrated activator with the increase in mixed water for the same amount of binder. Moreover, the flow value of the mixture with a w/b of 0.35 was measured as 205 mm, whereas that with a w/b of 0.45 showed excessive fluidity, exceeding the flow table.

Table 3. Setting time and flow value of the mixture without ZP (23 ± 2 °C).

w/b	Setting Time (min)		Flow Value (mm)
	Initial	Final	
0.35	20	25	205
0.45	55	65	overflow

Water absorption was measured following the ASTM C1403 [42] method using a 50 × 50 × 50 mm³ sample. Dry density was calculated as follows.

$$\rho_d = \frac{m_d}{V}$$

where ρ_d is the dry density, m_d is the mass of the sample after oven-drying at 105 ± 5 °C for 24 h, and V is the volume of the sample.

The UPV was measured on a 40 × 40 × 160 mm³ prismatic mold sample. First, the receiver was contacted on the right and the oscillator on the left. Next, the measurement was conducted again with the oscillator on the right and the receiver on the left. The average value of both measurements for a sample was taken as one measurement value. UPV measurement was performed on three samples, and the average value was used as the UPV value.

3. Results and Discussion

3.1. Compressive Strength

Figure 1 shows the compressive strength measurement results according to the w/b ratio, curing temperature, and ZP concentration. Regardless of the w/b value and curing temperature, the ZP-containing samples showed low compressive strength values compared with the free-ZP samples. Previous studies have reported that aerated concrete has low compressive strength due to the foamed pores [24].

For the samples with a w/b of 0.35 (Figure 1a,b) and 0.45 (Figure 1c,d), the compressive strength values of the samples with a high w/b were low, with or without ZP. Herein, an increase in the w/b ratio was designed as a mixture in which mixed water increases in the same amount of binder. The concentration of the activator is 10% of the binder weight. Therefore, the mixed-water increment dilutes the concentration of the alkali solution. Consequently, the hydration reaction of the slag is reduced, affecting the decrease in strength. Figure 1a,b shows the results of the compressive strength of the samples with a w/b of 0.35. For the free-ZP samples, the compressive strength at all the measured ages increases as the curing temperature increases, i.e., the strength of the sample cured at 40 °C is more than that cured at 23 °C. However, for the ZP-containing samples, the compressive strength at 40 °C decreased slightly compared with that at 23 °C. The samples with a w/b of 0.45 showed a strong tendency to decrease the compressive strength with an increasing curing temperature (Figure 1c,d). This trend is more significant in the samples with a w/b of 0.45 than those with a w/b of 0.35 as the curing temperature increases. In previous studies of AASC, an increase in the curing temperature promoted the hydration reaction of slag and thus improved the strength. However, by using ZP in this study, no strength improvement was observed.

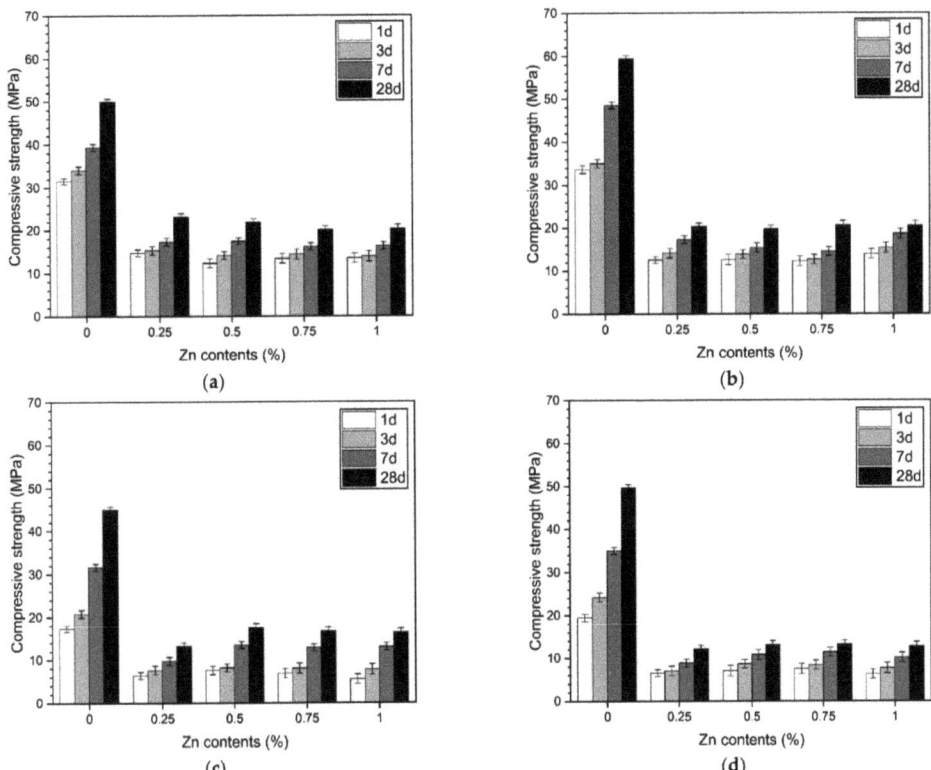

Figure 1. Compressive strength of samples with (**a**) w/b of 0.35, cured at 23 ± 2 °C, (**b**) w/b of 0.35, cured at 40 ± 2 °C, (**c**) w/b of 0.45, cured at 23 ± 2 °C, and (**d**) w/b of 0.45, cured at 40 ± 2 °C.

The decreasing tendency of the compressive strength, even with an increase in the curing temperature, can be considered as follows. ZP reacts with water to generate $Zn(OH)_2$ and H_2 inside the paste. At a curing temperature of 40 °C rather than 23 °C, the samples are set and hardened rapidly. Therefore, the time required for the expansion and movement of the H_2 is insufficient owing to the rapid setting. However, as shown in Table 3, the sample with a w/b of 0.45 has a longer setting time than the sample with a w/b of 0.35. Therefore, the samples with a w/b of 0.45 have enough time for the bubbles generated by the H2 gas in the paste to move and expand compared to the 0.35 samples. Therefore, the sample with a w/b of 0.45 has reduced strength due to the large diameter pores generated by the ZP. The high w/b ratio and the curing temperature at room temperature increase the setting time of the paste and increase the expansion and movement of the H_2, acting as a factor to increase the size of the pores.

Figure 2 shows the appearance of each demolded sample after a 24-h curing. Regardless of the w/b and curing temperature, the top surfaces of the ZP-mixed samples are expanded and swollen. To measure the compressive strength, the expanded part at the top was cut to a size of 50 × 50 × 50 mm^3 with a low-speed precision cutter. The upper parts of the samples in Figure 2c,d with a w/b of 0.45 are more expanded and swollen than those in Figure 2a,b with a w/b of 0.35. This trend suggests that, as described above, the expansion and movement of the H$_2$ caused by the reaction of ZP and water were more active in the samples with a w/b of 0.45 than those with a w/b of 0.35. The expansion height increases as the ZP concentration increases (Figure 2), indicating that, if the ZP amount is doubled (Equation (1)), the expansion height or volume will double as well. However, the ZP amount and the amount of expansion are not proportional. Thus, not all the generated gases are trapped inside the sample as some foams are destroyed or gases escape out of the sample. Such a quantity is difficult to precisely calculate. Similarly, in the study of aerated concrete using H$_2$O$_2$ as a foaming agent, it was reported that no linear proportional relationship existed between the amount of the foaming agent and the expansion rate [20].

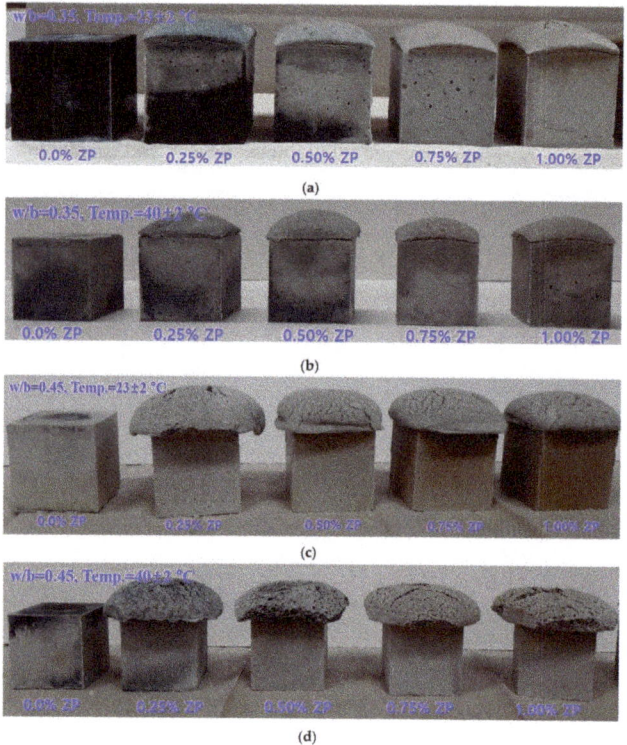

Figure 2. Sample appearances (after 24-h curing) of specimens with (**a**) w/b of 0.35, cured at 23 ± 2 °C, (**b**) w/b of 0.35, cured at 40 ± 2 °C, (**c**) w/b of 0.45, cured at 23 ± 2 °C, and (**d**) w/b of 0.45, cured at 40 ± 2 °C.

3.2. Reaction Products

Figures 3 and 4 show the XRD analysis results for determining the hydration reactants according to the w/b ratio, curing temperature, and ZP concentration. In the ZP-free samples, hydrotalcite, stratlingite, hydroganet, monocarboaluminate, C–S–H(I), C–S–H gel, calcite, and katoite were observed regardless of the w/b [9,28,43–45]. Table 4 summarizes the classification and types of the hydration reactants shown in Figures 3 and 4.

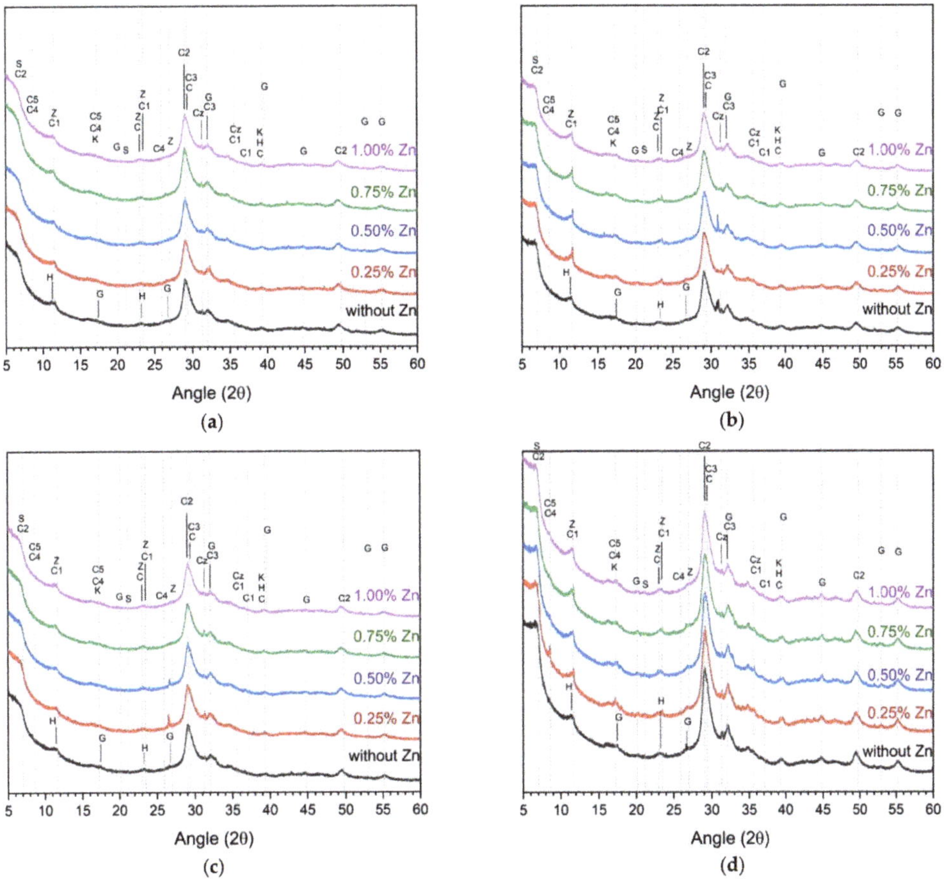

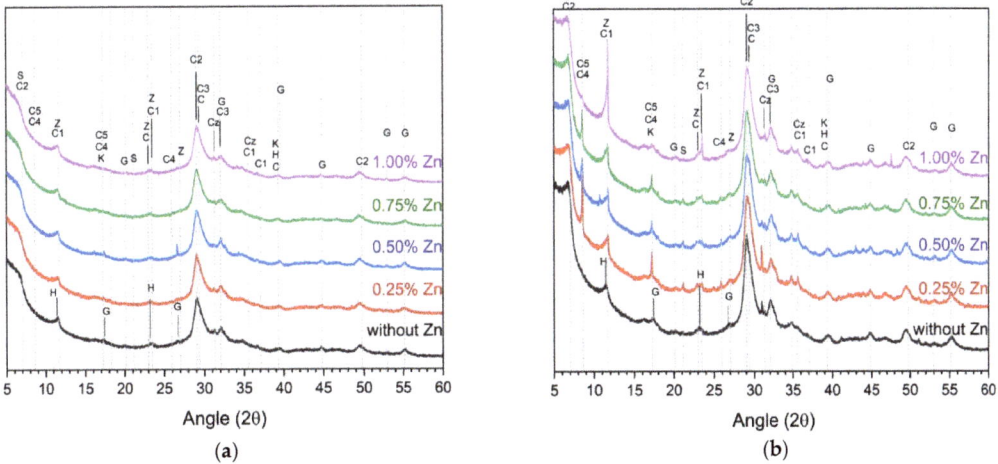

Figure 3. XRD analysis of samples with a w/b of 0.35 (**a**) cured at 23 ± 2 °C after 1 day, (**b**) cured at 23 ± 2 °C after 28 d, (**c**) cured at 40 ± 2 °C after 1 day, and (**d**) cured at 40 ± 2 °C after 28 d.

Figure 4. *Cont.*

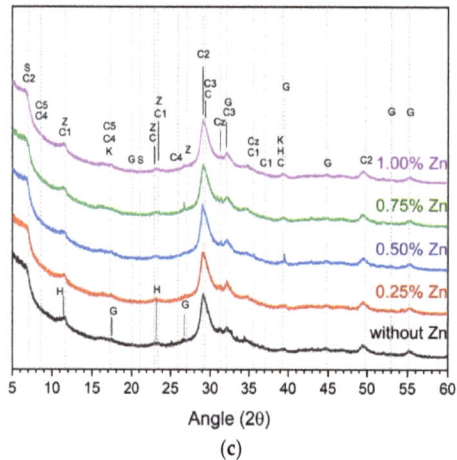
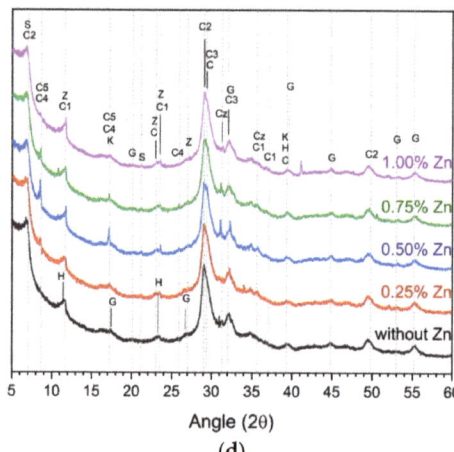

Figure 4. XRD analysis of samples with a w/b of 0.45 (**a**) cured at 23 ± 2 °C after 1 day, (**b**) cured at 23 ± 2 °C after 28 d, (**c**) cured at 40 ± 2 °C after 1 day, and (**d**) cured at 40 ± 2 °C after 28 d.

Table 4. Hydration reaction summary (in Figures 3 and 4).

Label	Hydration Reactant
C	calcite
C1	monocarboaluminate
C2	C–S–H(I)
C3	C–S–H gel
C4	calcium aluminum oxide sulfate hydrate
C5	calcium iron sulfate hydrate
H	Hydrotalcite
K	Katoite
S	Stratlingite
G	Hydrogarnet
Cz	Calcium zinc oxide (Wurzite type)
Z	Zinc hydroxide

Moreover, after 28 d as compared to one day, the slag hydration reaction increased and the peak heights of the stratlingite, hydrogarnet, C–S–H(I), and C–S–H increased. In the ZP-containing samples, calcium aluminum oxide sulfate hydrate, calcium iron sulfate hydrate, calcium ZnO, and zinc hydroxide were additionally observed. Precisely, calcium zinc oxide and zinc hydroxide are hydration reactants found in previous studies involving ZnO [36,46,47].

It has been reported that the crystalline phase of calcium zincate ($CaZn_2(OH)_6 \cdot 2H_2O$, wurtzite type) reduces the pozzolanic reaction by consuming the $Ca(OH)_2$ [48,49]. However, since there is no $Ca(OH)_2$ in AASC, it can be considered that the calcium zincate is formed by the presence of calcium eluted from the slag and OH-ions supplied from the activator. Other studies have reported that the ZnO interferes with the C–S–H-gel formation. ZP reacts with water to generate ZnO and H_2 gas (Equation (1)). ZnO forms Zn-based hydrates through the following reactions (Equations (2)–(4)). In the reactions up to Equations (2)–(4), the reaction proceeds by OH^- supplied from the alkali activator and calcium ions eluted from the slag [34,46,50,51]. Precisely, the environment in which OH^- ions are sufficiently supplied by the activator induces the reactions up to Equations (2)–(4) quickly.

$$ZnO + H_2O \rightarrow Zn^{2+} + 2OH^- \rightarrow Zn(OH)_2 \qquad (2)$$

$$ZnO + H_2O + 2OH^- \rightarrow Zn(OH)_4^{2-} \tag{3}$$

$$2Zn(OH)_4^{2-} + Ca^{2+} + H_2O \rightarrow Ca(Zn(OH)_3)_2 \cdot H_2O + 2OH^- \tag{4}$$

The addition of ZP slightly affected the hydration reaction product of AASC. No remarkable peak change of the hydration reaction product nor the formation of a new hydration reaction product was observed due to the change in the ZP concentration.

3.3. Microstructures

Figures 5 and 6 show SEM images according to the w/b and curing temperature. The shapes of the foams observed in Figures 5 and 6 are not completely spherical but elliptical, formed by aeration [52,53]. Figure 5a–d shows the cross-sectional SEM images of the samples cured at 23 °C, with a ZP concentration of 0.25–1.00%. Several voids were observed in the cross-sectional view, and some were connected. Even with increased ZP concentration, the diameters of the pores did not change significantly, showing similar sizes. Figure 5e–f shows the cross-sectional SEM images of the samples cured at 40 °C, with similar pore sizes and distributions regardless of the ZP concentration. The number of pores of such samples slightly increased compared with those cured at 23 °C.

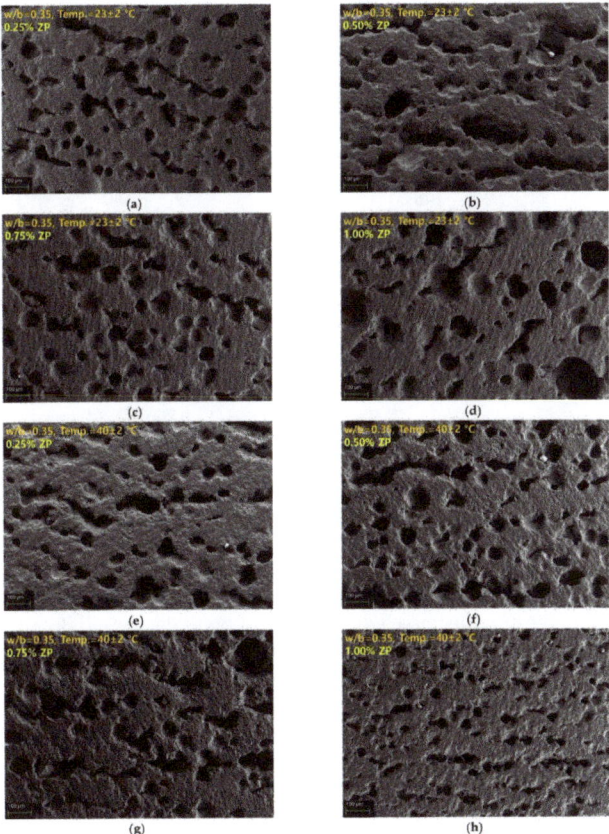

Figure 5. SEM images of samples with a w/b of 0.35 (**a**) cured at 23 ± 2 °C, 0.25% ZP, (**b**) cured at 23 ± 2 °C, 0.50% ZP, (**c**) cured at 23 ± 2 °C, 0.75% ZP, (**d**) cured at 23 ± 2 °C, 1.00% ZP, (**e**) cured at 40 ± 2 °C, 0.25% ZP, (**f**) cured at 40 ± 2 °C, 0.50% ZP, (**g**) cured at 40 ± 2 °C, 0.75% ZP, and (**h**) cured at 40 ± 2 °C, 1.00% ZP.

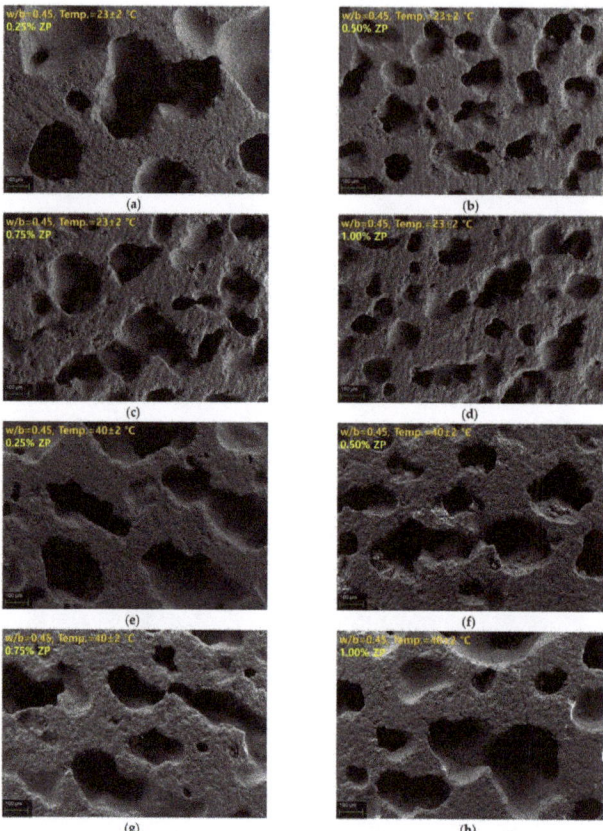

Figure 6. SEM images of samples with a w/b of 0.45 (**a**) cured at 23 ± 2 °C, 0.25% ZP, (**b**) cured at 23 ± 2 °C, 0.50% ZP, (**c**) cured at 23 ± 2 °C, 0.75% ZP, (**d**) cured at 23 ± 2 °C, 1.00% ZP, (**e**) cured at 40 ± 2 °C, 0.25% ZP, (**f**) cured at 40 ± 2 °C, 0.50% ZP, (**g**) cured at 40 ± 2 °C, 0.75% ZP, and (**h**) cured at 40 ± 2 °C, 1.00% ZP.

Figure 6 shows the cross-sectional SEM images of the samples with a w/b of 0.45 according to the curing temperature. Figure 6a–h show the cross-sectional SEM images of the samples cured at 25 °C and 40 °C, respectively. The diameters of the pores observed in the latter increased more than those observed in the former. Although the diameters of the pores increased, the number of pores decreased. Besides, the diameters of the samples with a w/b of 0.45 are 2–3 times larger than those with a w/b of 0.35, which agrees with the results of previous studies where the foam content increased as the w/b increased [26].

The pore diameter and distribution characteristics observed in the cross-sectional SEM images of the samples seem to be significantly affected by the w/b and curing temperature rather than the ZP concentration. A high w/b and low curing temperatures delay the slag hydration reaction and increase the setting time. This phenomenon gives sufficient time for the H_2 generated by the reaction of ZP and water to expand and move. Consequently, the pores present in the samples with a w/b of 0.45 are larger than the pores present in the samples with a w/b of 0.35. Besides, with a w/b ratio of 0.45, the pore diameter was larger for the samples cured at 40 °C than at 25 °C despite an increase in the curing temperature. Thus, the increase in the temperature affects the expansion of the H_2. However, for the samples with a w/b of 0.35, the pore diameter change according to the curing temperature was different from those with a w/b of 0.45. The samples with a w/b of 0.35 showed small-sized pores due to the fast-setting time, even at 25 °C. Increasing the curing temperature to 40 °C further inhibited the expansion and movement of the voids due to a faster setting. Consequently, with a w/b of 0.35, the difference in the pore diameter based on curing temperatures of 25 °C and 40 °C was insignificant. Despite increasing the curing temperature, the setting speed was faster than the expansion speed of the H_2, indicating that the pore diameter did not increase. To change the pore diameter by ZP, an appropriate setting time for smooth H_2 expansion and movement is necessary.

It has been reported in previous studies that an increase in the alkali activator concentration reduces the curing time, consequently reducing the foam pores' size [29]. Therefore, the rapid setting and hardening of AASC suppresses gas expansion. As a result, the size of the pores formed inside the specimen becomes smaller. The amount of the air foams increases as the amount of ZP increases, although the difference is insignificant. The difference in the amount and size of the air foam according to the amount of ZP is insignificant (Figure 5), indicating that, as the ZP increases, the number of foams increases. As the foams merge and the liquid film thickens, the size of the small foams increases, and, simultaneously, the breakdown of the foams increases [26,28]. However, the consolidation and destruction of these foams were difficult to observe in insufficiently grown foams due to the rapid setting time of the AASC. Therefore, when the w/b ratio of 0.45 is compared with that of 0.35, the consolidation of the air foams and the size increase are observed due to higher fluidity and longer setting time. This phenomenon becomes clear by comparing Figures 5 and 6. The increase in the foam content affects the decrease in compressive strength [26].

3.4. Water Absorption and Dry Density

Figure 7 shows the results of measuring the water absorption and ultrasonic pulse rates. The water absorption rate (Figure 7a) increased as the ZP concentration increased regardless of the w/b ratio and curing temperature. At a w/b of 0.35, the sample without ZP decreased from 25.58% to 23.78% when the curing temperature increased from 23 °C to 40 °C. Moreover, even with a w/b of 0.45, the water absorption rate of the sample without ZP decreased from 31.02% to 29.38% when the curing temperature increased from 23 °C to 40 °C.

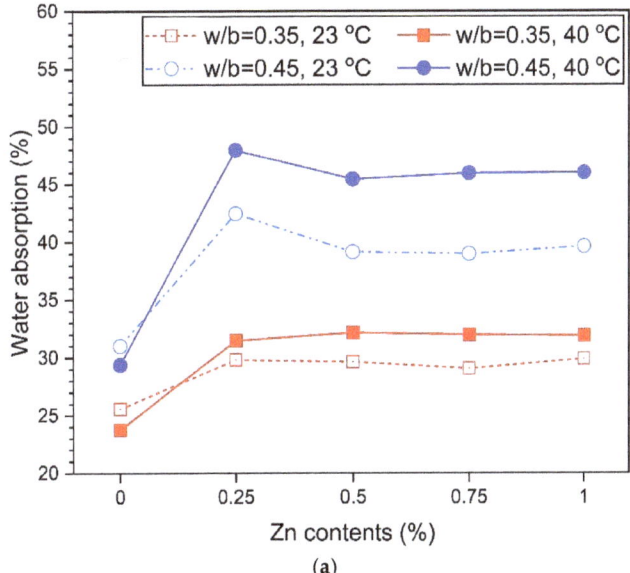

Figure 7. Cont.

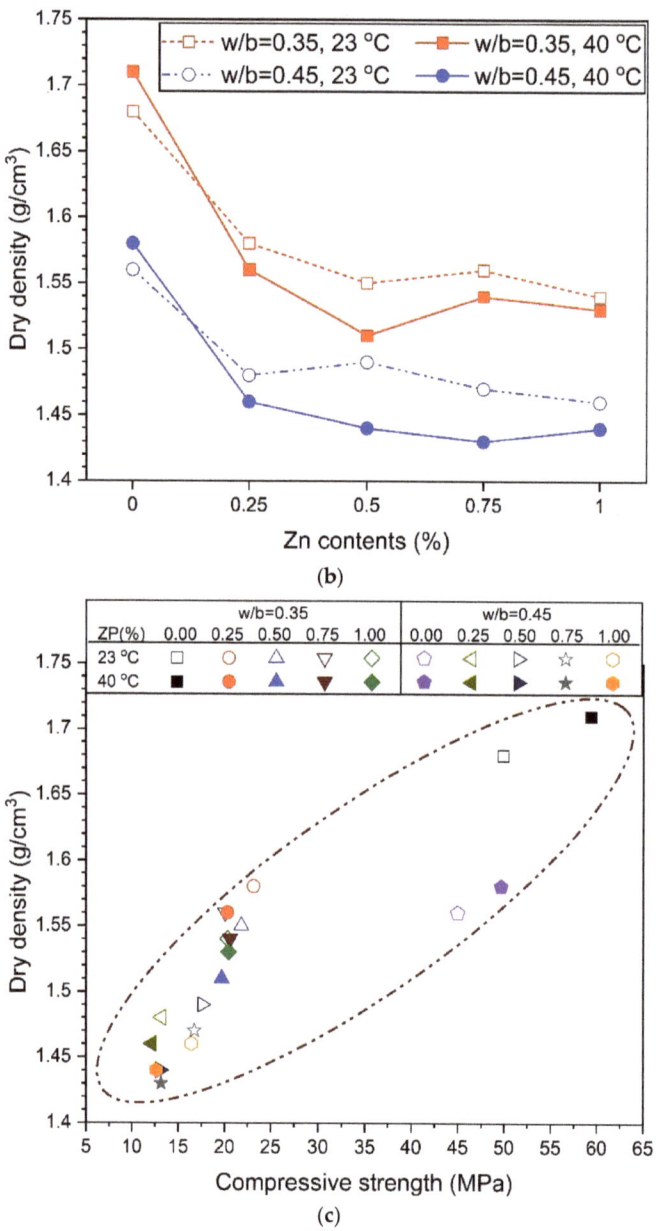

Figure 7. Water absorption and UPV variation with Zn content—(**a**) water absorption, (**b**) dry density, and (**c**) dry density vs. compressive strength.

With a w/b of 0.35, the difference between the absorption rates between the curing temperatures of 23 °C and 40 °C was <3%. The pore diameters and distribution were similar regardless of the curing temperature (Figure 5). Consequently, the difference in the absorption rate is also small. However, with a w/b of 0.45, the absorption rate was 5.5–7.0% at curing temperatures of 23 °C and 40 °C, showing a greater difference than with a w/b of 0.35. This trend shows the difference in the pore diameter based on the curing temperature (Figure 6). Thus, the sample cured at 40 °C has a larger pore diameter than that cured at 23 °C, indicating an increase in the water absorption space.

Figure 7b shows that the dry density is higher in the samples with a w/b of 0.35 than those with a w/b of 0.45. As shown in Figures 5 and 6, since the inner pore size of the samples with a w/b of 0.45 is more than that with a w/b of 0.35, the density is relatively smaller. According to previous studies, as the amount of the foaming agent increases, the dry density decreases. In Figure 7b, the samples with a w/b of 0.45 show similar trends to the results of previous studies. However, the sample with a w/b of 0.35 has the lowest density at 0.5% ZP, which increases slightly again at 0.75% and 1.0% ZP. This increment may be attributed to the merging and collapsing effects of the void structures [22,54]. In previous studies, it was reported that the compressive strength increases as the dry density increases [22,26]. Figure 7c shows the relationship between the dry density and compressive strength. The dry density and compressive strength were linearly proportional. That is, the higher the dry density, the higher the compressive strength. The results of Figure 7c are consistent with those of previous studies on aerated concrete that reported a linearly proportional relationship between the density and compressive strength [15,24].

3.5. Ultrasonic Pulse Velocity (UPV)

Figure 8 shows the measurement results of the UPV. For the ZP-free samples, the UPV also increases as the curing temperature increases from 23 °C to 40 °C, regardless of the w/b. However, as the ZP content increases, the UPV decreases, which may be attributed to the decrease in the voids in the matrix. Just as the difference in the absorption rate based on the curing temperature was insignificant and large with a w/b of 0.35 and 0.45, respectively, the difference in the UPV between the curing temperatures was larger with a w/b of 0.45 than that with a w/b of 0.35.

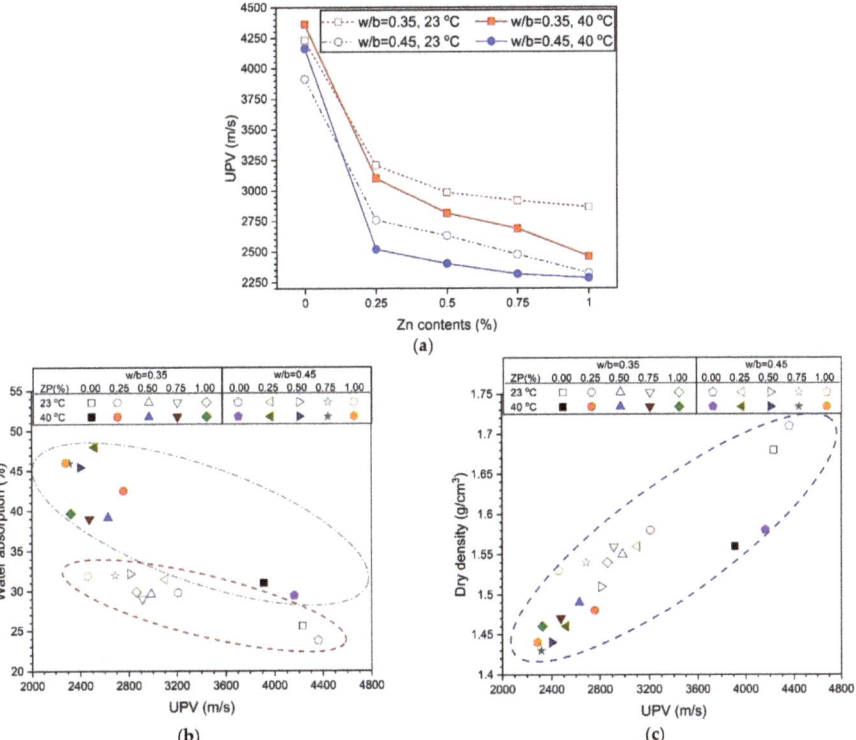

Figure 8. (a) Variations of UPV with Zn content, (b) water absorption with UPV, and (c) dry density with UPV.

The similarity between the water absorption rate and the UPV based on the w/b ratio and the curing temperature is examined in Figure 8b. A linear inverse relationship was established between the absorption rate and UPV. With a w/b of 0.35 and 0.45, the absorption rate and UPV are located in the upper left and lower right corners, respectively. An increase in the w/b indicates a high absorption rate and a low UPV. Figure 8c shows the correlation between the dry density and UPV. The dry density-UPV shows the relationship between the water absorption and UPV. The UPV increases as dry density increases, indicating a small number of voids. Therefore, increasing the ZP concentration decreases the dry density and UPV regardless of the w/b because voids are formed by H_2 generation due to the ZP mixing. Furthermore, as mentioned in the SEM of Figures 5 and 6, the samples with a w/b of 0.45 had larger pores than those with a w/b of 0.35. The density and UPV of the samples with a w/b of 0.45 were smaller than those with a w/b of 0.35.

4. Conclusions

The results of examining the strength and microstructural properties of AASC mixed with zinc powder (ZP) at a concentration of 0.25–1.0% are summarized as follows.

- ZP reduces the compressive strength regardless of the w/b and curing temperature. ZP reacts with water to generate H2 to form several pores inside the sample, causing a sharp decrease in the strength. Compared to the sample without ZP, a strength decrease of about 50% is observed. The larger the w/b ratio (0.45 > 0.35), the curing temperature (40 ± 2 °C > 23 ± 2 °C), and the concentration of ZP (1.0 > 0.75 > 0.5 > 0.25 > 0.0), the more the diameter of the bubbles formed inside the sample increases, which is a direct factor in the strength.
- The new hydration reactants by ZP are calcium zincate (Wurtzite type) and zinc hydroxide. No new hydration reactants were observed according to the concentration of ZP, w/b ratio, and curing temperature. In all the samples, the increase and sharpening of the C-S-H peaks over time were most pronounced. The decrease in the mechanical performance due to the formation of bubbles is a greater influence factor than the change in the hydration reactant due to the mixing of ZP.
- Increasing the ZP content increases the water absorption and decreases the dry density and UPV. Moreover, the difference in characteristics according to the w/b ratio and the curing temperature was clearly shown. Compared with a w/b = 0.35, at 0.45, the difference in the water absorption, dry density, and UPV according to the curing temperature was relatively large. The increase in the w/b ratio shows the effect of the diameter and distribution of the pores formed by ZP on the absorption rate, density, and UPV of the sample.
- The bubble formation characteristics according to the curing temperature show different aspects according to the w/b. The w/b = 0.35 had little effect on the pore size and distribution of the samples cured at 23 ± 2 °C and 40 ± 2 °C. However, in the case of W / B in 0.45, the difference in the pore diameter according to the curing temperature was significant. That is, it was observed that the pore diameter becomes larger at 40 ± 2 °C than 23 ± 2 °C. This tendency is that, the smaller the w/b ratio, the shorter the setting time, which is considered to be a factor limiting the expansion and movement of the pores by the H_2 gas.
- Zinc powder suggested the possibility of being used as a gas for the production of aerated concrete instead of aluminum powder. To control the diameters and distribution of the pores by ZP, it is necessary to increase the curing time of the AASC paste. The setting time or curing time is considered to be one of the important factors in controlling the movement and expansion of the H_2 generated by ZP. In addition, the construction site construction at room temperature (23 ± 2 °C) and at a high temperature (40 ± 2 °C) are expected to be fully utilized for manufacturing concrete products in factories.

Author Contributions: Conceptualization, C.K. and T.K.; methodology, T.K.; validation, K.S., C.K. and T.K.; formal analysis, T.K.; investigation, C.K. and K.S.; resources, C.K.; data curation, C.K. and T.K.; writing—original draft preparation, C.K. and T.K.; writing—review and editing, T.K.; visualization, T.K.; supervision, T.K.; project administration, C.K. and T.K.; funding acquisition, K.S. and T.K. All authors have read and agreed to the published version of the manuscript.

Funding: This work was supported by the National Research Foundation of Korea (NRF) grant funded by the Korea Government (MOE) (NRF-2020R1I1A1A01056497).

Institutional Review Board Statement: Not applicable.

Informed Consent Statement: Not applicable.

Data Availability Statement: Data sharing is not applicable to this article.

Acknowledgments: The authors thank the Core Research Facility of Pusan National University funded by the Korea Ministry of Education for the technical support on XRD and SEM analysis.

Conflicts of Interest: The authors declare no conflict of interest.

References

1. Cheah, C.B.; Tan, L.E.; Ramli, M. Recent advances in slag-based binder and chemical activators derived from industrial by-products—A review. *Constr. Build. Mater.* **2021**, *272*, 12167. [CrossRef]
2. Elahi, M.M.A.; Hossain, M.M.; Karim, M.R.; Zain, M.F.M. A review on alkali-activated binders: Materials composition and fresh properties of concrete. *Constr. Build. Mater.* **2020**, *260*, 19788. [CrossRef]
3. Athira, V.S.; Bahurudeen, A.; Saljas, M.; Jayachandran, K. Influence of different curing methods on mechanical and durability properties of alkali activated binders. *Constr. Build. Mater.* **2021**, *299*, 123963. [CrossRef]
4. Gökçe, H.S.; Tuyan, M.; Nehdi, M.L. Alkali-activated and geopolymer materials developed using innovative manufacturing techniques: A critical review. *Constr. Build. Mater.* **2021**, *303*, 124483. [CrossRef]
5. Ibrahim, M.; Maslehuddin, M. An overview of factors influencing the properties of alkali-activated binders. *J. Cleaner Prod.* **2021**, *286*, 124972. [CrossRef]
6. Mendes, B.C.; Pedroti, L.G.; Vieira, C.M.F.; Marvila, M.; Azevedo, A.R.G.; Franco de Carvalho, J.M.; Ribeiro, J.C.L. Application of eco-friendly alternative activators in alkali-activated materials: A review. *J. Build. Eng.* **2021**, *35*, 102010. [CrossRef]
7. Shi, C.; Roy, D.; Krivenko, P. *Alkali-Activated Cements and Concretes*; CRC Press: Boca Raton, FL, USA, 2003.
8. Wang, W.; Noguchi, T. Alkali-silica reaction (ASR) in the alkali-activated cement (AAC) system: A state-of-the-art review. *Constr. Build. Mater.* **2020**, *252*, 119105. [CrossRef]
9. Ruan, S.; Zhu, W.; Yang, E.-H.; Weng, Y.; Unluer, C. Improvement of the performance and microstructural development of alkali-activated slag blends. *Constr. Build. Mater.* **2020**, *261*, 120017. [CrossRef]
10. Adesanya, E.; Perumal, P.; Luukkonen, T.; Yliniemi, J.; Ohenoja, K.; Kinnunen, P.; Illikainen, M. Opportunities to improve sustainability of alkali-activated materials: A review of side-stream based activators. *J. Cleaner Prod.* **2021**, *286*, 125558. [CrossRef]
11. Gu, G.; Xu, F.; Ruan, S.; Huang, X.; Zhu, J.; Peng, C. Influence of precast foam on the pore structure and properties of fly ash-based geopolymer foams. *Constr. Build. Mater.* **2020**, *256*, 119410. [CrossRef]
12. Amran, M.; Fediok, R.; Vatin, N.; Lee, Y.H.; Murali, G.; Ozbakkaloglu, T.; Klyuev, S.; Alabduljabber, H. Fiber-Reinforced Foamed Concrete: A Review. *Materials* **2020**, *13*, 4323. [CrossRef]
13. Hou, L.; Li, J.; Lu, Z.; Niu, Y. Influence of foaming agent on cement and foam concrete. *Constr. Build. Mater.* **2021**, *280*, 122399. [CrossRef]
14. Pasupathy, K.; Ramakrishnan, S.; Sanjayan, J. Enhancing the mechanical and thermal properties of aerated geopolymer concrete using porous lightweight aggregates. *Constr. Build. Mater.* **2020**, *264*, 120713. [CrossRef]
15. Fu, X.; Lai, Z.; Lai, X.; Lu, Z.; Lv, S. Preparation and characteristics of magnesium phosphate cement based porous materials. *Constr. Build. Mater.* **2016**, *127*, 712–723. [CrossRef]
16. Novais, R.M.; Ascensão, G.; Ferreira, N.; Seabra, M.P.; Labrincha, J.A. Influence of water and aluminium powder content on the properties of waste-containing geopolymer foams. *Ceram. Int.* **2018**, *44*, 6242–6249. [CrossRef]
17. Kioupis, D.; Zisimopoulou, A.; Tsivilis, S.; Kakali, G. Development of porous geopolymers foamed by aluminum and zinc powders. *Ceram. Internet* **2021**, *47*, 26280–26292. [CrossRef]
18. Shuai, Q.; Xu, Z.; Yao, Z.; Chen, X.; Jiang, Z.; Peng, X.; An, R.; Li, Y.; Jiang, X.; Li, H. Fire resistance of phosphoric acid-based geopolymer foams fabricated from metakaolin and hydrogen peroxide. *Mater. Lett.* **2020**, *263*, 127228. [CrossRef]
19. Yan, S.; Zhang, F.; Liu, J.; Ren, B.; He, P.; Jia, D.; Yang, J. Green synthesis of high porosity waste gangue microsphere/geopolymer composite foams via hydrogen peroxide modification. *J. Cleaner Prod.* **2019**, *227*, 483–494. [CrossRef]
20. Shi, J.; Liu, B.; Liu, Y.; Wang, E.; He, Z.; Xu, H.; Ren, X. Preparation and characterization of lightweight aggregate foamed geopolymer concretes aerated using hydrogen peroxide. *Constr. Build. Mater.* **2020**, *256*, 119442. [CrossRef]
21. Yang, Y.; Zhou, Q.; Deng, Y.; Lin, J. Reinforcement effects of multi-scale hybrid fiber on flexural and fracture behaviors of ultra-low-weight foamed cement-based composites. *Cement Concrete Comp.* **2020**, *108*, 103509. [CrossRef]

22. Ducman, V.; Korat, L. Characterization of geopolymer fly-ash based foams obtained with the addition of Al powder or H_2O_2 as foaming agents. *Mater. Charact.* **2016**, *113*, 207–213. [CrossRef]
23. Li, T.; Huang, F.; Zhu, J.; Tang, J.; Liu, J. Effect of foaming gas and cement type on the thermal conductivity of foamed concrete. *Constr. Build. Mater.* **2020**, *231*, 117197. [CrossRef]
24. Ji, Z.; Li, M.; Su, L.; Pei, Y. Porosity, mechanical strength and structure of waste-based geopolymer foams by different stabilizing agents. *Constr. Build. Mater.* **2020**, *258*, 119555. [CrossRef]
25. Falliano, D.; De Domenico, D.; Ricciardi, G.; Gugliandolo, E. Experimental investigation on the compressive strength of foamed concrete: Effect of curing conditions, cement type, foaming agent and dry density. *Constr. Build. Mater.* **2018**, *165*, 735–749. [CrossRef]
26. He, J.; Gao, Q.; Song, X.; Bu, X.; He, J. Effect of foaming agent on physical and mechanical properties of alkali-activated slag foamed concrete. *Constr. Build. Mater.* **2019**, *226*, 280–287. [CrossRef]
27. Pasupathy, K.; Ramakrishnan, S.; Sanjayan, J. influence of recycled concrete aggregate on the foam stability of aerated geopolymer concrete. *Constr. Build. Mater.* **2021**, *271*, 121850. [CrossRef]
28. Hajimohammadi, A.; Ngo, T.; Mendis, P.; Kashani, K.; van Deventer, J.S.J. Alkali activated slag foams: The effect of the alkali reaction on foam characteristics. *J. Clean. Prod.* **2017**, *147*, 330–339. [CrossRef]
29. Kränzlein, E.; Pollmann, H.; Krcmar, W. Metal powders as foaming agents in fly ash based geopolymer synthesis and their impact on the structure depending on the Na/Al ratio. *Cem. Concr. Comp.* **2018**, *90*, 161–168. [CrossRef]
30. Klapiszewska, I.; Parus, A.; Ławniczak, Ł.; Jesionowski, T.; Klapiszewski, Ł.; Ślosarczyk, A. Production of antibacterial cement composites containing ZnO/lignin and ZnO-SiO$_2$/lignin hybrid admixtures. *Cem. Concr. Comp.* **2021**, *124*, 104250. [CrossRef]
31. Noeiaghaei, T.; Dhami, N.; Mukherjeem, A. Nanoparticles surface treatment on cemented materials for inhibition of bacterial growth. *Constr. Build. Mater.* **2017**, *150*, 880–891. [CrossRef]
32. Abo-El-Enein, S.A.; El-Hosiny, F.I.; El-Gamal, S.M.A.; Amin, M.S.; Ramadan, M. Gamma radiation shielding, fire resistance and physicochemical characteristics of Portland cement pastes modified with synthesized Fe_2O_3 and ZnO nanoparticles. *Constr. Build. Mater.* **2018**, *173*, 687–706. [CrossRef]
33. Le Pivert, M.; Zerelli, B.; Martin, N.; Capochichi-Gnambodoe, M.; Leprince-Wang, Y. Smart ZnO decorated optimized engineering materials for water purification under natural sunlight. *Constr. Build. Mater.* **2020**, *257*, 119592. [CrossRef]
34. Troconis de Rincón, O.; Pérez, O.; Paredes, E.; Caldera, Y.; Urdaneta, C.; Sandoval, I. Long-term performance of ZnO as a rebar corrosion inhibitor. *Cem. Concr. Comp.* **2002**, *24*, 79–87. [CrossRef]
35. Loh, K.; Gaylarde, C.C.; Shirakawa, M.A. Photocatalytic activity of ZnO and TiO_2 'nanoparticles' for use in cement Mixes. *Constr. Build. Mater.* **2018**, *167*, 853–859. [CrossRef]
36. Bica, B.O.; Staub de Melo, J.V. Concrete blocks nano-modified with zinc oxide (ZnO) for photocatalytic paving: Performance comparison with titanium dioxide (TiO_2). *Constr. Build. Mater.* **2020**, *252*, 119120. [CrossRef]
37. Reichlek, R.; Mccurdy, E.; Heple, L. Zinc Hydroxide: Solubility Product and Hydroxy-597 complex Stability Constants from 12.5–75 °C. *Can. J. Chem.* **1975**, *53*, 3841–3845. [CrossRef]
38. Degen, A.; Kosec, M. Effect of pH and impurities on the surface charge of zinc oxide in 599 aqueous solution. *J. Eur. Ceram. Soc.* **2000**, *20*, 667–673. [CrossRef]
39. ASTM International. *Standard Practice for Mechanical Mixing of Hydraulic Cement Pastes and Mortars of Plastic Consistency*; ASTM C305; ASTM International: West Conshohocken, PA, USA, 2014.
40. ASTM International. *Standard Test Method for Time of Setting of Hydraulic-Cement Paste by Gillmore Needles*; ASTM C266; ASTM International: West Conshohocken, PA, USA, 2015.
41. ASTM International. *Standard Specification for Flow Table for Use in Tests of Hydraulic Cement*; ASTM C230; ASTM International: West Conshohocken, PA, USA, 2008.
42. ASTM International. *Standard Test Method for Rate of Water Absorption of Masonry Mortars*; ASTM C1403; ASTM International: West Conshohocken, PA, USA, 2015.
43. Kim, T.; Kang, C. The Mechanical Properties of Alkali-Activated Slag-Silica Fume Cement Pastes by Mixing Method. *Int. J. Concr. Struct. Mater.* **2020**, *14*, 41. [CrossRef]
44. Jun, Y.; Kim, T.; Kim, J.H. Chloride-bearing characteristics of alkali-activated slag mixed with seawater: Effect of different salinity levels. *Cement Concrete Comp.* **2020**, *112*, 103680. [CrossRef]
45. Yum, W.S.; Jeong, Y.; Yoon, S.; Jeon, D.; Jun, Y.; Oh, J.E. Effects of $CaCl_2$ on hydration and properties of lime(CaO)-activated slag/fly ash binder. *Cement Concrete Comp.* **2017**, *84*, 111–123. [CrossRef]
46. Garg, N.; White, C.E. Mechanism of zinc oxide retardation in alkali activated materials: An in situ X-ray pair distribution function investigation. *J. Mater. Chem. A* **2017**, *5*, 11794–11804. [CrossRef]
47. Mohsen, A.; Abdel-Gawwad, H.A.; Ramadan, M. Performance, radiation shielding, and anti-fungal activity of alkali-activated slag individually modified with zinc oxide and zinc ferrite nano-particles. *Constr. Build. Mater.* **2020**, *257*, 119584. [CrossRef]
48. Taylor-Lange, S.C.; Riding, K.A.; Juenger, M.C.G. Increasing the reactivity of metakaolin-cement blends using zinc oxide. *Cem. Concr. Comp.* **2012**, *34*, 835–847. [CrossRef]
49. Amer, M.W.; Fawwaz, I.K.; Akl, M.A. Adsorption of lead, zinc and cadmium ions on polyphosphate-modified kaolinite clay. *J. Environ. Chem. Ecotoxicol.* **2010**, *2*, 1–8.

50. Nochaiya, T.; Sekine, Y.; Choopun, S.; Chaipanich, A. Microstructure, characterizations, functionality and compressive strength of cement-based materials using zinc oxide nanoparticles as an additive. *J. Alloys Compd.* **2015**, *630*, 1–10. [CrossRef]
51. Šiler, P.; Kolářová, I.; Novotný, R.; Másilko, J.; Pořízka, J.; Bednárek, J.; Švec, J.; Opravil, T. Application of isothermal and isoperibolic calorimetry to assess the effect of zinc on cement hydration. *J. Therm. Anal. Calorim.* **2018**, *133*, 27–40. [CrossRef]
52. Nambiar, E.K.K.; Ramamurthy, K. Air–void characterisation of foam concrete. *Cem. Concr. Res.* **2007**, *37*, 221–230. [CrossRef]
53. Cabrillac, R.; Fiorio, B.; Beaucour, A.; Dumontet, H.; Ortola, S. Experimental study of the mechanical anisotropy of aerated concrete and of the adjustment parameters on the induced porosity. *Constr. Build. Mater.* **2006**, *20*, 286–295. [CrossRef]
54. Masi, G.; Rickard, W.D.A.; Bignozzi, M.C.; Riessen, A. The influence of short fibres and foaming agents on the physical and thermal behaviour of geopolymer composites. *Adv. Sci. Technol.* **2014**, *92*, 56–61. [CrossRef]

Article

Development of Geopolymers as Substitutes for Traditional Ceramics for Bricks with Chamotte and Biomass Bottom Ash

Juan María Terrones-Saeta *, Jorge Suárez-Macías, Francisco Javier Iglesias-Godino and Francisco Antonio Corpas-Iglesias

Department of Chemical, Environmental, and Materials Engineering, Higher Polytechnic School of Linares, University of Jaen, Scientific and Technological Campus of Linares, 23700 Linares, Jaen, Spain; jsuarez@ujaen.es (J.S.-M.); figodino@ujaen.es (F.J.I.-G.); facorpas@ujaen.es (F.A.C.-I.)
* Correspondence: terrones@ujaen.es

Citation: Terrones-Saeta, J.M.; Suárez-Macías, J.; Iglesias-Godino, F.J.; Corpas-Iglesias, F.A. Development of Geopolymers as Substitutes for Traditional Ceramics for Bricks with Chamotte and Biomass Bottom Ash. *Materials* 2021, 14, 199. https://doi.org/10.3390/ma14010199

Received: 25 November 2020
Accepted: 27 December 2020
Published: 4 January 2021

Publisher's Note: MDPI stays neutral with regard to jurisdictional claims in published maps and institutional affiliations.

Copyright: © 2021 by the authors. Licensee MDPI, Basel, Switzerland. This article is an open access article distributed under the terms and conditions of the Creative Commons Attribution (CC BY) license (https://creativecommons.org/licenses/by/4.0/).

Abstract: The greater environmental awareness, new environmental regulations and the optimization of resources make possible the development of sustainable materials as substitutes for the traditional materials used in construction. In this work, geopolymers were developed as substitutes to traditional ceramics for brick manufacture, using as raw materials: chamotte, as a source of aluminosilicate, and biomass bottom ashes from the combustion of almond shell and alpeorujo (by-product produced in the extraction of olive oil composed of solid parts of the olive and vegetable fats), as the alkaline activator. For the feasibility study, samples were made of all possible combinations of both residues from 100% chamotte to 100% biomass bottom ash. The tests carried out on these sample families were the usual physical tests for ceramic materials, notably the compression strength test, as well as colorimetric tests. The freezing test was also carried out to study the in-service behavior of the different sample groups. The families with acceptable results were subjected to Fourier transform infrared (FTIR) analysis. The results of the previous tests showed that the geopolymer was indeed created for the final families and that acceptable mechanical and aging properties were obtained according to European standards. Therefore, the possibility of creating geopolymers with chamotte and biomass bottom ashes as substitutes for conventional ceramics was confirmed, developing an economical, sustainable material, without major changes in equipment and of similar quality to those traditionally used for bricks.

Keywords: geopolymer; chamotte; biomass bottom ash; ceramic; circular economy; environment

1. Introduction

The construction sector is one of the most demanding sectors in terms raw materials and the one that causes the greatest greenhouse gas production [1,2]. This fact is mainly due to the high production of materials as well as their low cost. More specifically, the consumption of ceramic materials in the building sector, among others, causes the scarcity of natural resources such as clay [3–5] as well as generates significant CO_2 emissions due to poorly optimized industrial processes [6]. Moreover, the construction sector accounts for the largest percentage of the global energy consumption [7].

On this basis, and with new circular economy trends, in recent years research lines have been developed based on the creation of construction materials with the incorporation of waste [8–10]. Thus, it is possible to reduce the extraction of virgin materials and to take advantage of waste from other industries [11–14]. Therefore, the economic and ecological flows of the materials are closed [15]. Moreover, in the field of ceramic materials, manufacturing processes produce high CO_2 emissions mainly due to the high temperatures generated in their manufacture, around 950 °C for traditional ceramics.

It is therefore essential to search for new materials with more optimized manufacturing processes that use residues in their composition, have good qualities and that can serve at the end of their useful life for the creation of other materials [16]. Based on this, in recent years, different lines based on geopolymers have been developed as cement substitutes or [17–20], as in this research, geopolymers as substitutes for traditional ceramic materials such as construction bricks. The term geopolymer was coined in 1978 by Joseph Davidovits [21] and is one of the most promising materials for the construction sector [22,23].

The geopolymer is an inorganic polymer [24] formed by the reaction of a source of aluminosilicate (binder) with an alkaline solution (activator) [25]. In this process called geopolymerization, aluminate and silicate monomers are formed, then they become oligomers and finally geopolymers. The water in this process is depleted, so the drying conditions are very important for its end resistance [26].

The sources of aluminosilicates that have been used correspond mostly to waste. Among these wastes are coal fly ash [27–30], slag waste from metallurgical industries [31–34], metakaolin [35–37], glass wastes [38–40], bagasse [41–43] and even hazardous waste [44–47]. It is therefore a material that not only reduces the emissions of other materials such as cement [48] or the extraction of clay for ceramic materials, but also its manufacturing process emits less greenhouse gases [49,50] and uses waste from other industries as raw materials [51]. The geopolymer is therefore a green material for the environment [52–54] and it is framed within the new circular economy.

In turn, sodium hydroxide or potassium hydroxide in appropriate proportions have been used for activator or alkaline solutions. An increase in the concentration of both would cause the rapid precipitation of the aluminosilicate gel and lower its compressive strength [55]. In addition, a low concentration of activator would cause an incomplete geopolymerization process and lower compressive strength. Therefore, the concentration of the activator as well as the curing temperature [56] is essential for obtaining the best mechanical properties [57]. The properties of geopolymers are diverse, and include high temperature resistance [58–60], piezoelectric properties [61], and good behavior in contact with steel [62].

Based on what has been said and in order to develop substitutes for traditional ceramics that are less harmful to the environment [63], geopolymers with chamotte and biomass bottom ash were developed in this paper. The source of aluminosilicate is the chamotte and the activator is the biomass bottom ash, since the ashes have a high percentage of potash because they correspond to the combustion of almond husks and alpeorujo. It is therefore a material that uses waste as raw materials and more optimized manufacturing processes [64].

As mentioned, the building consumes huge amounts of virgin materials to obtain new materials [65]. The increasing construction of buildings and the renovation of existing ones cause more materials to be consumed and in turn to produce more waste [66]. Companies that manufacture bricks with red clay generate a large amount of waste, mainly, by bricks that do not have the right geometric and visual shapes or breaking in the transport [67]. These bricks are crushed to deposit them in the vicinity of the industries with the consequent environmental impact. This material derived from the manufacture of defective and crushed red clay bricks is called chamotte [68]. The chamotte has had a use in civil engineering as a filler material or in the ceramic industry itself as an additive for stoneware or new bricks [69,70], however, in these processes, their properties are not optimized. The composition of the chamotte with high proportions of silica and alumina make it ideal for use in geopolymers. Some authors have studied geopolymers with chamotte and sodium hydroxide as an activator [71–73], however, the literature is scarce and the development of timely studies in this area is still necessary.

On the other hand, the biomass bottom ash depends on their composition of the biomass used in combustion, so each case must be studied. This waste is a big problem since the global biomass production is 140 billion tons per year [74]. A lot of research is being carried out for biomass flying ashes [75] and very few studies for biomass bottom ashes.

Biomass bottom ashes are a residue with inorganic components and to a lesser extent with organic ones [76]. Its quality and composition depends on the biomass used and the combustion process [77], being characterized as a non-hazardous waste according to European legislation [78]. However, it is currently an environmental problem when produced in large quantities, not valorizing it and depositing it in landfill. Although, there are few studies on its valorization that have been successful [79–82].

Therefore, the scope of this research is to develop geopolymers with one hundred percent waste, chamotte and biomass bottom ash, as substitutes for traditional ceramics. For this, different groups of samples were made with different percentages of both materials, conformed with water after going through a drying process. Finally, its physical, compressive strength, and colorimetric properties were studied and its durability was evaluated with the freezing test traditionally used in ceramics. Samples manufactured with different percentages of waste combination that reflected suitable results in the previous tests were analyzed with Fourier transform infrared (FTIR).

In short, with the development of this new geopolymer material as a substitute for traditional ceramics for the manufacture of bricks, a series of obvious economic and environmental advantages are achieved. On the one hand, the cost of clay extraction is reduced, as well as the cost of the manufacturing process by preventing high-temperature sintering of the ceramics and in addition, the cost of waste is practically non-existent, as it is not used at present. From an environmental point of view, it can be said that the impact on the landscape and the environment is reduced, as well as greenhouse gas emissions, since it is not necessary to extract raw materials and avoid depositing the waste in dumps. In addition, the production process is optimized by reducing the emission of harmful gases without major modifications to the machinery used. Finally, a new life is given to the waste that is not currently used, closing the flow of materials and developing a new circular economy.

2. Materials and Methods

The materials used in this work as well as the methodology followed are detailed in the following paragraphs.

2.1. Materials

The materials used are entirely industrial by-products. On the one hand, (and as a source of aluminosilicates) the commonly called chamotte will be used. Chamotte is a by-product derived from the ceramic industry that is the basis of the material for the shaping of the geopolymer after activation. In turn, the biomass bottom ash of almond husk and alpeorujo combustions will be used as an alkaline activator, henceforth biomass bottom ash (BBA).

Therefore, since both by-products will be analyzed in depth in the methodology, successive paragraphs shall describe their origin and training.

2.1.1. Chamotte

Chamotte is an inherent industrial by-product of ceramics production. The samples taken belong to the companies of the province of Jaen, Spain. These companies are dedicated to the manufacture of bricks with red clay.

In the manufacturing process, bricks that are not accepted for commercialization are discarded mainly because of their dimensions or shapes. Given their volume, they are crushed in the plant to be able to store it more easily and if possible, its subsequent withdrawal for use in other activities, such as the filling of embankments, sports courts, etc.

Based on the above, the by-product used was almost entirely sintered according to an appropriate process, so it offers stable physical and chemical characteristics. Since the process is similar between the different brick manufacturing companies as well as the raw material used, there is a repeatability of the properties of the by-product over time.

This material after its process is easily found in very fine grading, so its use is immediate within the conformed of the geopolymer.

2.1.2. Biomass Bottom Ash from the Combustion of Almond Husk and Alpeorujo

The biomass bottom ash used in this project, hereinafter BBA, belongs to the companies located in Jaen, Spain. These biomass bottom ashes correspond to the by-product generated in the combustion of the almond husk and alpeorujo for the generation of electrical energy.

Using such a specific combustion material creates a by-product of similar physical and chemical properties over time. This material will be analyzed in the following sections and has the fundamental role of providing the alkaline activation of the chamotte for the formation of the geopolymer, and consequently, its mechanical properties.

It should be noted that before using the by-product, it was crushed to obtain a fine grain size. The process of crushing the biomass bottom ash, which has a maximum particle size of 16 mm, was carried out with the same equipment used for crushing clay in the ceramics industry. Furthermore, as the biomass bottom ash has low resistance, due to the materials from which it comes, the process is fast, economical and of high quality.

2.2. Methodology

The methodology to be followed in this work is clear and objective to evaluate the possibility of geopolymer conformation through the use of by-products of the ceramic industry and the generation of energy. The main purpose is to create a sustainable and economical material as a substitute for traditional ceramics.

First of all, both by-products were analyzed in order to determine their chemical composition. In this way, the present elements and compounds, capable of fulfilling the functions of aluminosilicate and alkaline activator, were evaluated, respectively. The physical properties were evaluated to determine the ease of the material for its treatment and its subsequent conformation in the successive processes.

Once both by-products were analyzed, different samples with different combination percentages were formed. Taking as a base material the chamotte, increasing percentages of biomass bottom ash were added from 10% to 100%, with increases of 10%. In this way, the variation could be observed of the physical properties of the geopolymer in all possible combinations of both elements.

The two residues were mixed in the corresponding percentage and conformed in a matrix with a pressure of 30 ± 1 MPa. Once the samples were formed, their dimensions were measured and they were dried at room temperature (20 ± 2 °C) for 24 h and at 90 ± 2 °C for another 24 h.

After the drying process was carried out, we proceeded to leach the elements that had not reacted and involved a useless load. This phase consisted of a continuous recirculation of water in a tank after submerging the samples. Once this process was carried out in the laboratory, the samples were dried again at a temperature of 90 ± 2 °C for 24 h, finally measuring their dimensions and weight.

The physical tests after the conformed samples are the typical tests performed on the ceramic elements to confirm the quality. Moreover, the aesthetic properties of the test sample families and the compressive strength were studied.

Finally, an accelerated ageing test was conducted to evaluate the behavior of different families over time. In this case, and because it is one of the most common tests used for ceramics, the freezing test was carried out. The result of this test was assessed visually.

In a final point, all the results obtained from the different families were analyzed to obtain a combination field of both residues that create a suitable geopolymer according to the European ceramics regulations. The combinations of chamotte and biomass bottom ash that showed acceptable results in the tests were finally analyzed with Fourier transform infrared (FTIR). In this way, the formation of the geopolymer in these combinations as well as the variations between them could be observed.

Based on the comments and according to the logical scheme, the following subsections will be divided into several groups, the initial tests of the by-products, the geopolymers conformed and ageing tests and Fourier transform infrared.

2.2.1. Initial Tests of the by-Products

Based on the comments, and as an initial and essential premise of this work, the industrial by-products, chamotte and biomass bottom ash were analyzed in detail.

First, both by-products were crushed and sieved by the 0.25 mm sieve and then dried at a temperature of $105 \pm 2\ °C$. The resulting material was the one that had been used in all the tests of this work and in the conform of geopolymers.

It should be taken into account that the humidity of the products under study for the conformation of the geopolymers would not be a problem in itself, as it could be in other materials. However, this moisture should be taken into account to subtract it from the water necessary for conforming.

The tests performed on the aforementioned samples can be classified into two sections, physical tests, intended to determine the particle density UNE-EN 1097-7 [83] and laser diffraction granulometry; as well as chemical tests, aimed at determining the different chemical elements in the samples, elemental analysis, loss on ignition and X-ray fluorescence. It is essential to detect those chemical elements that will help the geopolymerization process, as well as those harmful elements that must be monitored in the process.

2.2.2. Conformed of Geopolymers: Physical and Mechanical Tests of the Conformed Samples

Characterized the initial materials and studied their suitability for use in the realization of geopolymers, we proceeded to the conformation of the different families of test samples based on the combination of both industrial by-products, chamotte and biomass bottom ash (BBA).

The starting aluminosilicate, which will be activated later, is the chamotte. Therefore, it is the base element on which it was proceeded to add increasing amounts of the alkaline activator, biomass bottom ash.

This increase was made from 0% to 100%, reflecting all possible combinations of both materials for the further study of the geopolymer conformed. In this way, it is possible to analyze the optimal combination and possible cases in which they reflect the characteristics acceptable by the regulations in this regard. An analytical chemical study of the combination of both elements would be extremely difficult and unrepresentative of reality, since, being industrial by-products, the elements are not high in purity. The different sample groups are represented in Table 1, showing the percentage of each by-product for each group.

Table 1. Sample groups composed of geopolymers with different combination percentages of chamotte and biomass bottom ash (BBA).

Samples Groups	Chamotte, %	BBA, %
10C0A	100	0
9C1A	90	10
8C2A	80	20
7C3A	70	30
6C4A	60	40
5C5A	50	50
4C6A	40	60
3C7A	30	70
2C8A	20	80
1C9A	10	90
0C10A	0	100

It should be noted that samples groups 10C0A and 0C10A, made up of 100% chamotte and 100% biomass bottom ash, respectively, obviously do not produce geopolymers, since there is no activation of aluminosilicates. However, both families have been carried out to physically, mechanically and aesthetically check the variations that occur in the formation of the geopolymer, as well as to be certain that the geopolymer has been formed.

From each of the detailed families, six samples were formed in order to have statistically analytical results.

The samples were formed following the same process for all families, this being the one detailed below:

- The chamotte and the biomass bottom ash were mixed until the resulting mass was homogenized and according to the corresponding percentages of each family.
- Subsequently, 20% water was added to the previous mass, mixing again until obtaining the homogenization of the product.
- This resulting mixture was conformed in a steel matrix of internal dimensions of 60×30 mm, applying a gradual pressure through a piston until reaching 30 ± 1 MPa. This pressure was maintained for one minute.
- Once the mixture was compacted, the sample was removed, leaving the sample fully conformed.

It should be noted that the percentage of 20% water added to the mixture for conforming was determined empirically to optimize this process. Higher percentages of water caused an excess of water exudation during compression.

Once the samples were made, they were left at room temperature (20 ± 2 °C) for 24 h and at 90 ± 2 °C for another 24 h to remove excess water that has not reacted during the geopolymerization process. As mentioned above, the curing temperature of the geopolymer has a significant influence on the mechanical characteristics. However, if the geopolymer is to replace traditional ceramics, the production times must be similar. Therefore, first a curing at ambient temperature for 24 h is produced to increase the resistance and subsequently, a drying at higher temperature to decrease the production times once the resistance of the geopolymer has been reached.

Subsequently, and in order for this process to take place in full, once the different samples of the different families were dried, their geometric dimensions were measured and weighed to subsequently undergo a process of continuous recirculation of water (20 ± 2 °C). This process has two main objectives, the first of which is to eliminate possible excess elements that are properly diluted in the water and have not reacted, or have no utility within the geopolymer; on the other hand, to provide the water necessary for the geopolymerization reaction to occur if, at first, it could have been stopped due to a lack thereof. After this continuous water recirculation process, the conformed samples were dried again. Once dried for 24 h at a temperature of 90 ± 2 °C, the geometric dimensions and mass were measured, for the subsequent study of the variation of the physical properties in the geopolymerization process.

Once six samples were obtained for each of the families, the physical properties of the different sample groups were studied through the tests usually used for ceramic materials. These tests were the determination of mass loss, the determination of dimensions UNE-EN 772-16 [84], capillary water absorption UNE-EN 772-11 [85], cold water absorption UNE-EN 772-21 [86], boiling water absorption UNE-EN 772-7 [87], bulk density and open porosity UNE-EN 772-4 [88]. The purpose of carrying out the present tests is the study of the physical characteristics of the materials formed to compare them with traditional ceramics, since the main objective of the project was the replacement of the latter by geopolymers.

Subsequently, the color of the various samples of the families was objectively evaluated. For this, the colorimeter will be used, which will reflect the color of the different samples in combination with the primary colors.

Finally, the mechanical properties of the different families will be studied through the compression test UNE-EN 772-1 [89], that will be able to obtain the resistance of each of them. With this essay, families can be accepted or rejected based on European regulations and their compressive strength.

It should be noted that traditional ceramics formed with red clay for the manufacture of bricks are those that the present work seeks to replace with geopolymers. Therefore, the comparison ceramics were performed with the same forming conditions (water and compaction) and were sintered in the oven at a temperature of 950 ± 10 °C, with heating ramps of 4 °C/min and temperature maintenance for 1 h.

2.2.3. Ageing Tests (Freezing Test) and Fourier Transform Infrared (FTIR) of the Geopolymers

The main purpose of the freezing test was the study of the behavior of the different sample families after the effect produced by a continuous cycle of ice and melt UNE 67028 [90]. In this way, its durability could be evaluated over time and the quality of the geopolymer was obtained before the inclement weather.

There were taken six samples of each family for its development and they were introduced into the melting tank, progressively submerging them at a temperature of 15 ± 2 °C and in a minimum time of 3 h. Subsequently, they were removed and they were left to rest for a period of 1 min, in order to introduce them into a cold room without any contact between them. They were kept in the chamber for 18 h, remaining at least 11 h at the temperature of −15 ± 2 °C. They were subsequently removed from the chamber and introduced into the melting tank for at least 6 h. This process was repeated for a total of 25 cycles.

After performing the 25 test cycles for the samples of the families, the visual inspection was carried out. The objective of the visual inspection testing (VT) is to evaluate the appearance of breaks, spalling and chipping greater than 15 mm, according to the UNE 67028 standard [90]. If any of the defects mentioned in several of the samples of the different families appears, this would be classified as a freezing geopolymer, unsuitable for use.

Sample families that obtained acceptable results in the freezing test were analyzed with Fourier transform infrared. For this purpose, the samples of these families were manufactured again with the process detailed in the methodology. The 10C0A and 0C10A families corresponding to 100% chamotte and 100% biomass bottom ash, respectively, were also analyzed. In this way, it was possible to evaluate the differences that existed between the different spectra of the detailed families and the base materials, thus analyzing the formation of the geopolymer and chemically corroborating its existence.

To carry out this test, the samples were first crushed to a particle size of less than 0.063 mm. The detailed samples were analyzed with the Bruker Tensor20 spectrophotometer (Tensor20, Bruker, Billerica, MA, USA) which allowed the recording of the FTIR spectra of solid, liquid and gaseous samples in the mid and near infrared range. In addition, in this case, it was used in the attenuated total reflectance (ATR). The standard spectral resolution was 4 cm^{-1}, with a spectral range of 4000–400 cm^{-1}.

3. Results and Discussions

The successive sections describe the results of the different tests detailed above in the methodology, including in each of them the partial conclusions that may be derived from their analysis.

3.1. Initial Tests of by-Products

This section details the results obtained and the conclusions derived from the tests destined for the determination of the physical and chemical properties of the elements under study, chamotte and biomass bottom ashes.

First of all, and within the physical properties, it has been obtained that the particle density of the chamotte and the bottom ashes of biomass is 2.54 and 2.65 t/m^3, respectively. Both densities are adequate and similar, so no volume correction would be necessary for the geopolymer conforming process. The results given off are similar to the diversity of materials used in construction, established as the usual particle density 2.65 t/m^3.

In turn, Figure 1 shows the particle size distribution of the chamotte (sieved by the 0.25 mm sieve). It is distinguished that the highest percentage of particles have a size between 40 and 200 micrometers. This microscopic granulometry makes chamotte an ideal by-product for use as an aluminosilicate in the conformation of the geopolymers. This fact is due to the fact that its fineness as well as its amorphous form makes an excellent combination with the activator possible.

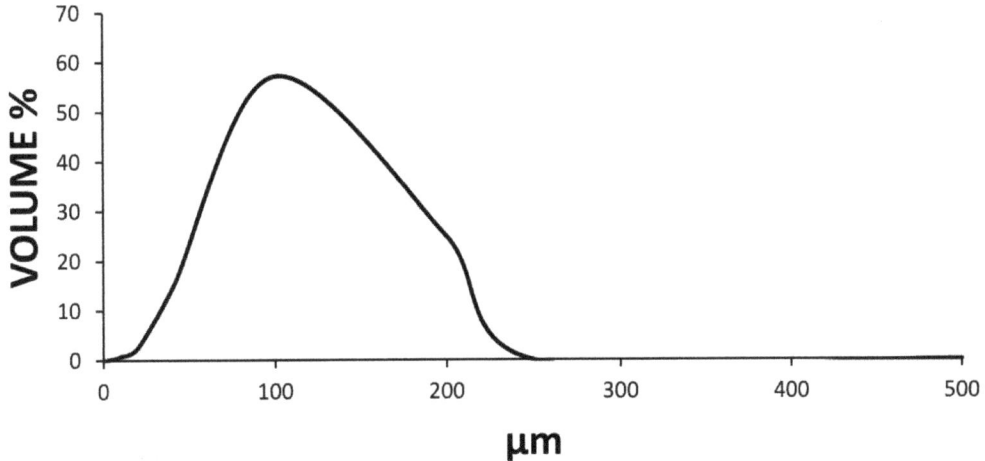

Figure 1. Graph of the laser granulometry of the sieved chamotte by a 0.25 mm sieve.

Similarly, Figure 2 shows the particle size distribution of the bottom ash of the biomass sieved by a 0.25 mm sieve after mashing and drying. The distribution of the particles was observed from 10 to 200 um, as a finer material than the chamotte and suitable for use as an activator in geopolymers. It should be noted that different authors have studied and corroborated that the fineness of the materials used in the conformation of the geopolymer greatly influences the final mechanical characteristics.

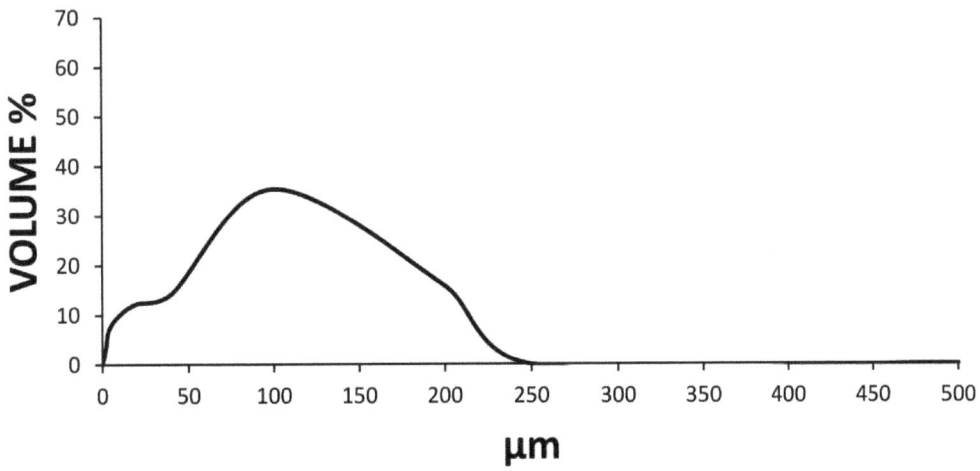

Figure 2. Graph of the laser granulometry of the sieved BBA by a 0.25 mm sieve.

Once the physical properties were evaluated and the results obtained for the conformation of geopolymers were acceptable, the chemical properties were studied. The first test performed was that of elementary analysis, to determine the percentage of carbon, nitrogen, hydrogen and sulfur present in both samples.

The results of the elemental analysis test for the chamotte and the biomass bottom ash, observable in Table 2, reflect a very low percentage of carbon. This fact is obvious, since they derive from sintering or combustion procedures that are carried out at very high temperatures. On the other hand, it should be highlighted that the value of sulfur in both samples was zero—without assuming a problem to be analyzed later. If, on the contrary, there was sulfur in one of the two samples, it should be studied later to prevent environmental pollution problems.

Table 2. Elemental analysis of the chamotte and the BBA.

Sample	Nitrogen, %	Carbon, %	Hydrogen, %	Sulfur, %
Chamotte	0.00 ± 0.00	0.24 ± 0.01	0.08 ± 0.00	0.00 ± 0.00
BBA	0.05 ± 0.00	4.64 ± 0.14	0.48 ± 0.02	0.00 ± 0.00

This test was complemented with that of loss on ignition, which is detailed in Table 3 for both materials.

Table 3. Loss on ignition for the chamotte and the BBA.

Sample	Loss on Ignition, %
Chamotte	1.74 ± 0.10
BBA	8.16 ± 0.19

As the results reflect, the loss on ignition in both samples was markedly reduced. This fact, as mentioned in the previous trial, is due to the production process of both by-products, which was produced at high temperatures. The loss on ignition of the chamotte is lower than that of the BBA, since, for its production through ceramic materials, a sintering temperature of the clay material is necessary. In the case of BBA, even when high temperatures are reached, the process is much faster so it can lead to unburning.

The X-ray fluorescence of chamotte is shown in Table 4 and reflects an elemental composition similar to that of any traditional ceramic. The silicon-aluminum ratio is suitable for the formation of geopolymers and in turn, the percentages of magnesium, calcium and iron are just right so as not to cause any problems. Therefore, it can be concluded that chamotte is a good source of aluminosilicate for the manufacture of geopolymers, without contaminating or containing harmful elements that could interfere with the process.

Table 4. X-ray fluorescence of the chamotte.

Element	wt, %
Si	27.32 ± 0.12
Al	8.16 ± 0.10
Ca	5.95 ± 0.10
Fe	4.57 ± 0.09
K	3.80 ± 0.09
Mg	1.92 ± 0.05
Ti	0.455 ± 0.023
Sx	0.119 ± 0.006
Na	0.201 ± 0.012
P	0.0965 ± 0.0048
Mn	0.0665 ± 0.0033

Table 4. Cont.

Element	wt, %
Sr	0.0523 ± 0.0030
Zr	0.0375 ± 0.0037
V	0.0209 ± 0.0018
Ni	0.0242 ± 0.0016
Rb	0.0208 ± 0.0043
Cr	0.0146 ± 0.0017
Pt	0.0162 ± 0.0039
Cl	0.0107 ± 0.0008
Ru	0.0070 ± 0.0026
Total weight % oxygen	45.39 ± 0.47

On the other hand, the X-ray fluorescence of biomass bottom ash, shown in Table 5, reflects a high percentage of potassium. This fact is very interesting and necessary for its use as an activator of the mentioned aluminosilicate. The other two majority elements are silicon and calcium which appear to a lesser extent and do not represent a problem for the geopolymer, as they even increase the silicon–aluminum ratio and can help to obtain resistance. The rest of the elements present in the biomass bottom ash sample do not represent a problem for the viability of the conformed geopolymer and there are no elements hazardous to the environment. However, they are a burden on the material unlike the use of pure potassium hydroxide, which is remedied for its lower price and a higher percentage of addition.

Table 5. X-ray fluorescence of the biomass bottom ash.

Element	wt, %
K	23.91 ± 0.19
Si	11.21 ± 0.10
Ca	11.10 ± 0.13
Px	3.58 ± 0.06
Mg	4.21 ± 0.08
Al	2.57 ± 0.06
Fe	1.33 ± 0.05
Sx	0.230 ± 0.011
Na	0.229 ± 0.019
Cl	0.255 ± 0.013
Ti	0.128 ± 0.006
Sr	0.0859 ± 0.0043
Mn	0.0442 ± 0.0022
Cu	0.0240 ± 0.0016
Ni	0.0221 ± 0.0012
Cr	0.0135 ± 0.0013
Zr	0.0106 ± 0.0027
Rb	0.0070 ± 0.0035
Zn	0.0047 ± 0.0016
V	0.0024 ± 0.0012
Total weight % oxygen	32.89 ± 0.36

3.2. Physical and Mechanical Tests of the Conformed Samples

The families made of samples with the different percentages of added chamotte and BBA were tested to study their viability. Figure 3 shows the results of the loss of weight, linear shrinkage, capillary water absorption and cold water absorption of the different samples of the families after the water recirculation process. That is, after conforming and subsequent drying process in an oven, the dry weight was determined and subsequently, they were submerged in a recirculated bath and dried after 24 h.

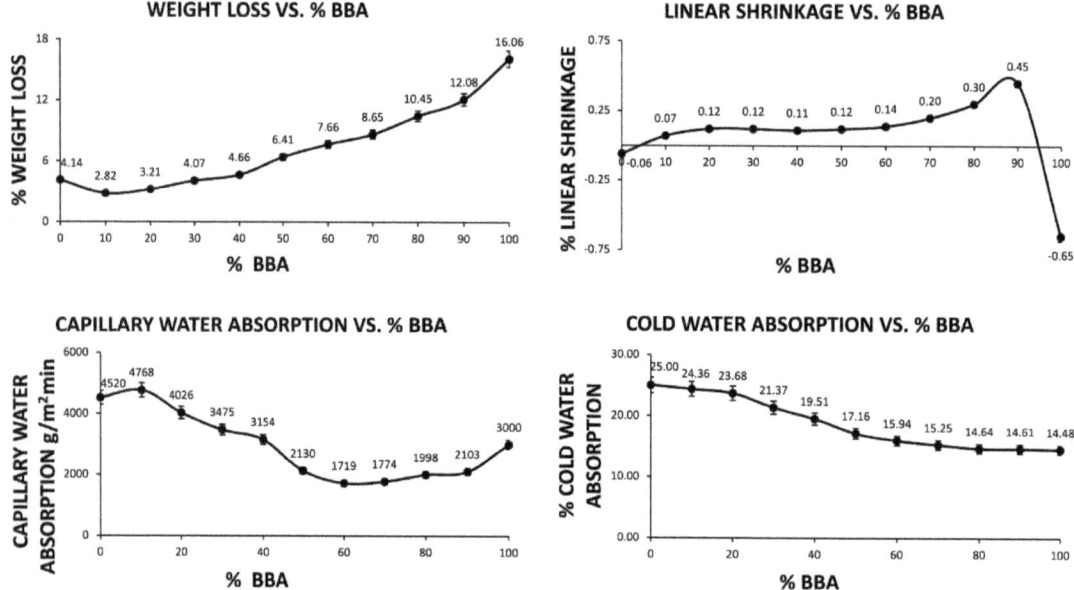

Figure 3. Graphs of the loss of weight, linear shrinkage, capillary water absorption and the cold water absorption of the different sample groups of geopolymers.

As can be seen in Figure 3, the percentage of weight loss is increasing with respect to the percentage of BBA about mixing in the geopolymer. This fact is mainly based on the ability of the geopolymerization process in the circulation of water to eliminate those superfluous elements that are not part of the geopolymer's structure. These elements are mostly present in the BBA, since the chamotte is a stable material because of its sintering process. Compared to a traditional ceramic after sintering, which has a weight loss of around 9.5%, it can be deduced that it is quite similar, in most cases being even lower.

The linear shrinkage values for the 100% chamotte and 100% BBA families are substantially different from the other families. This is because there is no combination of the two by-products and the geopolymer cannot be formed. On the other hand, the families formulated with a combination of both by-products have a greater linear to medium shrinkage that increases the percentage of BBA, which is not excessively high in any case. This fact is corroborated when comparing the linear shrinkages obtained with a traditional ceramic. A red clay ceramic after sintering has an average linear shrinkage of 2.7%.

As in the previous case, the water absorption rate of the families with 100% chamotte and 100% BBA is significantly different from the rest of the families. In the families composed of both wastes, a tendency towards a decrease in the absorption rate is observed with an increase in the addition of BBA. This fact implies the creation of a denser material and with less open porosity with the increase in the percentage of BBA. The reduction in the water absorption rate creates a material suitable for outdoor use, since contact with water would not cause great absorption and an increase in the weight of the material to be supported by the structure. In comparison, a traditional ceramic has a capillary water absorption of 1700 g/m^2·min, similar to the samples formed with 60% BBA and higher percentages.

As with the capillary water absorption, rate cold water absorption reflects a reduction in the absorption capacity of samples with a higher percentage of BBA in their formulation. This fact predicts a higher density of the materials created and a lower porosity, which in turn could lead to higher compressive strength. The traditional ceramics usually have a cold water absorption of 13%, something lower than the values obtained and derived from the sintering and conforming process.

At the same time, the results of the boiling water absorption, open porosity, bulk density and compressive strength tests for the different sample groups are detailed in Figure 4. The progressive addition of BBA causes a lower boiling water absorption, a higher bulk density of the test samples and a lower open porosity of the families. This fact will be directly related to the quality of the geopolymer.

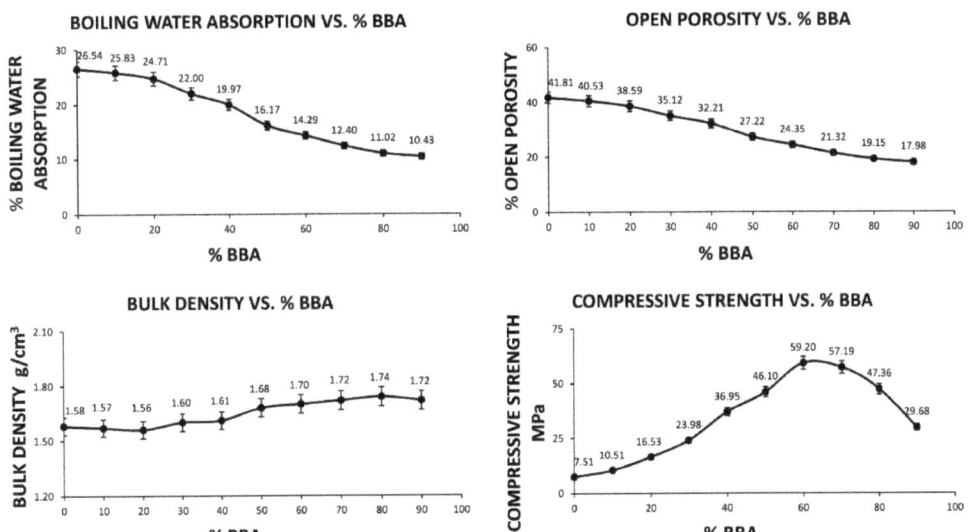

Figure 4. Graphs of boiling water absorption, open porosity, bulk density and compressive strength of the different samples groups of geopolymers.

It should be noted that the samples made of 100% BBA cracked and collapsed during the boiling water absorption test, therefore they were totally discarded from further interpretation. The unsuitability of this family does not suppose a problem, since it cannot be considered a geopolymer as it is only made up of BBA and there is no suitable aluminosilicate. In family 10C0A, a similar fact occurs, only the aluminosilicate exists and not the activator so it cannot be considered a geopolymer. However, both families show that the geopolymer has been formed as there are great differences in its physical and mechanical properties.

The boiling water absorption and the open porosity in a traditional ceramic is usually 12% and 24%, respectively. As can be seen in Figure 4, the families with percentages of 60% BBA addition in the mixture coincide approximately. On the other hand, lower values are obtained for higher percentages of addition of BBA at 60%.

On the other hand, a standard ceramic usually has a bulk density of around 2 t/m^3, while geopolymers in all families have lower densities. This decrease in density, far from being a problem, can become a strength, because if the adequate resistances prescribed by the regulations are achieved, having a lower density makes possible a lower thermal insulation, as well as a better acoustic insulation. On the other hand, its lower weight makes it possible to create lighter claddings that do not overload the structure of the building.

Finally, the compression strength test of the conformed samples of the different families of geopolymers is essential to evaluate the formation of the geopolymer. Moreover, with this test, the appropriate percentage of the combination of chamotte and BBA will be selected for production.

Firstly, the results clearly reflect the formation of a geopolymer structure, since it can be seen that the compressive strength of the samples with a combination of chamotte and BBA increases with respect to the samples conformed with chamotte alone. In turn, over 60% of BBA on the mix creates a material which, although it has a higher density and lower porosity, has a reduction in compressive strength. This is mainly due to the fact that there is no adequate proportion of chamotte and BBA, so that only a percentage of the BBA reacts with the chamotte and all the unmixed BBA remains in excess. This fact decreases the resistance notably as it is an inert load. It should be noted that in this research the material is composed entirely of waste, so that in these there are a number of chemical compounds that do not only not favor the geopolymerization reaction but lead to a decrease in resistance.

Both conclusions confirm the geopolymer structure formed, with maximum compressive strength around the combination of 60% BBA and 40% chamotte. However, there is a range of the combination of both by-products that comply with the regulations on the resistance of ceramic materials, more specifically beginning with family 9C1A. This European standard sets a minimum compression strength for red clay bricks of 10 MPa.

The ageing test will rule out those families of samples that do not have adequate behavior in service, even though they reflect adequate initial mechanical characteristics. In turn, the colorimetric test will classify the samples according to color, which is essential within ceramic materials. The ultimate aim is to create a resistant but pleasantly aesthetic material that is accepted by the market.

Figure 5 shows an orderly representation of a sample from each of the families of geopolymers. As can be seen, there is a darkening of the samples due to the increasing percentage of BBA. At the ends are the colors of the chamotte and the BBA, respectively. In the central zone are the intermediate combinations of both residues. Since the aesthetics of an element are a personal appreciation, and ultimately it is the market that must choose it, it can only be identified faithfully in order to establish a color scale that does not vary according to the photograph taken.

Figure 5. Image of the different families of samples from family 10C0A (left) to family 0C10A (right).

In order to determine the color of each sample accurately, the colorimeter was used, giving the following values for the primary colors red, green and blue, as detailed in Table 6.

Table 6. RGB color coordinates of the samples conformed by the chamotte and the BBA of the different groups.

Groups	Chamotte, %	BBA, %	Red	Green	Blue
10C0A	100	0	379 ± 19	182 ± 9	115 ± 7
9C1A	90	10	249 ± 12	119 ± 5	77 ± 3
8C2A	80	20	253 ± 8	126 ± 7	84 ± 4
7C3A	70	30	232 ± 10	126 ± 7	87 ± 4
6C4A	60	40	192 ± 10	109 ± 5	78 ± 3
5C5A	50	50	177 ± 10	107 ± 6	79 ± 3
4C6A	40	60	170 ± 6	112 ± 4	85 ± 3
3C7A	30	70	155 ± 8	115 ± 4	93 ± 3
2C8A	20	80	147 ± 9	115 ± 6	97 ± 5
1C9A	10	90	142 ± 8	122 ± 6	109 ± 5
0C10A	0	100	118 ± 5	115 ± 6	110 ± 5

The color coordinates of the different groups of samples are another characteristic of the material, not limited by the regulations but by the quality controls of the industry. It is usual that a ceramic material incorporating waste is not accepted by the industry because of the color it reflects, even if it has adequate physical and mechanical characteristics. The quality criteria established by the producing companies have maximum permissible variations in the color of the final material, and therefore the addition of waste that varies sharply in color is rejected. In this case, it can be seen that the variation in color is gradual and towards darker shades, which is important and easy to market.

3.3. Ageing Tests (Freezing Test) and Fourier Transform Infrared (FTIR) of the Geopolymers

Once all the parameters of the different families, physical, mechanical and aesthetic, had been determined, the freezing test was carried out.

The ultimate aim of this test was to study the behavior of the different families in service, i.e., to study the variation in the initial characteristics over time. To evaluate ageing, this test requires a visual inspection after 25 freezing and defrosting cycles, determining which families are affected and should be discarded. The affected geopolymers will be called freeze geopolymers and will be rejected.

Figure 6 shows the picture of the different sample families before and after the freezing test. For comparison, one sample is taken that has been tested and another that has not been tested for each family.

After performing the freezing test and the observation of the above pictures, it can be concluded that only sample families 6C4A, 5C5A, 4C6A and 3C7A are suitable. Families 10C0A, 9C1A, 8C2A, 7C3A, 2C8A, 1C9A and 0C10A are freezing geopolymers, as they present spalling and flaking of more than 15 mm. Freezing geopolymers are rejected because they may not represent a sufficient quality of service.

On the basis of this result, it can be commented that although the physical and mechanical characteristics of the previous families, except for families 10C0A and 0C10A, were within the current regulations, the freezing test revealed a subgroup of samples with better mechanical and physical behavior during their useful life. These families will definitely be the ones considered as possible solutions, corresponding to combination percentages of 40% BBA with 60% chamotte up to 70% BBA with 30% chamotte.

The acceptable sample families listed above were then subjected to Fourier transform infrared (FTIR) analysis. For this purpose, new samples were made with the process detailed in the methodology and analyzed in order to compare the spectra. In turn, the families 10C0A and 0C10A corresponding to 100% chamotte and 100% BBA, respectively, were analyzed. In this way, it is easy to observe in the comparison of the spectra the differences that exist between them as well as the modifications that are produced by the process of geopolymerization.

Figure 7 shows the spectra of all the detailed families as well as in the right margin the amplification of all the spectra between 850 and 1150 cm^{-1}. It can be seen how due to the geopolymerization process, the asymmetric stretching frequency changes to a lower value for the families 6C4A, 5C5A, 4C6A and 3C7A than the one of the band presenting the chamotte and the BBA around 1010 cm^{-1}. This is because AlO4 partially replaces SiO$_4$ and changes the chemical environment of the Si–O bond. On the other hand, the comparison of the spectra in the 850–1150 cm^{-1} zone shows that the intensity of the frequency band detailed above and that of 875 cm^{-1} increases for the 6C4A, 5C5A, 4C6A and 3C7A geopolymer families with respect to the chamotte and BBA bands. The increase in intensity indicates an increase in chain length and more aluminosilicate gel formed, i.e., a more complete geopolymerization process. It should be noted that the higher intensity is reflected in the 4C6A group of samples, which in turn has been the most resistant family. It can be concluded that the Fourier transform infrared (FTIR) analysis coincides with the results with the previous compression test.

Figure 6. Image of the samples before the freezing test (left) and after the freezing test (right) for each of the geopolymer families studied.

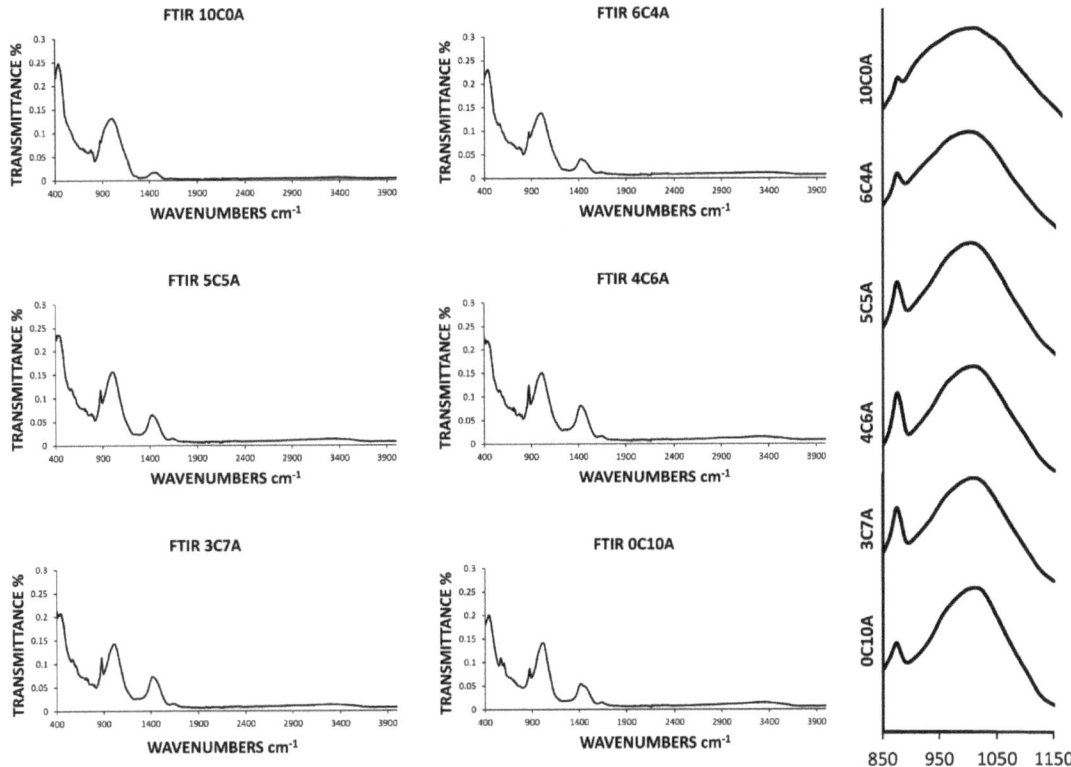

Figure 7. Fourier transform infrared (FTIR) analysis of the families with acceptable results (6C4A, 5C5A, 4C6A and 3C7A), as well as of the family with 100% chamotte (10C0A) and the family with 100% BBA (0C10A). On the right, a comparison of the intensity of the spectra for the families detailed in the region 850–1150 cm^{-1} is shown.

4. Conclusions

The development of the research methodology present in this work has led to a series of conclusions derived from each test. In order to obtain an objective and representative final conclusion of this investigation, the conclusions derived from each of the sub-sections of the methodology are set out below:

- The physical–chemical characterization of the chamotte and the biomass bottom ashes showed the suitability of both materials for the conformation of geopolymers. The elemental composition of the chamotte provides the perfect base of aluminosilicate, in combination with the high percentage of potassium present in the biomass bottom ashes. On the other hand, the similarity between the densities of both by-products and their microscopic granulometry facilitates the mixing process.
- The physical tests carried out on the families of samples conformed have reflected logical and statistically representative behavior. The loss of weight and linear shrinkage increased as the percentage of BBA in the mix increased. However, the bulk density is much lower than that of a traditional ceramic, which is of interest for other properties such as thermal or acoustic insulation. On the other hand, the rate of capillary water absorption, the cold water absorption and boiling water absorption, as well as the open porosity, decreased as the percentage of BBA in the mixture increased.

- The mechanical tests reflected a perfect quadratic curve, with a maximum of around 60% biomass bottom ashes in the mixture. However, all the families, except for 10C0A and 0C10A, showed adequate resistance behavior according to the regulations in force.
- The freezing tests determined that only the 6C4A, 5C5A, 4C6A and 3C7A families have adequate resistance to the ageing test.
- The Fourier transform infrared (FTIR) analysis reflected the formation of the geopolymer for the 6C4A, 5C5A, 4C6A and 3C7A sample groups.
- Geopolymers with acceptable results are formed with 40% BBA and 60% chamotte up to 70% BBA and 30% chamotte, the optimum combination being 60% BBA and 40% chamotte.

On the basis of the partial conclusions mentioned above and obtained from the detailed investigation methodology, it can be concluded that it is possible to produce geopolymers with physical, mechanical and aesthetic characteristics similar to those of traditional ceramics. Therefore, a sustainable material can be developed, thanks to the use of industrial by-products and to the low gas emissions of the manufacturing process, with appropriate characteristics and composed entirely of waste.

Author Contributions: Conceptualization, F.A.C.-I., F.J.I.-G., J.M.T.-S. and J.S.-M.; methodology, F.A.C.-I., F.J.I.-G., J.M.T.-S. and J.S.-M.; software, J.M.T.-S. and J.S.-M.; validation, F.A.C.-I. and F.J.I.-G.; formal analysis, F.A.C.-I. and F.J.I.-G.; investigation, J.M.T.-S. and J.S.-M.; resources, F.A.C.-I.; data curation, F.J.I.-G.; writing—original draft preparation, J.S.-M.; writing—review and editing, J.M.T.-S.; visualization, J.M.T.-S.; supervision, F.A.C.-I.; project administration, J.S.-M.; funding acquisition, F.A.C.-I. All authors have read and agreed to the published version of the manuscript.

Funding: This research received no external funding.

Institutional Review Board Statement: Not applicable.

Informed Consent Statement: Not applicable.

Data Availability Statement: Data is contained within the article.

Acknowledgments: Technical and human support provided by CICT of Universidad de Jaen (UJA, MINECO, Junta de Andalucía, FEDER) is gratefully acknowledged.

Conflicts of Interest: The authors declare no conflict of interest.

References

1. Zhang, L.; Liu, B.; Du, J.; Liu, C.; Wang, S. CO_2 emission linkage analysis in global construction sectors: Alarming trends from 1995 to 2009 and possible repercussions. *J. Clean. Prod.* **2019**, *221*, 863–877. [CrossRef]
2. Oti, J.E.; Kinuthia, J.M. Stabilised unfired clay bricks for environmental and sustainable use. *Appl. Clay Sci.* **2012**, *58*, 52–59. [CrossRef]
3. Kazmi, S.M.S.; Abbas, S.; Munir, M.J.; Khitab, A. Exploratory study on the effect of waste rice husk and sugarcane bagasse ashes in burnt clay bricks. *J. Build. Eng.* **2016**, *7*, 372–378. [CrossRef]
4. Kazmi, S.M.S.; Abbas, S.; Saleem, M.A.; Munir, M.J.; Khitab, A. Manufacturing of sustainable clay bricks: Utilization of waste sugarcane bagasse and rice husk ashes. *Constr. Build. Mater.* **2016**, *120*, 29–41. [CrossRef]
5. Subashi De Silva, G.H.M.J.; Mallwattha, M.P.D. Strength, durability, thermal and run-off properties of fired clay roof tiles incorporated with ceramic sludge. *Constr. Build. Mater.* **2018**, *179*, 390–399. [CrossRef]
6. Chatterjee, A.; Sui, T. Alternative fuels—Effects on clinker process and properties. *Cem. Concr. Res.* **2019**, *123*, 105777. [CrossRef]
7. Liu, B.; Zhang, L.; Sun, J.; Wang, D.; Liu, C.; Luther, M.; Xu, Y. Composition of energy outflows embodied in the gross exports of the construction sector. *J. Clean. Prod.* **2020**, *248*, 119296. [CrossRef]
8. Dondi, M.; Guarini, G.; Raimondo, M.; Zanelli, C. Recycling PC and TV waste glass in clay bricks and roof tiles. *Waste Manag.* **2009**, *29*, 1945–1951. [CrossRef]
9. Munir, M.J.; Kazmi, S.M.S.; Wu, Y.F.; Hanif, A.; Khan, M.U.A. Thermally efficient fired clay bricks incorporating waste marble sludge: An industrial-scale study. *J. Clean. Prod.* **2018**, *174*, 1122–1135. [CrossRef]
10. Thapa, V.B.; Waldmann, D.; Wagner, J.F.; Lecomte, A. Assessment of the suitability of gravel wash mud as raw material for the synthesis of an alkali-activated binder. *Appl. Clay Sci.* **2018**, *161*, 110–118. [CrossRef]
11. dos Reis, G.S.; Cazacliu, B.G.; Cothenet, A.; Poullain, P.; Wilhelm, M.; Sampaio, C.H.; Lima, E.C.; Ambros, W.; Torrenti, J.M. Fabrication, microstructure, and properties of fired clay bricks using construction and demolition waste sludge as the main additive. *J. Clean. Prod.* **2020**, *258*, 120733. [CrossRef]

12. Raut, S.P.; Ralegaonkar, R.V.; Mandavgane, S.A. Development of sustainable construction material using industrial and agricultural solid waste: A review of waste-create bricks. *Constr. Build. Mater.* **2011**, *25*, 4037–4042. [CrossRef]
13. Saboya, F.; Xavier, G.C.; Alexandre, J. The use of the powder marble by-product to enhance the properties of brick ceramic. *Constr. Build. Mater.* **2007**, *21*, 1950–1960. [CrossRef]
14. Zhang, L. Production of bricks from waste materials—A review. *Constr. Build. Mater.* **2013**, *47*, 643–655. [CrossRef]
15. Zhou, W.; Yan, C.; Duan, P.; Liu, Y.; Zhang, Z.; Qiu, X.; Li, D. A comparative study of high- and low-Al2O3 fly ash based-geopolymers: The role of mix proportion factors and curing temperature. *Mater. Des.* **2016**, *95*, 63–74. [CrossRef]
16. Gharzouni, A.; Vidal, L.; Essaidi, N.; Joussein, E.; Rossignol, S. Recycling of geopolymer waste: Influence on geopolymer formation and mechanical properties. *Mater. Des.* **2016**, *94*, 221–229. [CrossRef]
17. Zhang, Z.; Zhu, Y.; Yang, T.; Li, L.; Zhu, H.; Wang, H. Conversion of local industrial wastes into greener cement through geopolymer technology: A case study of high-magnesium nickel slag. *J. Clean. Prod.* **2017**, *141*, 463–471. [CrossRef]
18. Shang, J.; Dai, J.G.; Zhao, T.J.; Guo, S.Y.; Zhang, P.; Mu, B. Alternation of traditional cement mortars using fly ash-based geopolymer mortars modified by slag. *J. Clean. Prod.* **2018**, *203*, 746–756. [CrossRef]
19. Adesanya, E.; Ohenoja, K.; Luukkonen, T.; Kinnunen, P.; Illikainen, M. One-part geopolymer cement from slag and pretreated paper sludge. *J. Clean. Prod.* **2018**, *185*, 168–175. [CrossRef]
20. Tennakoon, C.; Shayan, A.; Sanjayan, J.G.; Xu, A. Chloride ingress and steel corrosion in geopolymer concrete based on long term tests. *Mater. Des.* **2017**, *116*, 287–299. [CrossRef]
21. Sabbatini, A.; Vidal, L.; Pettinari, C.; Sobrados, I.; Rossignol, S. Control of shaping and thermal resistance of metakaolin-based geopolymers. *Mater. Des.* **2017**, *116*, 374–385. [CrossRef]
22. Shi, C.; Jiménez, A.F. New cements for the 21st century: The pursuit of an alternative to Portland cement. *Cem. Concr. Res.* **2011**, *41*, 750–763. [CrossRef]
23. Davidovits, J. Geopolymers and geopolymeric materials. *J. Therm. Anal.* **1989**, *35*, 429–441. [CrossRef]
24. Palomo, A.; Grutzeck, M.W.; Blanco, M.T. Alkali-activated fly ashes: A cement for the future. *Cem. Concr. Res.* **1999**, *29*, 1323–1329. [CrossRef]
25. Zhang, M.; Deskins, N.A.; Zhang, G.; Cygan, R.T.; Tao, M. Modeling the Polymerization Process for Geopolymer Synthesis through Reactive Molecular Dynamics Simulations. *J. Phys. Chem. C* **2018**, *122*, 6760–6773. [CrossRef]
26. Xu, H.; Van Deventer, J.S.J. The geopolymerisation of alumino-silicate minerals. *Int. J. Miner. Process.* **2000**, *59*, 247–266. [CrossRef]
27. Cai, J.; Li, X.; Tan, J.; Vandevyvere, B. Fly ash-based geopolymer with self-heating capacity for accelerated curing. *J. Clean. Prod.* **2020**, *261*, 121119. [CrossRef]
28. Qian, L.-P.; Wang, Y.-S.; Alrefaei, Y.; Dai, J.-G. Experimental study on full-volume fly ash geopolymer mortars: Sintered fly ash versus sand as fine aggregates. *J. Clean. Prod.* **2020**, *263*, 121445. [CrossRef]
29. Chuah, S.; Duan, W.H.; Pan, Z.; Hunter, E.; Korayem, A.H.; Zhao, X.L.; Collins, F.; Sanjayan, J.G. The properties of fly ash based geopolymer mortars made with dune sand. *Mater. Des.* **2016**, *92*, 571–578. [CrossRef]
30. Zhang, Z.; Wang, H.; Zhu, Y.; Reid, A.; Provis, J.L.; Bullen, F. Using fly ash to partially substitute metakaolin in geopolymer synthesis. *Appl. Clay Sci.* **2014**, *88–89*, 194–201. [CrossRef]
31. Hertel, T.; Pontikes, Y. Geopolymers, inorganic polymers, alkali-activated materials and hybrid binders from bauxite residue (red mud)—Putting things in perspective. *J. Clean. Prod.* **2020**, *258*, 120610. [CrossRef]
32. Nazari, A.; Sanjayan, J.G. Synthesis of geopolymer from industrial wastes. *J. Clean. Prod.* **2015**, *99*, 297–304. [CrossRef]
33. Pontikes, Y.; Machiels, L.; Onisei, S.; Pandelaers, L.; Geysen, D.; Jones, P.T.; Blanpain, B. Slags with a high Al and Fe content as precursors for inorganic polymers. *Appl. Clay Sci.* **2013**, *73*, 93–102. [CrossRef]
34. Bignozzi, M.C.; Manzi, S.; Lancellotti, I.; Kamseu, E.; Barbieri, L.; Leonelli, C. Mix-design and characterization of alkali activated materials based on metakaolin and ladle slag. *Appl. Clay Sci.* **2013**, *73*, 78–85. [CrossRef]
35. Liang, G.; Zhu, H.; Zhang, Z.; Wu, Q.; Du, J. Investigation of the waterproof property of alkali-activated metakaolin geopolymer added with rice husk ash. *J. Clean. Prod.* **2019**, *230*, 603–612. [CrossRef]
36. Medri, V.; Papa, E.; Lizion, J.; Landi, E. Metakaolin-based geopolymer beads: Production methods and characterization. *J. Clean. Prod.* **2020**, *244*, 118844. [CrossRef]
37. Kuenzel, C.; Neville, T.P.; Donatello, S.; Vandeperre, L.; Boccaccini, A.R.; Cheeseman, C.R. Influence of metakaolin characteristics on the mechanical properties of geopolymers. *Appl. Clay Sci.* **2013**, *83–84*, 308–314. [CrossRef]
38. Si, R.; Dai, Q.; Guo, S.; Wang, J. Mechanical property, nanopore structure and drying shrinkage of metakaolin-based geopolymer with waste glass powder. *J. Clean. Prod.* **2020**, *242*, 118502. [CrossRef]
39. Xiao, R.; Ma, Y.; Jiang, X.; Zhang, M.; Zhang, Y.; Wang, Y.; Huang, B.; He, Q. Strength, microstructure, efflorescence behavior and environmental impacts of waste glass geopolymers cured at ambient temperature. *J. Clean. Prod.* **2020**, *252*, 119610. [CrossRef]
40. Novais, R.M.; Ascensão, G.; Seabra, M.P.; Labrincha, J.A. Waste glass from end-of-life fluorescent lamps as raw material in geopolymers. *Waste Manag.* **2016**, *52*, 245–255. [CrossRef]
41. Faisal, M.; Muhammad, K. Synthesis and characterization of geopolymer from bagasse bottom ash, waste of sugar industries and naturally available china clay. *J. Clean. Prod.* **2016**, *129*, 491–495.
42. Arulrajah, A.; Kua, T.A.; Suksiripattanapong, C.; Horpibulsuk, S.; Shen, J.S. Compressive strength and microstructural properties of spent coffee grounds-bagasse ash based geopolymers with slag supplements. *J. Clean. Prod.* **2017**, *162*, 1491–1501. [CrossRef]

43. Nkwaju, R.Y.; Djobo, J.N.Y.; Nouping, J.N.F.; Huisken, P.W.M.; Deutou, J.G.N.; Courard, L. Iron-rich laterite-bagasse fibers based geopolymer composite: Mechanical, durability and insulating properties. *Appl. Clay Sci.* **2019**, *183*, 105333. [CrossRef]
44. Zhuang, X.Y.; Chen, L.; Komarneni, S.; Zhou, C.H.; Tong, D.S.; Yang, H.M.; Yu, W.H.; Wang, H. Fly ash-based geopolymer: Clean production, properties and applications. *J. Clean. Prod.* **2016**, *125*, 253–267. [CrossRef]
45. Fu, S.; He, P.; Wang, M.; Cui, J.; Wang, M.; Duan, X.; Yang, Z.; Jia, D.; Zhou, Y. Hydrothermal synthesis of pollucite from metakaolin-based geopolymer for hazardous wastes storage. *J. Clean. Prod.* **2020**, *248*, 119240. [CrossRef]
46. Capasso, I.; Lirer, S.; Flora, A.; Ferone, C.; Cioffi, R.; Caputo, D.; Liguori, B. Reuse of mining waste as aggregates in fly ash-based geopolymers. *J. Clean. Prod.* **2019**, *220*, 65–73. [CrossRef]
47. Nath, S.K.; Kumar, S. Reaction kinetics of fly ash geopolymerization: Role of particle size controlled by using ball mill. *Adv. Powder Technol.* **2019**, *30*, 1079–1088. [CrossRef]
48. Hajimohammadi, A.; Ngo, T.; Mendis, P.; Sanjayan, J. Regulating the chemical foaming reaction to control the porosity of geopolymer foams. *Mater. Des.* **2017**, *120*, 255–265. [CrossRef]
49. Habert, G.; d'Espinose de Lacaillerie, J.B.; Roussel, N. An environmental evaluation of geopolymer based concrete production: Reviewing current research trends. *J. Clean. Prod.* **2011**, *19*, 1229–1238. [CrossRef]
50. Turner, L.K.; Collins, F.G. Carbon dioxide equivalent (CO_2-e) emissions: A comparison between geopolymer and OPC cement concrete. *Constr. Build. Mater.* **2013**, *43*, 125–130. [CrossRef]
51. Lahoti, M.; Wong, K.K.; Tan, K.H.; Yang, E.-H. Effect of alkali cation type on strength endurance of fly ash geopolymers subject to high temperature exposure. *Mater. Des.* **2018**, *154*, 8–19. [CrossRef]
52. Part, W.K.; Ramli, M.; Cheah, C.B. An overview on the influence of various factors on the properties of geopolymer concrete derived from industrial by-products. *Constr. Build. Mater.* **2015**, *77*, 370–395. [CrossRef]
53. Peyne, J.; Gautron, J.; Doudeau, J.; Rossignol, S. Development of low temperature lightweight geopolymer aggregate, from industrial Waste, in comparison with high temperature processed aggregates. *J. Clean. Prod.* **2018**, *189*, 47–58. [CrossRef]
54. Singh, B.; Ishwarya, G.; Gupta, M.; Bhattacharyya, S.K. Geopolymer concrete: A review of some recent developments. *Constr. Build. Mater.* **2015**, *85*, 78–90. [CrossRef]
55. Sékou, T.; Siné, D.; Lanciné, T.D.; Bakaridjan, C. Synthesis and Characterization of a Red Mud and Rice Husk Based Geopolymer for Engineering Applications. *Macromol. Symp.* **2017**, *373*, 1600090. [CrossRef]
56. Huiskes, D.M.A.; Keulen, A.; Yu, Q.L.; Brouwers, H.J.H. Design and performance evaluation of ultra-lightweight geopolymer concrete. *Mater. Des.* **2016**, *89*, 516–526. [CrossRef]
57. Hu, Z.; Wyrzykowski, M.; Lura, P. Estimation of reaction kinetics of geopolymers at early ages. *Cem. Concr. Res.* **2020**, *129*, 105971. [CrossRef]
58. Kong, D.L.Y.; Sanjayan, J.G.; Sagoe-Crentsil, K. Comparative performance of geopolymers made with metakaolin and fly ash after exposure to elevated temperatures. *Cem. Concr. Res.* **2007**, *37*, 1583–1589. [CrossRef]
59. Kong, D.L.Y.; Sanjayan, J.G. Effect of elevated temperatures on geopolymer paste, mortar and concrete. *Cem. Concr. Res.* **2010**, *40*, 334–339. [CrossRef]
60. Sellami, M.; Barre, M.; Toumi, M. Synthesis, thermal properties and electrical conductivity of phosphoric acid-based geopolymer with metakaolin. *Appl. Clay Sci.* **2019**, *180*, 105192. [CrossRef]
61. Bi, S.; Liu, M.; Shen, J.; Hu, X.M.; Zhang, L. Ultrahigh Self-Sensing Performance of Geopolymer Nanocomposites via Unique Interface Engineering. *ACS Appl. Mater. Interfaces* **2017**, *9*, 12851–12858. [CrossRef] [PubMed]
62. Aguirre-Guerrero, A.M.; Robayo-Salazar, R.A.; de Gutiérrez, R.M. A novel geopolymer application: Coatings to protect reinforced concrete against corrosion. *Appl. Clay Sci.* **2017**, *135*, 437–446. [CrossRef]
63. Yan, S.; He, P.; Jia, D.; Duan, X.; Yang, Z.; Wang, S.; Zhou, Y. In-situ preparation of fully stabilized graphene/cubic-leucite composite through graphene oxide/geopolymer. *Mater. Des.* **2016**, *101*, 301–308. [CrossRef]
64. Yan, D.; Chen, S.; Zeng, Q.; Xu, S.; Li, H. Correlating the elastic properties of metakaolin-based geopolymer with its composition. *Mater. Des.* **2016**, *95*, 306–318. [CrossRef]
65. Roviello, G.; Menna, C.; Tarallo, O.; Ricciotti, L.; Ferone, C.; Colangelo, F.; Asprone, D.; di Maggio, R.; Cappelletto, E.; Prota, A.; et al. Preparation, structure and properties of hybrid materials based on geopolymers and polysiloxanes. *Mater. Des.* **2015**, *87*, 82–94. [CrossRef]
66. Medri, V.; Papa, E.; Mazzocchi, M.; Laghi, L.; Morganti, M.; Francisconi, J.; Landi, E. Production and characterization of lightweight vermiculite/geopolymer-based panels. *Mater. Des.* **2015**, *85*, 266–274. [CrossRef]
67. Azevedo, A.R.G.; França, B.R.; Alexandre, J.; Marvila, M.T.; Zanelato, E.B.; Xavier, G.C. Influence of sintering temperature of a ceramic substrate in mortar adhesion for civil construction. *J. Build. Eng.* **2018**, *19*, 342–348. [CrossRef]
68. Kittl, P.; Diaz, G.; Alarcón, H. Dosification of a cement-talc-chamotte refractory mortar subjected to thermal shock. *Cem. Concr. Res.* **1992**, *22*, 736–742. [CrossRef]
69. Fiala, L.; Konrád, P.; Fořt, J.; Keppert, M.; Černý, R. Application of ceramic waste in brick blocks with enhanced acoustic properties. *J. Clean. Prod.* **2020**, *261*, 121185. [CrossRef]
70. Nayana, A.M.; Rakesh, P. Strength and durability study on cement mortar with ceramic waste and micro-silica. *Mater. Today Proc.* **2018**, *5*, 24780–24791. [CrossRef]
71. Amin, S.K.; El–Sherbiny, S.A.; El–Magd, A.A.M.A.; Belal, A.; Abadir, M.F. Fabrication of geopolymer bricks using ceramic dust waste. *Constr. Build. Mater.* **2017**, *157*, 610–620. [CrossRef]

72. Huseien, G.F.; Sam, A.R.M.; Shah, K.W.; Asaad, M.A.; Tahir, M.M.; Mirza, J. Properties of ceramic tile waste based alkali-activated mortars incorporating GBFS and fly ash. *Constr. Build. Mater.* **2019**, *214*, 355–368. [CrossRef]
73. Keppert, M.; Vejmelková, E.; Bezdička, P.; Doleželová, M.; Čáchová, M.; Scheinherrová, L.; Pokorný, J.; Vyšvařil, M.; Rovnaníková, P.; Černý, R. Red-clay ceramic powders as geopolymer precursors: Consideration of amorphous portion and CaO content. *Appl. Clay Sci.* **2018**, *161*, 82–89. [CrossRef]
74. Martirena, F.; Monzó, J. Vegetable ashes as Supplementary Cementitious Materials. *Cem. Concr. Res.* **2018**, *114*, 57–64. [CrossRef]
75. Nalbantoglu, Z.; Gucbilmez, E. Improvement of calcareous expansive soils in semi-arid environments. *J. Arid Environ.* **2001**, *47*, 453–463. [CrossRef]
76. Vassilev, S.V.; Baxter, D.; Andersen, L.K.; Vassileva, C.G. An overview of the composition and application of biomass ash. Part 1. Phase–mineral and chemical composition and classification. *Fuel* **2013**, *105*, 40–76. [CrossRef]
77. Tortosa Masiá, A.A.; Buhre, B.J.P.; Gupta, R.P.; Wall, T.F. Characterising ash of biomass and waste. *Fuel Process. Technol.* **2007**, *88*, 1071–1081. [CrossRef]
78. EUR-Lex—32000D0532—ES. Available online: https://eur-lex.europa.eu/LexUriServ/LexUriServ.do?uri=CELEX:32000D0532:ES:HTML (accessed on 3 March 2020).
79. Rosales, J.; Cabrera, M.; Beltrán, M.G.; López, M.; Agrela, F. Effects of treatments on biomass bottom ash applied to the manufacture of cement mortars. *J. Clean. Prod.* **2017**, *154*, 424–435. [CrossRef]
80. Giro-Paloma, J.; Mañosa, J.; Maldonado-Alameda, A.; Quina, M.J.; Chimenos, J.M. Rapid sintering of weathered municipal solid waste incinerator bottom ash and rice husk for lightweight aggregate manufacturing and product properties. *J. Clean. Prod.* **2019**, *232*, 713–721. [CrossRef]
81. Alam, Q.; Hendrix, Y.; Thijs, L.; Lazaro, A.; Schollbach, K.; Brouwers, H.J.H. Novel low temperature synthesis of sodium silicate and ordered mesoporous silica from incineration bottom ash. *J. Clean. Prod.* **2019**, *211*, 874–883. [CrossRef]
82. James, K.A.; Thring, W.R.; Helle, S.; Ghuman, S.H. Ash Management Review: Applications of Biomass Bottom Ash. *Energies* **2012**, *5*, 3856. [CrossRef]
83. UNE-EN 1097-7:2009 Tests for Mechanical and Physical Properties of Aggregates—Part 3: Determination of Loose Bulk Density and Voids. Available online: https://www.une.org/encuentra-tu-norma/busca-tu-norma/norma?c=N0042553 (accessed on 16 September 2020).
84. UNE-EN 772-16:2011 Methods of Test for Masonry Units—Part 16: Determination of Dimensions. Available online: https://www.une.org/encuentra-tu-norma/busca-tu-norma/norma/?c=N0047875 (accessed on 30 September 2020).
85. UNE-EN 772-11:2011 Methods of Test for Masonry Units—Part 11: Determination of Water Absorption of Aggregate Concrete, Autoclaved Aerated Concrete, Manufactured Stone and Natural Stone Masonry Units due to Capillary Action and the Initial Rate of Water Absorption of Clay Masonry Units. Available online: https://www.une.org/encuentra-tu-norma/busca-tu-norma/norma?c=N0047874 (accessed on 30 September 2020).
86. UNE-EN 772-21:2011 Methods of Test for Masonry Units—Part 21: Determination of Water Absorption of Clay and Calcium Silicate Masonry Units by Cold Water Absorption. Available online: https://www.une.org/encuentra-tu-norma/busca-tu-norma/norma?c=N0047877 (accessed on 30 September 2020).
87. UNE-EN 772-7:1999 Methods of Test for Masonry Units—Part 7: Determination of Water Absorption of Clay Masonry Damp Proof Course Units by Boiling in Water. Available online: https://www.une.org/encuentra-tu-norma/busca-tu-norma/norma?c=N0009121 (accessed on 30 September 2020).
88. UNE-EN 772-4:1999 Methods of Test for Masonry Units—Part 4: Determination of Real and Bulk Density and of Total and Open Porosity for Natural Stone Masonry Units. Available online: https://www.une.org/encuentra-tu-norma/busca-tu-norma/norma/?c=N0009120 (accessed on 30 September 2020).
89. UNE-EN 772-1:2011+A1:2016 Methods of Test for Masonry Units—Part 1: Determination of Compressive Strength. Available online: https://www.une.org/encuentra-tu-norma/busca-tu-norma/norma/?Tipo=N&c=N0056681 (accessed on 30 September 2020).
90. UNE 67028:1997 EX Clay Bricks. Freezing Test. Available online: https://www.une.org/encuentra-tu-norma/busca-tu-norma/norma/?c=N0006752 (accessed on 30 September 2020).

Article

Immobilization of Heavy Metals in Boroaluminosilicate Geopolymers

Piotr Rożek *, Paulina Florek, Magdalena Król and Włodzimierz Mozgawa

Faculty of Materials Science and Ceramics, AGH University of Science and Technology, 30 Mickiewicza Av., 30-059 Krakow, Poland; paulina@agh.edu.pl (P.F.); mkrol@agh.edu.pl (M.K.); mozgawa@agh.edu.pl (W.M.)
* Correspondence: prozek@agh.edu.pl

Abstract: Boroaluminosilicate geopolymers were used for the immobilization of heavy metals. Then, their mechanical properties, phase composition, structure, and microstructure were investigated. The addition of borax and boric acid did not induce the formation of any crystalline phases. Boron was incorporated into the geopolymeric network and caused the formation of N–B–A–S–H (hydrated sodium boroaluminosilicate) gel. In the range of a B/Al molar ratio of 0.015–0.075, the compressive strength slightly increased (from 16.1 to 18.7 MPa), while at a ratio of 0.150, the compressive strength decreased (to 12 MPa). Heavy metals (lead and nickel) were added as nitrate salts. The loss of the strength of the geopolymers induced by heavy metals was limited by the presence of boron. However, it caused an increase in heavy metal leaching. Despite this, heavy metals were almost entirely immobilized (with immobilization rates of >99.8% in the case of lead and >99.99% in the case of nickel). The lower immobilization rate of lead was due to the formation of macroscopic crystalline inclusions of $PbO \cdot xH_2O$, which was vulnerable to leaching.

Keywords: alkali-activation; fly ash; boron; lead; nickel; leaching

---

Citation: Rożek, P.; Florek, P.; Król, M.; Mozgawa, W. Immobilization of Heavy Metals in Boroaluminosilicate Geopolymers. *Materials* **2021**, *14*, 214. https://doi.org/10.3390/ma14010214

Received: 30 November 2020
Accepted: 29 December 2020
Published: 4 January 2021

Publisher's Note: MDPI stays neutral with regard to jurisdictional claims in published maps and institutional affiliations.

Copyright: © 2021 by the authors. Licensee MDPI, Basel, Switzerland. This article is an open access article distributed under the terms and conditions of the Creative Commons Attribution (CC BY) license (https://creativecommons.org/licenses/by/4.0/).

1. Introduction

Rapidly growing industries have given humanity ever higher standards of living, but improper management can have serious consequences for the environment. One of the most important impacts is environmental pollution, such as greenhouse gas emissions to the atmosphere, the formation of heaps from the deposit of solid wastes, and the contamination of water and soil with heavy metals. In order to protect the environment, some restrictions have been imposed and some goals have been established in order to create a "recycling society" that seeks to eliminate waste generation and use it as secondary raw material. Ordinary Portland cement (OPC) is one of the most used materials in the world, but it has a large carbon footprint, so research into obtaining its ecological alternative is obviously needed.

Geopolymers, like alkali-activated aluminosilicates, are defined as binder materials built from a tri-dimensional network structure of –Si–O–Si(Al)– bonds and synthesized by treating the reactive solid source of SiO_2 and Al_2O_3 with an alkaline solution [1]. Due to their similarity in chemical structure and physical properties to hardened ordinary Portland cement, geopolymers are considered a green alternative of the most commonly used building material in the world [2]. The relatively low alkali activation temperature of aluminosilicates (<90 °C), along with the lack of need for calcine substrates and fuse clinkers, results in reduced energy consumption and a lower amount of CO_2 released into the atmosphere in the case of geopolymers in comparison with OPC. The final solid products of alkali-activated aluminosilicates are highly chemical and fire-resistant, and they exhibit relatively high mechanical strengths [3]. Therefore, the physical properties of geopolymers show their potential to be environmentally-friendly Portland cement alternatives that can be used as construction and building materials. There are various known reactive solid sources of SiO_2 and Al_2O_3, starting from synthetic pure chemical reagents to natural minerals and industrial by-products. According to the literature, geopolymers have been successfully obtained from coal fly ash [4], bottom ash [5], red mud [6,7], slag [8], and biomass fly ash [9]. This fact suggests an excellent opportunity to reuse materials that are currently waste.

The composition of ashes, generated both in power plants and waste incinerators, depends on the source of solid fuels used in the combustion process, but it usually contains heavy metals. Without treatment, hazardous ions can be washed out and pose a threat to the environment. Heavy metals can get into the soil and water, enter plants, and then pass through the food chain to animal and human bodies. Immobilization is the chemical process of material transformation in such a way that soluble compounds are not eluted. One of the most effective materials for deactivating compounds is glass due to its high chemical resistance [10]. However, such a process is rarely used due to the fact that it is energy-consuming. In the literature, one can find more cost-effective methods with the use of cement pastes [11], slag-alkali binders [12], and red ceramics [13].

Another important aspect of alkali-activated aluminosilicates is their ability to immobilize hazardous ions, which has been widely investigated in the past few years, e.g., the immobilization mechanism of heavy metal cations (Cd^{2+}, Pb^{2+}, and Zn^{2+}) and anions (AsO_4^{3-} and $Cr_2O_7^{2-}$) in the composite geopolymers based on granulated blast furnace slag and drinking water treatment residue [14] or the immobilization of hexavalent chromium in fly ash-based geopolymers [15]. Additionally, the results described in [16] showed a matrix based on alkali-activated aluminosilicates that could be an alternative to cementitious binders. The proposed mechanisms of the heavy metal immobilization in geopolymers involve the ion-exchange of charge-balancing cations (Na^+) with heavy metal cations; covalent bonding of heavy metals to the aluminosilicate network of a geopolymer; the precipitation of hydroxides, carbonates, silicates of heavy metals; and the physical encapsulation of heavy metals in a low-permeable geopolymeric matrix [17]. Heavy metals may be trapped in closed micropores created in a geopolymeric matrix [18]. When the heavy metal leaching from geopolymers is below the standard limits, the geopolymers can be used in some construction applications [19].

In order to make geopolymers more cost-effective and eco-friendly, some attempts have been made to introduce other elements, such as boron, which could totally or partially substitute alumina or silica. Boron is used in different industries, such as, energy, agriculture, medicine, and glass, detergent, and ceramics manufacturing [20]. It is present in nature in many kinds of minerals, and its main salts (borates) are not considered toxic substances [21]. There have been some studies regarding geopolymers with the addition of various boron compounds, such as borax, anhydrous borax, boric acid, amorphous and crystalline lithium tetraborate, and colemanite [22–30]. Boron may also be introduced to a geopolymer matrix when the raw aluminosilicate material contains this element. Celik et al. [20] made geopolymers based on metakaolin and colemanite waste, while Taveri et al. investigated the utilization of recycled borosilicate glass for geopolymer production [31]. The impact of using borax as a component of the alkalizing solution was undertaken in the mentioned literature, and the chemical was considered to be an environmentally-friendly additive that could influence mechanical properties. In Portland cement systems, boron acts as a hydration retarder, and it has a strong impact on hydration kinetics—it slows down the dissolution of aluminate phases [32]. Boron negatively affects the hydration and mechanical strength of Portland cement, but it has no influence on the alkaline activation of fly ashes [21]. However, in geopolymers activated with NaOH and sodium silicate, a decrease in the content of sodium silicate in the activator solution was found to lead to a steep decrease in compressive strength, while the presence of borax allowed for the obtainment of a much slighter decrease in compressive strength [24].

To the best of our knowledge, there have been descriptions of boroaluminosilicate geopolymers utilized as matrixes for the disposal of ashes containing heavy metal ions. Taking into account the nature of boron, its ability to complex various elements, and the commonly known immobilizing properties of borosilicate glasses [33], the immobilizing properties of boroaluminosilicate geopolymers were investigated in the present work. The aim of this research was to utilize alkali-activated fly ash as a matrix for the immobilization of heavy metals and to assess the role of boron in immobilization efficiency. The results of structural studies of boroaluminosilicate geopolymers, obtained via an alkali-activation method of fly ash with two different sources of boron, are also presented. The results show the possibility of the utilization of industrial by-products bearing high amounts of heavy metals and boron compounds for the production of geopolymers that could be used, e.g., in roads basements construction.

2. Materials and Methods

Coal fly ash (FA) obtained from a Polish power plant was used as an aluminosilicate source for alkali-activation, which was conducted with a sodium hydroxide solution. Additionally, borax and boric acid were used to introduce boron to the structure of the geopolymers. Lead nitrate, $Pb(NO_3)_2$, and nickel nitrate, $Ni(NO_3)_2 \cdot 6H_2O$, were utilized to determine the immobilization properties of the geopolymers.

Geopolymers were obtained by mixing fly ash with an alkali activator (NaOH) solution (AA). Two series with boron were prepared—one with borax (BX), $Na_2[B_4O_5(OH)_4] \cdot 8H_2O$ and one with boric acid (BA), H_3BO_3—as sources of boron. The amount of boric compound represented four molar ratios of boron to aluminum: B/Al = 0.015, 0.030, 0.075, and 0.150 (Table 1). The amount of aluminum in fly ash was determined with the X-ray fluorescence (XRF) method.

Table 1. Geopolymer composition. AA: alkali activator (NaOH) solution; FA: fly ash; BX: borax; BA: boric acid.

B/Al (mol/mol)	AA (mol/dm^3)	AA/FA (g/g)	BX (wt.%)	BA (wt.%)
0			0	0
0.015			1	0.625
0.030	10	0.4	2	1.250
0.075			5	3.125
0.150			10	6.250

The samples with heavy metals were prepared by adding a metal salt to the geopolymeric slurry. The amount of the metal salt represented the amount of metal cations in relation to fly ash, and it was 2 and 4 wt.%. After mixing, the pastes were cast in silicone molds, thus obtaining cubic samples (20 × 20 × 20 mm^3) that were cured at 80 °C for one day. The samples were tested after another 27 days.

XRF was used for the quantitative analysis of fly ash and geopolymers (with an Axios PANalytical Max 4 kW spectrometer (PANalytical, Malvern, UK) with a wavelength dispersive and Rh source). The standardless analysis package Omnian (PANalytical) was used for the semiquantitative analysis of the spectra. The results are presented on a percentage scale (normalized to 100%). XRD (X-ray diffraction) was used to analyze the phase composition of the samples; a diffractometer (Empyrean, PANalytical) was with CuKα radiation and a graphite monochromator, and the measurements were carried out in the 2theta angle range of 5–60° for 3 h with a step of 0.007. Phases were identified with the use of an X'Pert HighScore Plus (PANalytical) application and the International Centre for Diffraction Data. SEM (scanning electron microscope) was used to observe the microstructure of the samples (with an FEI Nova NanoSEM 200 microscope; samples were sprayed with graphite), to observe the microstructure, and prepare EDS (energy-dispersive X-ray spectroscopy) maps (with ThermoFisher Scientific Phenom XL (ThermoFisher Scientific, Waltham, MA, USA); samples were sprayed with gold). FT-IR (Fourier transform infrared spectroscopy) was used to study the structure of the samples (with a Bruker VERTEX 70v vacuum spectrometer (Bruker, Billerica, MA, USA); samples were prepared as KBr pellets and spectra were measured in the mid infrared range (4000–400 cm^{-1}) with 128 scans and a resolution of 4 cm^{-1}). The bulk densities were calculated by dividing the mass of the sample by its volume. The compressive strength of each sample was measured with a ZwickRoell (Ulm, Germany) machine working on the principle of a hydraulic press.

The leaching test was prepared according to the procedure described in the EN 12457-2:2002 (Characterization of Waste—Leaching) standard. The geopolymer samples were crushed, and the size of the particles for the leaching test was below 4 mm. They were poured with distilled water and shaken for 24 h. The ratio of the leaching solution (L) to the sample (S) was: L/S = 10 L/kg. The concentrations of Pb^{2+} and Ni^{2+} cations in the leachates were analyzed by atomic absorption spectrometry (AAS). The analyses were conducted on a Philips PU 9100× spectrometer (Philips, Amsterdam, The Netherlands) with a calibration curve.

3. Results

3.1. Characterization of Fly Ash

The chemical composition of the FA is presented in Table 2. It was an aluminosilicate material (SiO_2 and Al_2O_3 > 80 wt.%) of class F (according to ASTM C 618-05). Some heavy metals—cobalt, zirconium, chromium, zinc, nickel, copper, and lead—were present in trace amounts. An 'amorphous halo' can be seen in Figure 1, which suggests the high presence of glassy content. Moreover, the crystalline phases of quartz and mullite were identified.

Table 2. The chemical composition of fly ash (FA).

Oxide Composition (wt.%):									
P_2O_5	SiO_2	TiO_2	Al_2O_3	Fe_2O_3	MgO	CaO	Na_2O	K_2O	LOI [1]
1.1	51.4	1.2	33.8	4.5	1.6	1.6	1.1	2.3	4.7
Trace Composition (mg/kg):									
Co	Sr	Ba	Zr	Cr	Zn	Ni	Cu	Pb	As
2038	774	510	367	229	134	111	103	97	65

[1] LOI = loss on ignition at 1000 °C.

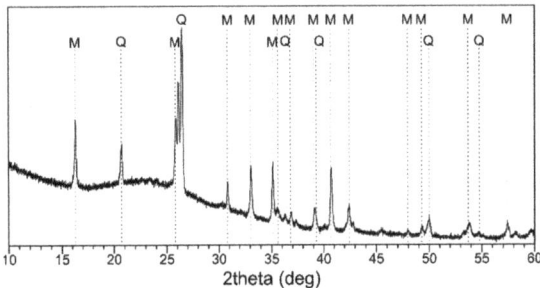

Figure 1. The XRD pattern of fly ash (M—mullite; Q—quartz).

3.2. Boroaluminosilicate Geopolymers

Mechanical properties: The compressive strength and bulk density of the prepared geopolymers are presented in Figure 2. The compressive strength of the reference sample (without boron) was 16.1 MPa, and the addition of boron to the extent expressed in a B/Al molar ratio of 0.015–0.075 caused a slight increase in strength (16.4–17.3 MPa in the case of BX as the source of B and 17.7–18.7 MPa in the case of BA as the source of B). The values of the compressive strength of the samples with borax and boric acid were quite similar, so the effect of borax "water" on strength could be excluded. A higher content of boron (B/Al = 0.150) induced a decrease in the compressive strength to 13.7 MPa for BX and 12.0 MPa for BA (the relative changes of compressive strength due to the presence of B are shown in Figure 3). In [30], it was observed that replacing sodium silicate with borax led to a compressive strength drop from 45 to 30 MPa.

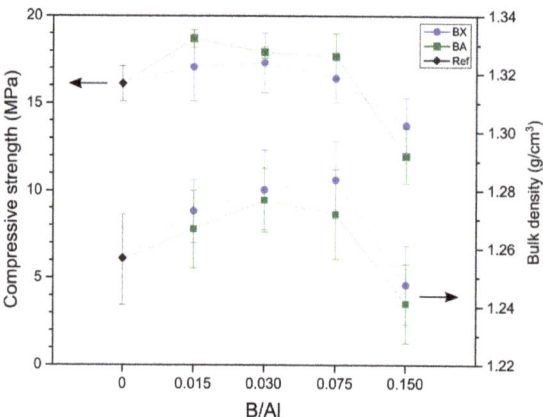

Figure 2. The compressive strength and bulk density of geopolymers with borax (BX) and boric acid (BA) as the source of boron (Ref—reference sample without boron).

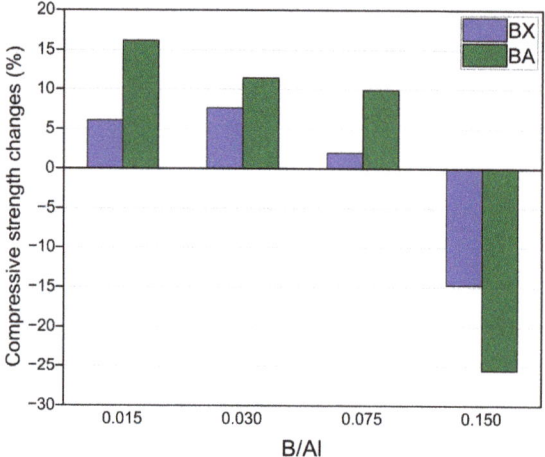

Figure 3. The relative changes in the compressive strength of geopolymers—the effect of the increasing boron content (in relation to geopolymer without boron).

Phase composition: As can be seen in Figure 4, some crystalline phases, mainly quartz and mullite, were identified in the geopolymers; these were the unreacted components of the fly ash. The newly formed phases were sodium carbonates and zeolite-like feldspathoids, which could have been hydroxysodalite or hydroxycancrinite. Zeolite-like phases and zeolites are common phases that form during the geopolymerization process [34]. Sodium carbonates (e.g., trona, $Na_3H(CO_3)_2·2H_2O$, and thermonatrite, $Na_2CO_3·H_2O$) formed due to the reaction between atmospheric CO_2 and free sodium in the pore solutions [35,36]. There were no visible differences between the patterns of the samples with and without boron. This is consistent with [37], in which the presence of borax and boric acid did not lead to the appearance of new phases. It can be then concluded that boron was built into the structure of a geopolymeric N–A–S–H gel (N–A–S–H = $Na_2O·Al_2O_3·SiO_2·H_2O$), thus forming N–B–A–S–H gel.

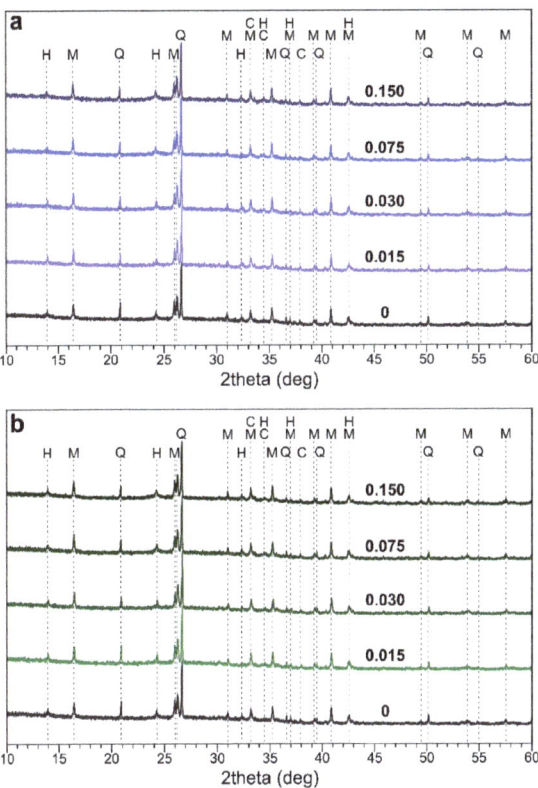

Figure 4. The XRD patterns of geopolymers with borax (**a**) and boric acid (**b**). 0—reference sample without boron (H—hydroxysodalite; M—mullite; Q—quartz; and C—sodium carbonates).

Structural studies: Figure 5 presents the infrared spectra of geopolymers with and without boron. In all of them, some groups of bands could be distinguished. The bands in the region of 3500–3400 cm^{-1} (stretching vibrations of O–H) and at around 1650 cm^{-1} (bending vibrations of H–O–H) were related to OH–groups and structural water molecules that were present in geopolymeric gel [38]. The bands at about 1450 and 860 cm^{-1} appeared due to the carbonation process, namely the stretching vibrations of C–O bond in the CO$_3$ groups. The main band at 1000 cm^{-1} could be assigned to the vibrations of Si–O–Si and Si–O–Al bonds. There were no distinct changes in the spectra caused by the addition of boron compounds.

Generally, boron may occur in the structure of oxide glasses in both triangular and tetrahedral coordination. In the presence of alkali, it is more likely to change from three- to four-fold coordination without non-bridging oxygen formation [39]. Additionally, the introduction of aluminum reduces the amount of [BO$_3$] groups in the glass [40]. Such a situation, namely the presence of alkalis and aluminum, existed in these geopolymeric systems and suggested that boron was present in their structure in a tetrahedral coordination. This was confirmed by the absence of a band at 1350 cm^{-1}, which could have been related to the vibrations of boron in three-fold coordination.

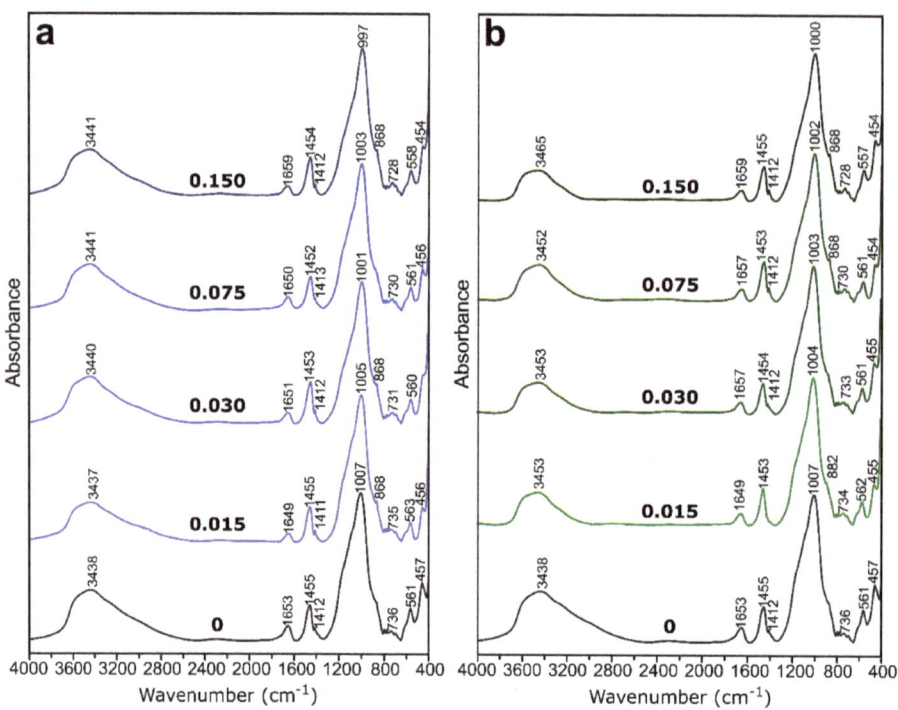

Figure 5. The FT-IR spectra of the geopolymers with borax (**a**) and boric acid (**b**). 0—reference sample without boron.

Williams and van Riesen [27] stated that boron in borax, which is both three- and four-fold coordinated, is likely to reorganize during the reorientation stage of the reaction, and it results in a four-fold coordinated boron in a geopolymer. They had similar observations that the trigonal boron dissolved and rearranged into tetrahedral boron, which was introduced in the geopolymeric cross-linked network that was made by Taveri et al. [31]. The tetrahedral boron in the mixture could directly participate in the geopolymerization, resulting in enhanced polycondensation [37]. It should be noted that tetrahedra such as [AlO$_4$] and [BO$_4$] introduce additional negative charges to a structure that should play an important role in the heavy metal immobilization properties of boroaluminosilicate geopolymers.

The only response of the spectra to the increasing content of boron was a slight shifting of the main band towards lower wavenumbers from 1007 to 997 cm^{-1} in the case of BX and from 1007 to 1000 cm^{-1} in the case of BA. This may indicate that boron in the tetrahedral coordination was incorporated into the aluminosilicate network of geopolymers. In glass, B–O vibrations in [BO$_4$] units were assigned to the band at 930 cm^{-1} [41], which should explain the shifting of geopolymeric band of 1007 cm^{-1} to lower wavenumbers in the presence of boron.

Microstructure: SEM images of the geopolymers with and without boron are presented in Figure 6. The amorphous geopolymeric gel was visible in all samples. The spherical shapes in Figure 6a,e are fly ash particles that dissolved to a greater or lesser extent and that were covered with the condensed gel. Some concave voids of halved cenospheres (hollow particles of fly ash), partially filled with the geopolymerization products, are visible in Figure 6c. No crystalline products, as the result of boron addition, were observed. However, the microstructure of the samples with the highest boron content (Figure 6c,f) seemed less compact and more porous. Since the bulk density did not change much with the addition of boron compounds (remaining in the range of 1.24–1.28 g/cm^3), the additional atoms in the structure must have increased the density. As such, the effect of structure relaxation caused by boron was evident.

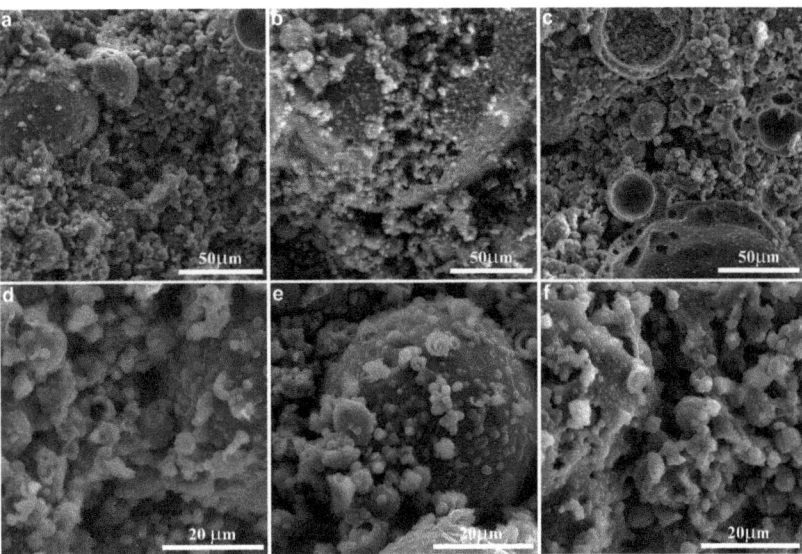

Figure 6. SEM images of the geopolymers reference sample (**a**,**d**), with BX (**b**,**c**) and BA (**e**,**f**). B/Al = 0 (**a**,**d**), 0.030 (**b**,**e**), and 0.150 (**c**,**f**).

3.3. Geopolymers with Heavy Metals

Mechanical properties: Figure 7 presents the compressive strength and bulk density of the samples containing heavy metals, namely nitrate salts of lead and nickel. Their presence caused a decrease in the compressive strength of the geopolymers, both with and without boron. The reference samples without boron achieved about 14 MPa (2 wt.% of Pb^{2+}), 12 MPa (4 wt.% of Pb^{2+}), 8 MPa (2 wt.% of Ni^{2+}), and 5 MPa (4 wt.% of Ni^{2+}), while for sample without heavy metals achieved 16 MPa (Figure 2). As such, the relative strength loss was 12% and 18% caused by lead presence, and there was a much greater loss caused by nickel presence: 52% and 70% (Figure 8a,b). Lee et al. [42] observed a drop in compressive strength from 24 to 17 and 13 MPa, respectively for geopolymers with 0, 0.5, and 1 wt.% of Pb. This negative impact on compressive strength could have been related to not only heavy metal cations but also the nitrates that were introduced together with Pb^{2+} and Ni^{2+}. Komnitsas et al. [43] stated that both nitrate and heavy metal ions influence compressive strength. Nitrate ions, even in low quantities, may prevent the hardening of a geopolymeric gel and, therefore, the development of compressive strength. The presence of a large quantity of nitrate is known to suppress silica solubility [44]. However, Nikolić et al. [45] showed that the addition of $NaNO_3$ (2.4 wt.% of NO_3^-) had no effect on the compressive strength of geopolymers, while $PbNO_3$ (the same amount of NO_3^-) caused an almost 25% decrease in compressive strength.

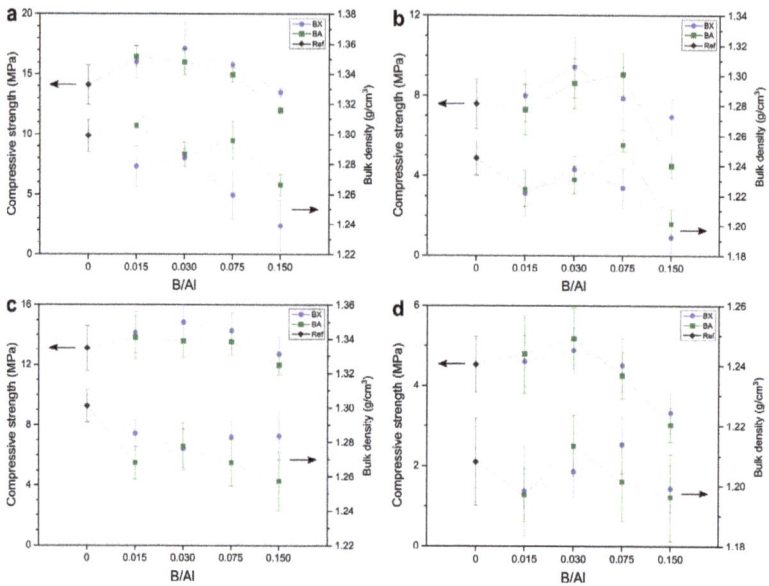

Figure 7. The compressive strength and bulk density of the geopolymers with borax (BX) and boric acid (BA) as the source of boron and (**a**) 2 wt.% of Pb^{2+}, (**b**) 2 wt.% Ni^{2+}, (**c**) 4 wt.% of Pb^{2+}, and (**d**) 4 wt.% Ni^{2+} (Ref—reference sample without boron).

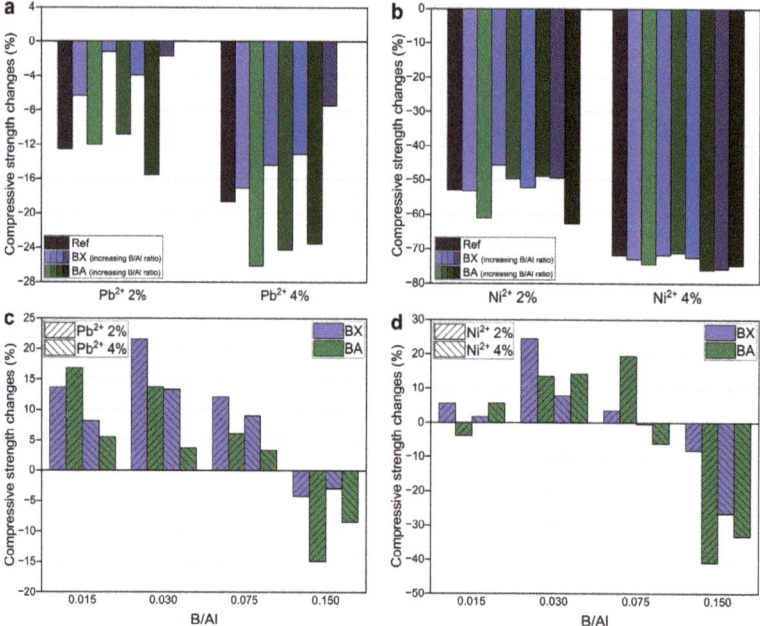

Figure 8. The relative changes in the compressive strength of geopolymers immobilizing heavy metals. The effect of the presence of (**a**) Pb^{2+} and (**b**) Ni^{2+} (in relation to the respective geopolymers without heavy metals). The effect of the boron presence (in relation to the geopolymer without boron) in geopolymers immobilizing (**c**) Pb^{2+} and (**d**) Ni^{2+}.

The loss of strength of the boroaluminosilicate geopolymers induced by Pb was significantly limited when B was introduced with borax (1–6% for 2 wt.% of Pb^{2+} and 7–17% for 4 wt.% of Pb^{2+}); in the case of boric acid, the strength loss was even higher. However, in the case of Ni, there was no enhancement in the strength loss of the boroaluminosilicate geopolymers in comparison to that of the reference sample.

In Figure 8c,d, one can see a similar effect of the boron presence, like in the case of the samples without heavy metals (Figure 3) that had a slight increase in the compressive strength in the range of the B/Al ratio of 0.015–0.075 (but 0.030 for 4 wt.% of Ni^{2+}). It can be stated that a certain amount of boron in the matrix of geopolymers has a positive impact on their compressive strength.

Phase composition: Several new peaks appeared in the diffraction patterns (Figure 9) as the result of heavy metal salt addition to the geopolymers. In the case of the samples with lead, the most matching phase was lead oxide, probably in the form of $PbO \cdot xH_2O$, which is likely to form in a highly alkaline environment. The addition of nickel nitrate caused the formation of sodium nitrate (nitratine) as the product of the reaction of NO_3^- anions with Na^+ cations from the alkaline activator. Minor amounts of some nickel compounds were also detected, most probably in the form of nickel hydroxide or silicate.

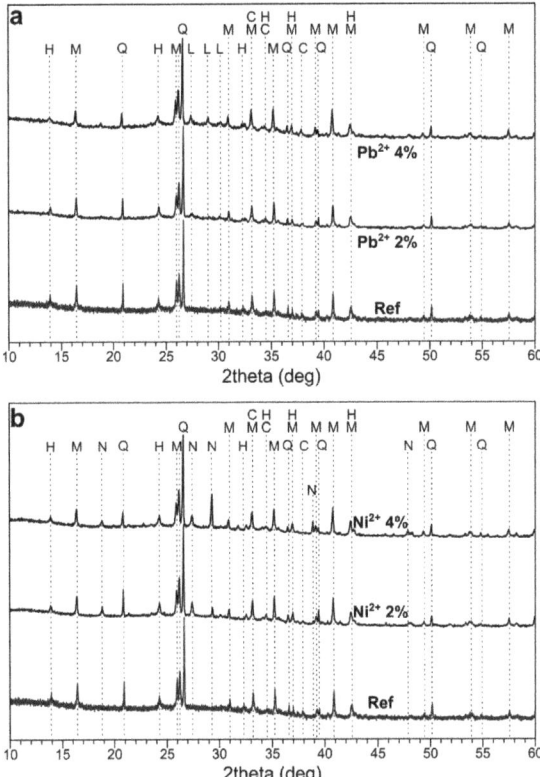

Figure 9. The XRD patterns of the geopolymers with heavy metals and without boron: (**a**) Pb^{2+} and (**b**) Ni^{2+} (H—hydroxysodalite; M—mullite; Q—quartz; C—sodium carbonate; L—lead oxide; N—nickel compounds; S—nitratine).

The formation of hydroxysodalite in the matrix of geopolymers may have had a positive impact on the immobilization of heavy metals. Ni^{2+} cations were immobilized in the center of a six-member ring of sodalite, while Pb^{2+} was six-fold coordinated by the oxygen atoms of the six-membered ring of the sodalite cage and three molecules of water [46].

Structural studies: The differences between the spectra of geopolymers with and without heavy metals (Figure 10) could be observed in three regions: about 1400, 1000, and 700 cm^{-1}. The bands at 1384 cm^{-1} were related to nitrate ions [43,47], and their intensities increased with the higher dosage of heavy metal salts. The main band at about 1000 cm^{-1} slightly shifted to the lower wavenumbers in the presence of heavy metals. This might have been related to the formation of connections between geopolymeric frameworks (cross-linking) by heavy metal cations [48,49], but it is more likely that it was related to the increase in the non-bridging oxygen because the introduced heavy metal cations bound with non-bridging oxygen [50]. The last changes were observed in the range of 710–650 cm^{-1} (small images in Figure 10) related to pseudolattice vibrations (of overtetrahedral fragments of the structure). The change in the band intensity in this range could have been related to the ion-exchange of Na$^+$ cations for heavy metal cations, since the ion-exchange induced an alteration in the ring's surrounding, changes in ionic radii and charges of the cations, and the deformations of rings [51]. A similarity of the geopolymer network and the zeolite framework in terms of the occurrence of ring structures was indicated by Bortnovsky et al. [52], proving that a geopolymeric network is built of deformed 6-, 8-, and 10-membered rings.

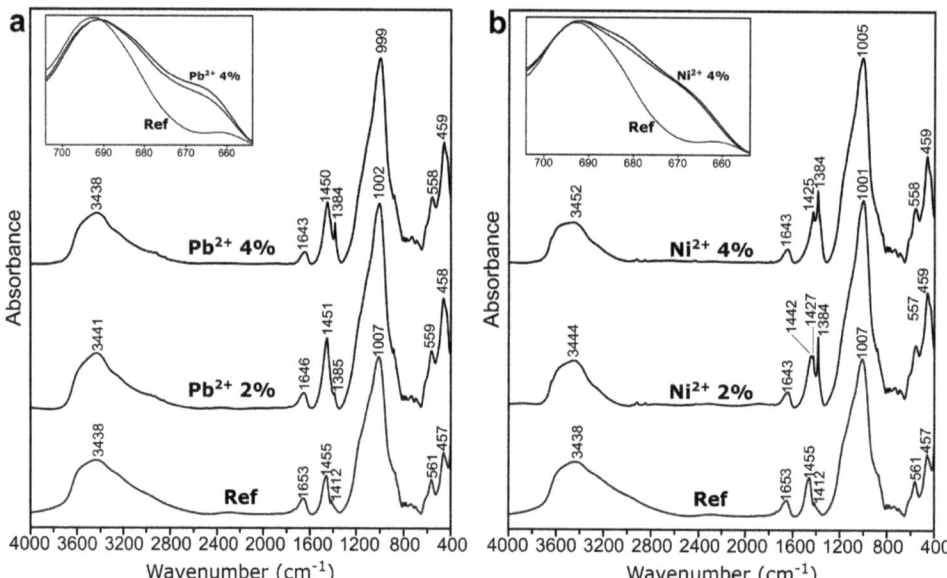

Figure 10. The FT-IR spectra of the geopolymers with heavy metals: (**a**) Pb^{2+} and (**b**) Ni^{2+}.

Microstructure: SEM images and EDS maps of the samples with heavy metals are presented in Figure 11. There were no significant changes in the general microstructure caused by the presence of Pb (Figure 11a), though some sporadically occurring inclusions could be observed (Figure 11b). The elemental mapping of the general structure of the sample with lead showed that Si, Al, O, Na, and Pb were distributed regularly throughout the sample, which indicates the bonding of lead cations to the geopolymeric network (both ion-exchange and covalent bonding mechanism could be involved here). The crystalline, macroscopic inclusion shown in Figure 11b was composed of Pb, clearly to a greater extent than the rest of the sample, proving the presence of precipitation mechanism of immobilization and probably being PbO·xH$_2$O. In the case of the sample with Ni, such macroscopic inclusions were not observed (Figure 11c), and only some microscopic regions with the higher nickel concentrations were visible.

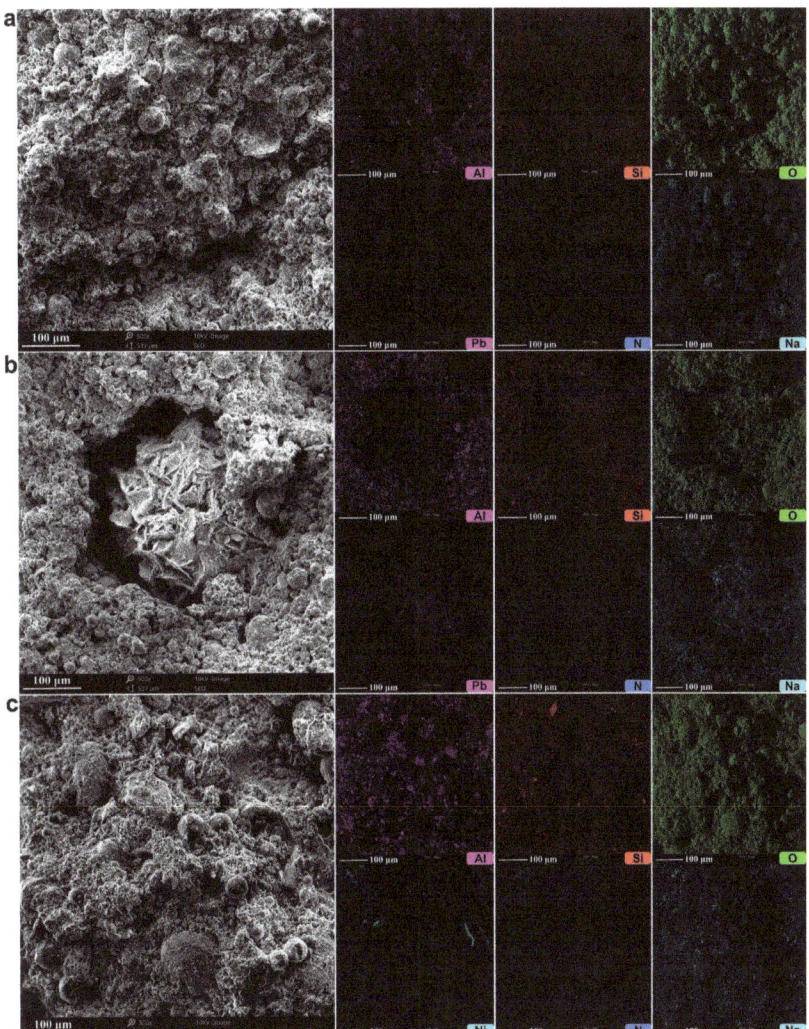

Figure 11. SEM images and EDS maps of the geopolymers with heavy metals: (**a**) the general image of the sample with Pb, (**b**) the inclusion in the sample with Pb, and (**c**) the general image of the sample with Ni.

Immobilization properties: The values of heavy metals leaching from the geopolymeric samples are presented in Figure 12. In the case of the samples with 2 and 4 wt.% of Ni^{2+}, the leaching was very low, namely < 0.05 mg/L for the reference sample (without boron). The leaching values were much higher for the reference samples with −0.6 mg/L (Pb^{2+} 2 wt.%) and 2.6 mg/L (Pb^{2+} 4 wt.%) of lead. The higher leachability of Pb than of Ni can be attributed to the occurrence of macroscopic crystalline inclusions of $PbO \cdot xH_2O$ with a significant solubility in water. Another reason for the higher immobilization efficiency of Ni than Pb may be related to their exchange energies, which are higher for Pb^{2+} than Ni^{2+}, and they are inversely related to solvation radii of heavy metals ions: Ni^{2+} (4.04 Å) >

Pb^{2+} (4.01 Å)—the larger the solvation radius is, the more favorable the immobilization is [46]. Nevertheless, the immobilization rates (Table 3) were excellent. For the reference sample, they were >99.9% for Pb^{2+} and >99.99% for Ni^{2+}. It should be noted that the heavy metal concentrations in the geopolymers were at extremely high levels, namely 20,000 and 40,000 mg/kg, but despite this, they were almost completely immobilized by the geopolymers. The typical contents in actual hazardous wastes, respectively, for Pb^{2+} and Ni^{2+} are: 398 and 90 mg/kg in municipal solid waste incineration fly ash [53], 1129 and 1420 mg/kg in municipal solid waste incineration bottom ash [54], 8966 and 50 mg/kg in waste glass from the spent fluorescent lamps [55], and 7060 and 93 mg/kg in sludge from wastewater treatment [56].

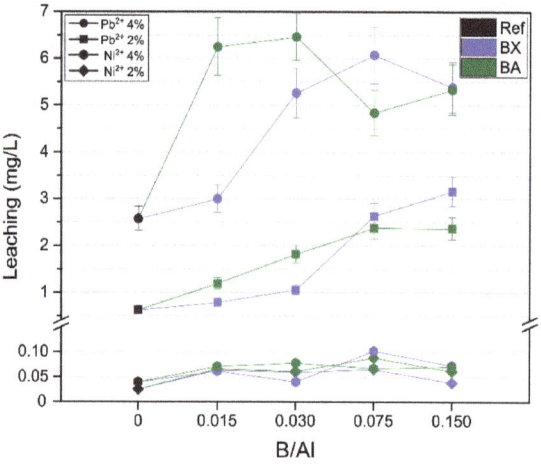

Figure 12. Heavy metal leaching from the geopolymers.

Table 3. Heavy metal immobilization in the geopolymers.

Sample	B/Al	Immobilization Rate (%)			
		Pb^{2+} 2%	Pb^{2+} 4%	Ni^{2+} 2%	Ni^{2+} 4%
Ref	0	99.97	99.94	>99.99	>99.99
BX	0.015	99.96	99.93	>99.99	>99.99
	0.030	99.95	99.87	>99.99	>99.99
	0.075	99.87	99.85	>99.99	>99.99
	0.150	99.84	99.87	>99.99	>99.99
BA	0.015	99.94	99.84	>99.99	>99.99
	0.030	99.91	99.84	>99.99	>99.99
	0.075	99.88	99.88	>99.99	>99.99
	0.150	99.88	99.87	>99.99	>99.99

The presence of boron in the structure led to an increase in heavy metal leaching. This was probably because the addition of a boron compound to the geopolymers led to the creation of additional immobilization sites due to the formation of [BO$_4$] tetrahedra, which provided more negative charges to the geopolymeric network and attracted heavy metal cations. However, the addition of boron caused an increase in geopolymer permeability, and this effect prevailed over the creation of additional immobilization sites, which resulted in an increased leachability. It should be noted that despite this, the immobilization rates of geopolymers with boron were still excellent: over 99.99% in the case of nickel and 99.84–99.96% in the case of lead. The presence of boron in real wastes subjected to alkali-activation can cause the increased leaching of some heavy metals, but its presence could also limit the compressive strength drop induced by heavy metal presence.

4. Conclusions

The addition of boron compounds to geopolymers slightly enhanced the compressive strength when the B/Al molar ratio was not higher than 0.075. A ratio of 0.150 caused a decrease in geopolymer strength. It was found that boron was incorporated into the geopolymeric structure and formed N-B-A-S-H gel. The boron presence did not induce the formation of any new crystalline phases. The immobilization properties of geopolymers were studied by introducing lead and nickel nitrates. A compressive strength drop was observed, and it was especially significant in the case of nickel addition. However, the boron presence in the matrix allowed us to limit the compressive strength decrease. Its presence had also a negative impact, because the heavy metal leaching increased with the increasing boron content. Nevertheless, the immobilization rates were excellent, especially when considering that heavy metals were added in extremal amounts of 20,000 and 40,000 mg/kg of fly ash. The immobilization rates in the case of the samples with lead were always higher than 99.8%, while in the case of the samples with nickel, the immobilization rates were higher than 99.99%.

Author Contributions: Conceptualization, P.R. and P.F.; investigation, P.R. and P.F.; data curation, P.R.; writing—original draft preparation, P.R. and P.F.; writing—review and editing, M.K.; visualization, P.R.; supervision, M.K. and W.M.; project administration, W.M.; funding acquisition, W.M. All authors have read and agreed to the published version of the manuscript.

Funding: This research was funded by National Science Centre in Poland, grant number 2018/31/B/ST8/03109.

Institutional Review Board Statement: Not-applicable.

Informed Consent Statement: Not-applicable.

Data Availability Statement: The data presented in this study are available on request from the corresponding author.

Conflicts of Interest: The authors declare no conflict of interest.

References

1. Provis, J.L.; Duxson, P.; Van Deventer, J.S.J. The role of particle technology in developing sustainable construction materials. *Adv. Powder Technol.* **2010**, *21*, 2–7. [CrossRef]
2. Duxson, P.; Provis, J.L.; Lukey, G.C.; van Deventer, J.S.J. The role of inorganic polymer technology in the development of "green concrete". *Cem. Concr. Res.* **2007**, *37*, 1590–1597. [CrossRef]
3. Zhang, P.; Zheng, Y.; Wang, K.; Zhang, J. A review on properties of fresh and hardened geopolymer mortar. *Compos. Part B Eng.* **2018**, *152*, 79–95. [CrossRef]
4. Król, M.; Rożek, P.; Mozgawa, W. Synthesis of the Sodalite by Geopolymerization Process Using Coal Fly Ash. *Pol. J. Environ. Stud.* **2017**, *26*, 2611–2617. [CrossRef]
5. Faisal, M.; Muhammad, K.; Gul, S. Synthesis and characterization of geopolymer from bagasse bottom ash, waste of sugar industries and naturally available China clay. *J. Clean. Prod.* **2016**, *129*, 491–495. [CrossRef]
6. Hu, W.; Nie, Q.; Huang, B.; Shu, X.; He, Q. Mechanical and microstructural characterization of geopolymers derived from red mud and fly ashes. *J. Clean. Prod.* **2018**, *186*, 799–806. [CrossRef]
7. Nie, Q.; Hu, W.; Huang, B.; Shu, X.; He, Q. Synergistic utilization of red mud for flue-gas desulfurization and fly ash-based geopolymer preparation. *J. Hazard. Mater.* **2019**, *369*, 503–511. [CrossRef]
8. Sun, Z.; Vollpracht, A. Isothermal calorimetry and in-situ XRD study of the NaOH activated fly ash, metakaolin and slag. *Cem. Concr. Res.* **2018**, *103*, 110–122. [CrossRef]
9. De Rossi, A.; Simão, L.; Ribeiro, M.J.; Novais, R.M.; Labrincha, J.A.; Hotza, D.; Moreira, R.F.P.M. In-situ synthesis of zeolites by geopolymerization of biomass fly ash and metakaolin. *Mater. Lett.* **2018**. [CrossRef]
10. Ojovan, M.I.; Lee, W.E. Glassy wasteforms for nuclear waste immobilization. *Metall. Mater. Trans. A Phys. Metall. Mater. Sci.* **2011**, *42*, 837–851. [CrossRef]
11. Gougar, M.L.D.; Scheetz, B.E.; Roy, D.M. Ettringite and C-S-H portland cement phases for waste ion immobilization: A review. *Waste Manag.* **1996**, *16*, 295–303. [CrossRef]
12. Lancellotti, I.; Ponzoni, C.; Barbieri, L.; Leonelli, C. Alkali activation processes for incinerator residues management. *Waste Manag.* **2013**, *33*, 1740–1749. [CrossRef] [PubMed]

13. Liguori, B.; Cassese, A.; Colella, C. Safe immobilization of Cr(III) in heat-treated zeolite tuff compacts. *J. Hazard. Mater.* **2006**, *137*, 1206–1210. [CrossRef]
14. Ji, Z.; Pei, Y. Immobilization efficiency and mechanism of metal cations (Cd^{2+}, Pb^{2+} and Zn^{2+}) and anions (AsO_4^{3-} and $Cr_2O_7^{2-}$) in wastes-based geopolymer. *J. Hazard. Mater.* **2020**, *384*, 121290. [CrossRef]
15. Nikolić, V.; Komljenović, M.; Džunuzović, N.; Ivanović, T.; Miladinović, Z. Immobilization of hexavalent chromium by fly ash-based geopolymers. *Compos. Part B Eng.* **2017**, *112*, 213–223. [CrossRef]
16. Rożek, P.; Król, M.; Knapik, A.; Mozgawa, W. Disposal of bottom ash from the incineration of hazardous waste in two different mineral matrixes. *Environ. Prog. Sustain. Energy* **2017**, *36*, 1074–1082. [CrossRef]
17. El-eswed, B.I. Chemical evaluation of immobilization of wastes containing Pb, Cd, Cu and Zn in alkali-activated materials: A critical review. *J. Environ. Chem. Eng.* **2020**, *8*, 104194. [CrossRef]
18. Khater, H.M.; Ghareib, M. Optimization of geopolymer mortar incorporating heavy metals in producing dense hybrid composites. *J. Build. Eng.* **2020**, *32*, 101684. [CrossRef]
19. Zhao, S.; Xia, M.; Yu, L.; Huang, X.; Jiao, B.; Li, D. Optimization for the preparation of composite geopolymer using response surface methodology and its application in lead-zinc tailings solidification. *Constr. Build. Mater.* **2021**, *266*, 120969. [CrossRef]
20. Celik, A.; Yilmaz, K.; Canpolat, O.; Al-mashhadani, M.M.; Aygörmez, Y.; Uysal, M. High-temperature behavior and mechanical characteristics of boron waste additive metakaolin based geopolymer composites reinforced with synthetic fibers. *Constr. Build. Mater.* **2018**, *187*, 1190–1203. [CrossRef]
21. Palomo, A.; López de la Fuente, J.I. Alkali-activated cementitous materials: Alternative matrices for the immobilisation of hazardous wastes—Part I. Stabilisation of boron. *Cem. Concr. Res.* **2003**, *33*, 281–288. [CrossRef]
22. Nicholson, C.L.; Murray, B.J.; Fletcher, R.A.; Brew, D.R.M.; MacKenzie, K.J.D.; Schmücker, M. Novel geopolymer materials containing borate structural units. *World Congr. Geopolymer* **2005**, *2005*, 31–33.
23. Nazari, A.; Maghsoudpour, A.; Sanjayan, J.G. Characteristics of boroaluminosilicate geopolymers. *Constr. Build. Mater.* **2014**, *70*, 262–268. [CrossRef]
24. Bagheri, A.; Nazari, A.; Sanjayan, J.G.; Rajeev, P. Alkali activated materials vs geopolymers: Role of boron as an eco-friendly replacement. *Constr. Build. Mater.* **2017**, *146*, 297–302. [CrossRef]
25. Bagheri, A.; Nazari, A.; Hajimohammadi, A.; Sanjayan, J.G.; Rajeev, P.; Nikzad, M.; Ngo, T.; Mendis, P. Microstructural study of environmentally friendly boroaluminosilicate geopolymers. *J. Clean. Prod.* **2018**, *189*, 805–812. [CrossRef]
26. Dupuy, C.; Gharzouni, A.; Sobrados, I.; Texier-Mandoki, N.; Bourbon, X.; Rossignol, S. 29Si, 27Al, 31P and 11B magic angle spinning nuclear magnetic resonance study of the structural evolutions induced by the use of phosphor- and boron–based additives in geopolymer mixtures. *J. Non-Cryst. Solids* **2019**, *521*, 119541. [CrossRef]
27. Williams, R.P.; van Riessen, A. Development of alkali activated borosilicate inorganic polymers (AABSIP). *J. Eur. Ceram. Soc.* **2011**, *31*, 1513–1516. [CrossRef]
28. Khezrloo, A.; Aghaie, E.; Tayebi, M. Split tensile strength of slag-based boroaluminosilicate geopolymer. *J. Aust. Ceram. Soc.* **2018**, *54*, 65–70. [CrossRef]
29. Bagheri, A.; Nazari, A.; Sanjayan, J.G. Fibre-reinforced boroaluminosilicate geopolymers: A comparative study. *Ceram. Int.* **2018**, *44*, 16599–16605. [CrossRef]
30. Bagheri, A.; Nazari, A.; Sanjayan, J.G.; Rajeev, P.; Duan, W. Fly ash-based boroaluminosilicate geopolymers: Experimental and molecular simulations. *Ceram. Int.* **2017**, *43*, 4119–4126. [CrossRef]
31. Taveri, G.; Tousek, J.; Bernardo, E.; Toniolo, N.; Boccaccini, A.R.; Dlouhy, I. Proving the role of boron in the structure of fly-ash/borosilicate glass based geopolymers. *Mater. Lett.* **2017**, *200*, 105–108. [CrossRef]
32. Bullerjahn, F.; Zajac, M.; Skocek, J.; Ben Haha, M. The role of boron during the early hydration of belite ye'elimite ferrite cements. *Constr. Build. Mater.* **2019**, *215*, 252–263. [CrossRef]
33. Farid, O.M.; Abdel Rahman, R.O. Preliminary assessment of modified borosilicate glasses for chromium and ruthenium immobilization. *Mater. Chem. Phys.* **2017**, *186*, 462–469. [CrossRef]
34. Rożek, P.; Król, M.; Mozgawa, W. Geopolymer-zeolite composites: A review. *J. Clean. Prod.* **2019**, *230*, 557–579. [CrossRef]
35. Nath, S.K.; Maitra, S.; Mukherjee, S.; Kumar, S. Microstructural and morphological evolution of fly ash based geopolymers. *Constr. Build. Mater.* **2016**, *111*, 758–765. [CrossRef]
36. Hajimohammadi, A.; van Deventer, J.S.J. Solid Reactant-Based Geopolymers from Rice Hull Ash and Sodium Aluminate. *Waste Biomass Valorization* **2017**, *8*, 2131–2140. [CrossRef]
37. Dupuy, C.; Havette, J.; Gharzouni, A.; Texier-Mandoki, N.; Bourbon, X.; Rossignol, S. Metakaolin-based geopolymer: Formation of new phases influencing the setting time with the use of additives. *Constr. Build. Mater.* **2019**, *200*, 272–281. [CrossRef]
38. Król, M.; Minkiewicz, J.; Mozgawa, W. IR spectroscopy studies of zeolites in geopolymeric materials derived from kaolinite. *J. Mol. Struct.* **2016**, *1126*, 200–206. [CrossRef]
39. Stoch, P.; Stoch, A. Structure and properties of Cs containing borosilicate glasses studied by molecular dynamics simulations. *J. Non-Cryst. Solids* **2015**, *411*, 106–114. [CrossRef]
40. Stoch, L.; Środa, M. Infrared spectroscopy in the investigation of oxide glasses structure. *J. Mol. Struct.* **1999**, *511–512*, 77–84. [CrossRef]
41. Adamczyk, A.; Handke, M.; Mozgawa, W. FTIR studies of $BPO_4 \cdot 2SiO_2$, $BPO_4 \cdot SiO_2$ and $2BPO_4 \cdot SiO_2$ joints in amorphous and crystalline forms. *J. Mol. Struct.* **1999**, *511–512*, 141–144. [CrossRef]

42. Lee, S.; van Riessen, A.; Chon, C.M.; Kang, N.H.; Jou, H.T.; Kim, Y.J. Impact of activator type on the immobilisation of lead in fly ash-based geopolymer. *J. Hazard. Mater.* **2016**, *305*, 59–66. [CrossRef] [PubMed]
43. Komnitsas, K.; Zaharaki, D.; Bartzas, G. Effect of sulphate and nitrate anions on heavy metal immobilisation in ferronickel slag geopolymers. *Appl. Clay Sci.* **2013**, *73*, 103–109. [CrossRef]
44. Zhang, J.; Provis, J.L.; Feng, D.; van Deventer, J.S.J. Geopolymers for immobilization of Cr^{6+}, Cd^{2+}, and Pb^{2+}. *J. Hazard. Mater.* **2008**, *157*, 587–598. [CrossRef] [PubMed]
45. Nikolić, V.; Komljenović, M.; Džunuzović, N.; Miladinović, Z. The influence of Pb addition on the properties of fly ash-based geopolymers. *J. Hazard. Mater.* **2018**, *350*, 98–107. [CrossRef]
46. Liu, J.; Luo, W.; Cao, H.; Weng, L.; Feng, G.; Fu, X.-Z.; Luo, J.-L. Understanding the immobilization mechanisms of hazardous heavy metal ions in the cage of sodalite at molecular level: A DFT study. *Microporous Mesoporous Mater.* **2020**, *306*, 110409. [CrossRef]
47. Ji, Z.; Pei, Y. Geopolymers produced from drinking water treatment residue and bottom ash for the immobilization of heavy metals. *Chemosphere* **2019**, *225*, 579–587. [CrossRef]
48. El-Eswed, B.I.; Aldagag, O.M.; Khalili, F.I. Efficiency and mechanism of stabilization/solidification of Pb(II), Cd(II), Cu(II), Th(IV) and U(VI) in metakaolin based geopolymers. *Appl. Clay Sci.* **2017**, *140*, 148–156. [CrossRef]
49. Guo, B.; Pan, D.; Liu, B.; Volinsky, A.A.; Fincan, M.; Du, J.; Zhang, S. Immobilization mechanism of Pb in fly ash-based geopolymer. *Constr. Build. Mater.* **2017**, *134*, 123–130. [CrossRef]
50. Hu, S.; Zhong, L.; Yang, X.; Bai, H.; Ren, B.; Zhao, Y.; Zhang, W.; Ju, X.; Wen, H.; Mao, S.; et al. Synthesis of rare earth tailing-based geopolymer for efficiently immobilizing heavy metals. *Constr. Build. Mater.* **2020**, *254*, 119273. [CrossRef]
51. Mozgawa, W.; Król, M.; Bajda, T. Application of IR spectra in the studies of heavy metal cations immobilization on natural sorbents. *J. Mol. Struct.* **2009**, *924–926*, 427–433. [CrossRef]
52. Bortnovsky, O.; Dedecek, J.; Tvaružková, Z.; Sobalík, Z.; Subrt, J. Metal Ions as Probes for Characterization of Geopolymer Materials. *J. Am. Ceram. Soc.* **2008**, *91*, 3052–3057. [CrossRef]
53. Luna Galiano, Y.; Fernández Pereira, C.; Vale, J. Stabilization/solidification of a municipal solid waste incineration residue using fly ash-based geopolymers. *J. Hazard. Mater.* **2011**, *185*, 373–381. [CrossRef]
54. Rożek, P.; Król, M.; Mozgawa, W. Solidification/stabilization of municipal solid waste incineration bottom ash via autoclave treatment: Structural and mechanical properties. *Constr. Build. Mater.* **2019**, *202*, 603–613. [CrossRef]
55. Bobirică, C.; Shim, J.H.; Park, J.Y. Leaching behavior of fly ash-waste glass and fly ash-slag-waste glass-based geopolymers. *Ceram. Int.* **2018**, *44*, 5886–5893. [CrossRef]
56. Sun, S.; Lin, J.; Zhang, P.; Fang, L.; Ma, R.; Quan, Z.; Song, X. Geopolymer synthetized from sludge residue pretreated by the wet alkalinizing method: Compressive strength and immobilization efficiency of heavy metal. *Constr. Build. Mater.* **2018**, *170*, 619–626. [CrossRef]

MDPI
St. Alban-Anlage 66
4052 Basel
Switzerland
www.mdpi.com

Materials Editorial Office
E-mail: materials@mdpi.com
www.mdpi.com/journal/materials

Disclaimer/Publisher's Note: The statements, opinions and data contained in all publications are solely those of the individual author(s) and contributor(s) and not of MDPI and/or the editor(s). MDPI and/or the editor(s) disclaim responsibility for any injury to people or property resulting from any ideas, methods, instructions or products referred to in the content.

www.ingramcontent.com/pod-product-compliance
Lightning Source LLC
LaVergne TN
LVHW070712100526
838202LV00013B/1076